The Montanans' FISHING GUIDE

DICK KONIZESKI

Volume I

MONTANA WATERS WEST OF CONTINENTAL DIVIDE

Mountain Press Publishing Company
279 West Front Street
Missoula, Montana 59801

Library of Congress Catalog Card Number 75-13141
Copyright 1975, Richard L. Konizeski

3rd EDITION
ISBN 0-87842-053-3

Second Printing, 1978

MOUNTAIN PRESS PUBLISHING COMPANY
279 West Front Street
Missoula, Montana 59801

Dedication

THIS BOOK IS DEDICATED
TO THE PRESERVATION OF
OUR WILDERNESS HERITAGE.

USE IT!
CHERISH IT!
PROTECT IT!

Dick Konizeski

Introduction

Montana, *Vacationers' Paradise, Land of the Shining Mountains, Untrammeled Forests* and *Unspoiled Waters* beneath *The Big Sky* high on the *Great Divide*.[1] Any way you look at it, it's right next door to Heaven and the two volumes (East and West) of *The Guide* are your passport to the recreational delights of more than 200 miles of truly "wild"[2] free-flowing rivers inaccessible except by boat or trail, 450 miles of unspoiled "blue ribbon"[3] rivers, 1600 tributaries ranging from foaming mountain cataracts to meandering valley brooks, with a combined total of 11,000 miles of top-flight fishing waters, and 1500 lakes ranging from crystal clear mountain jewels, to small ponds, to inland seas 150 miles long. Many lakes and streams are on the doorstep of civilization, readily accessible by public and private transportation. These lakes and streams offer a wide variety of accommodations and recreational opportunities ranging from launch trips to hydroplane racing, water skiing, sailing, canoeing, swimming, fishing or just plain loafing on the beach in the clear mountain air. Most of the lakes, however, are available only by horse or backpack deep in the rugged northern Rockies, home of the nation's most feared and respected carnivore, *Ursus horribilis* or the Grizzly Bear. There are also couger, bobcat, lynx, wolf, wolverene and popular big game species such as moose, elk, whitetailed and mule deer, caribou, big horn sheep, mountain goat, and in the prairies, antelope and buffalo. Truly a hunter's paradise second only in recreational potential to the finest trout fishing in the world. Here are caught brown trout weighing up to a state record of 29 pounds. rainbow, cutthroat, and Dolly Varden trout are common in the 5 to 15 pound class, and lake trout in the 20 to 35 pound

---

[1]Continental Divide
[2]Set aside by federal law as national recreational features.
[3]Defined by state law as of national recreational importance.

Acknowledgements

Many of the lakes and streams in Montana are readily accessible by auto, but most are in primitive, inaccessable regions. To survey them all the first time around was obviously a time consuming though pleasant task of many years duration. However, the accuracy of the original descriptions was not only the result of literally thousands of miles of back packing and endless hours of fishing, but also of the enthusiastic cooperation of various state and federal agencies, and a host of sportsmen including both native Montanans and many from out of state. Furthermore the fishing potentials and access routes to high mountain lakes and streams can and often do change radically in a single season due to a variety of causes such as exceptionally deep winter snow packs, poisoning out of undesirable and or replanting with more desirable species of fish, construction of new or abandonment of old roads and trails etc. The updating of the fishing data recorded herein is primarily the result of an ongoing program of helicopter surveys by the Montana State Fish and Game Department, plus thousands of tips on local waters by professional guides, and experienced anglers. Access changes were taken most from United States Forest Service records, and by word of mouth from state and federal forest service personnel. More than 500 corrections or additions are included. The lead articles have been rewritten and are authored mostly by fishery biologists who are eminently qualified by their professional training and on-the-job experience. Many new photographs have been included, contributed by the Montana State Fish and Game Department, and various local photographers, and amateurs who could be professionals if they wanted to.

It is through the cooperation of all these agencies and individuals that the usefulness, dependability and general overall excellence of this, the 4th edition of the MONTANANS' FISHING GUIDE, has been made possible.

R. L. KONIZESKI

Contents

MONTANA GAME FISH 1
FISHING THE BITTERROOT 9
   Bitterroot Drainage 11
THE BLACKFOOT IS BEAUTIFUL 41
   Blackfoot River Drainage 43
CLARK FORK EXPOSE' 69
   Clark Fork of the Columbia – Lower Drainage 72
   Clark Fork of the Columbia – Upper Drainage 107
BEHOLD THE FLATHEAD 139
   Flathead River Drainage 141
MIDDLE FORK OF THE FLATHEAD
— MONTANA'S WILDEST RIVER 193
   Middle Fork Drainage 195
NORTH FORK OF THE FLATHEAD
— "A FLOAT FISHERMAN'S PARADISE 200
   North Fork Drainage 201
SOUTH FORK OF THE FLATHEAD
— A WILDERNESS PARADISE 207
   South Fork Drainage 209
GLACIER NATIONAL PARK 227
   Glacier Park Lakes 229
THE KOOTENAI — "DAMMED FOREVER" 233
   Kootenai River Drainage 235
SWAN RIVER — GATEWAY TO THE MISSION MOUNTAIN WILD AREA AND
BOB MARSHALL WILDERNESS 287
   Swan River Drainage 289

Montana Game Fish

The following seven pages reproduce Montana fishing paintings done by wildlife artist Ron Jenkins. Many of the individual paintings are available as prints suitable for framing. For information write Mountain Press Publishing Company, 279 West Front Street, Missoula, Montana 59801.

BROOK TROUT. Range up to 9 pounds; are distinguished by variable dark brown-to-olive-to-scarlet red (in some alpine lakes) color, round red spots with blue margins on the sides, wavy lines across the back, and white margins along the ventral fins. This species is especially common in the Big Hole drainage and the lakes of the Beartooth Plateau (Clarks Fork of the Yellowstone drainage).

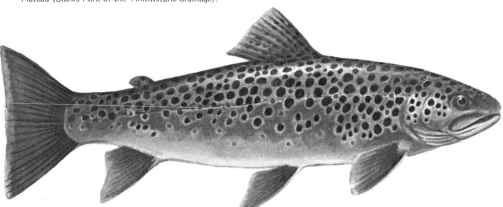

BROWN or GERMAN BROWN or LOCH LEVEN TROUT. Range up to about 20 pounds and are common to 5 pounds; are distinguished by their yellowish-brown color, relatively large dark spots on the back and a few red spots (sometimes with a narrow bluish band) on the sides; see especially the Yellowstone, Madison, Judith, Musselshell and upper Clark Fork drainages.

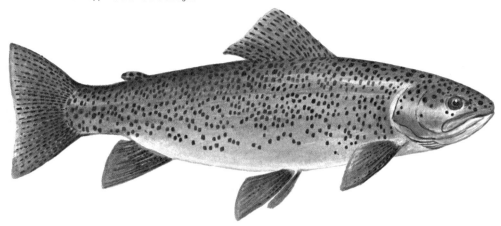

RAINBOW TROUT. Range up to 20 pounds and are common to 5 pounds; are distinguished by usually light sides and somewhat darker back with numerous irregularly-shaped spots and a broad red band along the midline of each side. They are abundant in all of the drainages and frequently cross with Cutthroat with hard-to-identify results.

CUTTHROAT TROUT. Range up to 15 pounds but are generally under 12 inches in the small headwaters of most Montana drainages, and up to 5 pounds in many lakes; are distinguished by a dark red or orange slash along the bottom of each jaw, and numerous dark spots on the posterior parts of the body.

GOLDEN TROUT. Generally average less than 1½ pounds; are distinguished by their beautiful coloration, generally yellowish color with bright carmine strips along the belly from the throat to the anal fin, small scales; have been introduced in suitable (alpine) lakes in most of the major drainages.

DOLLY VARDEN or BULL TROUT. Range up to 30 pounds and are common to 15 pounds; are distinguished by their large mouth, olive color, with round orange to yellowish spots on the sides, and a white border on the ventral fins, are generally slenderer than the Brook trout; see especially the South Fork of the Flathead drainage.

ARCTIC GRAYLING. Mostly average between 6 and 12 inches but range up to 2 pounds or 20 inches; are distinguished by their large beautifully colored dorsal fin, large eyes, small mouth, grayish-silver color and irregularly-shaped black dots; are found in several lakes in most of the major drainages, and in many lakes in the Big Hole and Clarks Fork of the Yellowstone drainages.

LAKE or MACKINAW TROUT. Range up to 40 pounds are common to 15 pounds; are distinguished by their light or greenish-gray color, many irregular light spots on the sides and back, large mouth and deeply forked caudal fin; see especially Whitefish and Flathead Lakes in the Flathead drainage, Spar Lake in the Kootenai drainage, and Twin Lakes in the Big Hole drainage.

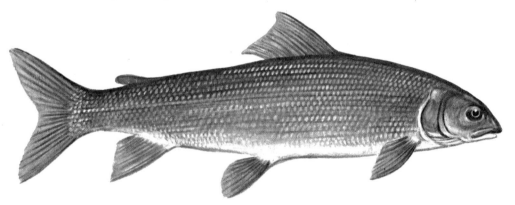

WHITEFISH. Range up to 5 pounds or 24 inches; are distinguished by their small mouth, silvery sides and belly, and olive drab back; see especially Rock Creek in the upper Clark Fork drainage, and the Kootenai and Yellowstone Rivers.

COHO or SILVER SALMON. The sea-going variety ranges up to maybe 30 pounds or so, but they've only recently (1969) been introduced to Montana waters so it's still too early to say how big they'll grow here. Otherwise they need no description except perhaps to say that they're not only large, they're fighters from the moment they feel the hook, and are truly an Epicurean's delight.

KOKANEE or SOCKEYE SALMON. Range up to 5 pounds but average between 12 and 14 inches; are distinguished by their dark-greenish color on the back, silver on the sides and belly, height of the anal fin is shorter than the base line, small teeth, and the mature males are a bloody red with no spots and a hooked nose; see Flathead Lake and the Clearwater on the Blackfoot drainage.

STURGEON. Several species are found in Montana. Some range up to several hundred pounds. They are distinguished by their flat belly, ventrally located mouth, cartilagenous nobby plates along the sides and back, and 4 barbels or wiskers directly in front of the mouth, see Kootenai and Yellowstone Rivers.

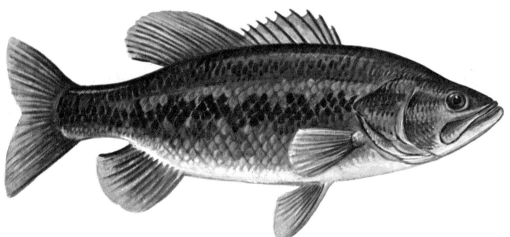

LARGEMOUTH BASS. Range up to 14 pounds; are distinguished by their dark metallic green back and greenish-yellow sides, a broken dark greenish band from the eye to the caudal fin, a large mouth that extends to well behind the eye in adults, and large scales; see Blanchard and many of the pothole lakes in the Flathead drainage, and sloughs along the upper Clark Fork drainage and many reservoirs in the eastern part of the state.

SMALLMOUTH BASS., Range up to 4 pounds but average around 1½ to 2½ pounds; are usually distinguished from Largemouth Bass by a shorter jaw that ends below the reddish colored eye, and a dorsal fin that is hardly emarginate; see Horseshoe Lake in the Swan River drainage.

YELLOW PERCH. Are reported up to 21 inches but mostly range between 5 and 15 inches; are distinguished by their deep greenish back with indentations extending down across their yellowish sides, and a dirty white belly, coarse scales, and an arched back leading to the head which has a more or less concave adult profile; they are common in many lakes in all drainages.

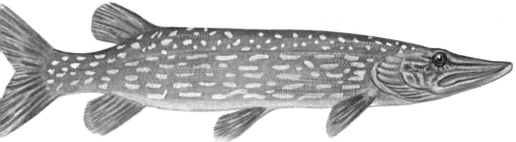

NORTHERN PIKE. Range up to 27 pounds but average between 3 and 8 pounds; are distinguished by their long narrow body and equally elongate head with its deep mouth and large vicious teeth, forked tail, and dirty greenish-yellow mottled body; see especially Pike Lake in the St. Mary drainage, and Dry Fork Reservoir in the Flathead drainage.

WALLEYE. The average angler cannot tell these from Sauger. Both are large members of the perch family. The average size creeled is 12 to 14 inches. Walleye and sauger are common in the reservoirs and large rivers in the eastern part of the state.

SUNFISH or PUMPKINSEEDS. Green sunfish and Bluegills have all been introduced into Montana. Mostly about the size of your hand (4 to 6 inches long), flat-bodied with coarse, iridescent scales. Bluegills have been widely planted in ranch ponds in eastern Montana. Pumpkinseeds alone are found west of the Continental Divide.

PADDLEFISH or SPOON-BILLED CATFISH. Are usually from 2 to 4 feet long (two in the 120 pound class have been taken in Montana). Smooth and very large gill slits; are caught (snagged) in the Lower Yellowstone and Missouri river drainages.

CHANNEL CAT. This hard looking but good eating fellow has a slender, smooth skinned body with a small head, whiskers on his chin, spines on his dorsal and pectoral fins, a long anal fin and a forked tail. He's bluish to greenish-gray on top and white below and you'll find him in sizes ranging up to 30 pounds or so in the lower Missouri, Yellowstone and Musselshell rivers.

BULLHEAD. Black bodied, smooth skinned, up to maybe 5 pounds in weight, big headed, wide mouthed with whiskers, more or less square tailed and plenty attractive in a frying pan. You'll find them here and there in the Lower Missouri, Yellowstone and Musselshell drainages and a few over the Divide in the lower Clarks Fork of the Columbia River and the Flathead drainage.

LING. If your wife took one horrified look, screamed and threw rod and all in the drink — that, brother, was a ling — a gruesome-looking half brother to the sea-going cod but resembling more closely the slippery eel. Many are those who claim its flesh is a toothsome as its looks are revolting — especially in the Kootenai River area although it's not bothered much elsewhere in the state.

BLODGETT CANYON
Courtesy USFS

Fishing The Bitterroot

The main stem of the Bitterroot River, approximately 100 yards wide, flows generally north through the beautiful Bitterroot Valley in western Montana. The main river originates near Conner at the confluence of its East and West Forks. As the river follows its meandering course (approximately 70 miles) through the valley, its flows are augmented by numerous creeks draining from the rugged Bitterroot Range on the west and from the Sapphire Range on the east. The Bitterroot drops 800 feet in elevation from Conner to its confluence with the Clark Fork below Missoula. The river was named by the Salish Indians after the Bitterroot flower (Lewisia rediviva), the state flower of Montana.

The Bitterroot Valley is one of the oldest agricultural valleys in the state, and the effects of land clearing are reflected in the water quality, discharge fluctuations, and fish populations. Spring runoff is very rapid with general flooding, creating many new channels every year. Home building and subdividing in the flood plain have led to many futile attempts to keep the river within its old established banks. Other attempts have been made to straighten its course and to rip rap its banks with rock. The projects speed up the river flow and cause increased sedimentation and stream bank destruction. River bank cover is mostly cottonwood, although there are many ponderosa pine, aspen, birch and willow, which provide bank protection and fish cover as long as they are protected from over-grazing.

During the summer irrigation season, farmers and ranchers completely dewater the lower portions of side tributaries and much of the main river. Areas such as "Victor Crossing" and "Tucker Crossing" are almost devoid of water every summer. Irrigation return makes up most of the summer flows below "Victor Crossing." Realizing that low flows are a weak link in any trout production picture, one would do well to avoid fishing these areas.

The Bitterroot was originally an excellent West slope cutthroat trout producer. Rainbow, brook and brown trout have since been introduced. More and more demands for irrigation water, increased sedimentation, and habitat destruction have contributed to declining numbers of all fish, but particularly the cutthroat and rainbow that spawn in the Spring of the year.

Brown trout and whitefish are the main species in the lower river and provide good fishing. The upper river near Darby provides excellent fishing for rainbow. The Bitterroot in its upper portions is a fabulous stream for floating and fishing. This type of trip is heartily recommended for all, beginners and experts alike.

BOYD R. OPHEIM
Fisheries Biologist
Montana State Fish & Game Department

RAVALLI COUNTY BIRD REFUGE
Courtesy Ernst Peterson

Bitterroot River Drainage

Ambrose Creek. A small stream flowing west to the Bitterroot River 4 miles north of Stevensville. The lower 3 miles are used entirely for irrigation, the upper reaches are followed by road for 7 miles and are heavily fished for small cutthroat.

Bailey Lake. A high alpine lake at the head of Lost Horse Creek reached by foot trail 1½ miles from the Lost Horse road at the USFS Guard Station. Bailey Lake covers about 12 acres and drops off gradually to a maximum depth of 32 feet at the northwest end. It provides fair fishing for 9 to 10 inch skinny rainbow plus rainbow-cutthroat hybrids.

Baker Creek. A small stream which flows for 4 miles from Baker Lake to the Nez Perce Fork of the Bitterroot River 6 miles above Conner. The lower mile is fair fishing for small cutthroat and rainbow.

Baker Lake. Take the West Fork (of the Bitterroot) highway for one mile past the Trapper Creek Guard Station, then a gravel road for 2½ miles up Pierce Creek where you bear off to the right onto a muchly switchbacked logging road for 6½ miles to Baker Point and thence a rough hunter's trail for another 1½ miles to the lake. A small (8.8 acres) 52 foot deep lake in sparsely timbered alpine country, it is seldom fished (probably because of its isolation) but produces good catches of 7 to 12 inch cutthroat.

Bass Creek. Drains Bass Lake 12 miles eastward to the Bitterroot River, 5 miles north of Stevensville. The lower 3 miles are mostly used for irrigation; from here there is a road for a couple of miles and then 7 miles of trail to the lake. The creek is poor-to-fair fishing for small cutthroat, rainbow and brook trout.

Bass Lake. Reached by the Bass Creek trail 7 miles up from the end of the road. It's 100 acres by 140 feet deep (here's one that doesn't freeze out), in an alpine area that is rocky and brushy with a little scrub timber. The State Fish and Game Department air-dropped it with rainbow in the summer of 1967. There has been good fishing since and should still be.

Bear Creek. Flows eastward from the junction of its North and South Forks for 9 miles through a deeply glaciated canyon to the Bitterroot River 2 miles south of Victor. The lower 4 miles are used for irrigation, but the upper reaches are followed by a good trail and are heavily fished for small cutthroat, rainbow and brook trout.

Bear Creek. A small stream about 6 miles long, that flows south to Lolo Creek 12 miles above the town of Lolo. It's followed by trail for the lower 1½ miles and is good fishing in early season for small cutthroat.

BITTERROOT

BAKER LAKE
Courtesy Ernst Peterson

BASS LAKE
Courtesy USFS

Bear Creek Lake. Take the Bear Creek trail for 5 miles from the end of the USFS road at the mouth of the canyon, to the mouth of the North Fork, and then a poor trail for 4 miles upstream to the lake. This is a beautiful cirque lake in sparsely timbered-to-rocky country, about 20 acres in extent, with a maximum depth of 30 feet and perhaps 3 acres in aquatic vegetation. There is an old abandoned dam at the outlet. It used to be good fishing for 10 to 14 inch brook trout but is reported to be barren now.

Bear Run Creek. A very small tributary of Miller Creek containing a few small cutthroat and brookies. A fine kids' fishing stream.

Beaver Creek. A small tributary of Big Creek, crossed by the Big Creek trail 2 miles below the lake and excellent fishing for small cutthroat.

Beaver Creek. A short stream (about 5 miles long) flowing to the West Fork of the Bitterroot 12 miles above Painted Rocks Lake. It is followed by a gravel road for its entire length and is good fishing for small cutthroat and brook trout (especially in the numerous beaver dams).

Beavertail Creek. A small, eastward flowing tributary of the West Fork of the Bitterroot River. It is followed by a gravel road and is fair fishing for small rainbow and cutthroat 1½ miles above the mouth, which is 2 miles above the West Fork Ranger Station.

Big Creek. A clear, wadeable stream that drains Big Creek Lake and flows eastward for 16 miles to join the Bitterroot River about 5 miles south of Stevensville. The lower 3 miles are dewatered for irrigation, the upper reaches are followed by trail and are good fishing for 8 to 10 inch cutthroat, brook and Dolly Varden.

MIDDLE FORK BEAR CREEK
Courtesy W.E. Thamarus, Jr.

BLODGETT CREEK CANYON
Courtesy Ernst Peterson

Big Creek Lakes.
Take the Big Creek USFS trail 11 miles from the end of a county road at the mouth of the canyon to the lower end of the lake and another couple of miles to the upper end. The lakes occupy a rock-walled glacial canyon with conifer cover in the bottoms. They are separate lakes only during low water in late summer and fall. Together they include about 240 acres, and have a maximum depth of over 100 feet. They are overpopulated with 12 to 16 inch rainbow and cutthroat, plus a few lunkers that will go to 3 or 4 pounds. Fishing is generally excellent for both species, each of which tends to occupy different areas.

Bitterroot Irrigation Ditch.
A large ditch that flows only during the summer months and is paralleled by a maintainence road for about 80 miles along the east side of the valley. It is excellent fishing in spots for brook and rainbow trout that range up to 5 pounds.

Blodgett Creek.
A clear, wadeable stream flowing from Blodgett Lake for about 20 miles due east through forested mountains to the Bitterroot River a couple miles north of Hamilton. The lower 3 miles are dewatered for irrigation, the upper reaches are followed by road and trail and are generally good fishing for small brook and Dolly Varden trout. There's a public campground at the end of the road.

Blue Joint Creek.
A clear, dashing inlet of Painted Rocks Lake, about 20 miles long; reached by road 6.6 miles from the West Fork highway around the south end of the lake and followed by trail to headwaters. It's an excellent mountain stream for 8 to 14 inch cutthroat, brook, Dolly Varden and rainbow trout.

Boulder Creek.
A beautiful tributary of the Bitterroot River, crossed at the mouth by the West Fork highway and followed by a road to a USFS campground a mile upstream and then for 10 miles by trail to headwaters by Boulder Lake. An easily fished stream in a deeply glaciated canyon, it's excellent for small red-bellied, green backed cutthroat.

BIG CREEK LAKE
Courtesy USFS

UPPER BIG CREEK LAKE
Courtesy USFS

Boulder Lake. A high alpine lake in sparsely timbered, glaciated country north of Bare Peak; reached by a good trail 10 miles from the end of the Boulder Creek road. It's a deep lake, about 20 acres, with an 8 foot dam at the outlet, spotty and only poor-to-fair fishing for 7 to 17 inch cutthroat.

Bryan Lake. Take a good USFS trail for 9 miles from the end of a county road 3 miles above the mouth of Bear Creek canyon; first up Bear Creek and then up the Middle Fork, to the lake which covers about 20 acres, averages 20 feet deep and has a fair amount of aquatic vegetation around the upper end. It was planted with rainbow years ago and the few that are left will average a pound or better and are in excellent condition.

Buck Creek. Flows from Buck Lake for about 4 miles to the East Fork of the Bitterroot River. Take the East Fork road 16 miles from Sula to the East Fork Ranger Station, then 2 miles up the Little East Fork road to the wilderness boundary and finally a good trail for 4½ miles to the mouth of the creek and continue upstream to the lake. The fishing is good here for small cutthroat trout.

Burnt Fork Bitterroot River. A beautiful, easily waded stream flowing from Burnt Fork Lake for 30 miles through the Sapphire Mountains to join the Bitterroot River about 1 mile north of Stevensville, and paralleled by road and trail to headwaters. It is excellent fishing, with many pools and eddies, for small cutthroat in the upper reaches; brook trout and a few Dolly Varden below.

Burnt Fork Lake. Reached by a good gravel road 6 miles up the creek from the mouth of the canyon, and then by poor trail 7 miles to the lake, or by USFS trail 5 miles from the end of the Willow Creek road, or by a poor (closed) jeep trail 7 miles north from the Skalkaho-Crooked Creek road. Burnt Fork lies in heavily-timbered, limestone terrain, covers about 30 acres, is dammed and has a water level fluctuation of about 20 feet in accordance to irrigation requirements. It is heavily fished and excellent for 7 to 12 inch Dolly Varden, plus a few small rainbow trout.

Butterfly Creek. A small stream about 3 miles long, flowing to Willow Creek 2 miles above the Butterfly Guard Station, and followed by a dirt road for about 1 mile. This stream was once good fishing for small cutthroat and rainbow but is now only fair due to logging operations. It was planted with rainbow in the 1950's.

LOWER CAMAS LAKE
Courtesy Ernst Peterson

CAMERON CREEK FALLS
Courtesy Ernst Peterson

Calf Creek. A very small tributary of Willow Creek, about 1½ miles below the Butterfly Guard Station. There is no trail, but it is fair fishing for about 1 mile above the mouth for small cutthroat and rainbow trout.

Camas Creek. A small, eastward flowing tributary of the Bitterroot River 8 miles south of Hamilton. The lower 3 miles are used for irrigation, the upper reaches are fair fishing for small cutthroat, rainbow and brook trout. There is a good trail along the north bank; reached via the Lost Horse County road for 3 miles to the Roaring Lion-Camas Creek road and on up for another 3½ miles to the trail take off point.

Camas Lakes. Upper and Lower, 14.8 and 7 acres by 54 and 17 feet maximum depth, a mile apart on the Camas Creek drainage, reached by pack trail 3½ miles above the end ou the Camas Creek road, and then up the creek on foot a long, last mile. Lower Camas Lake is overpopulated with 6 to 8 inch rainbow. Upper Camas is no better than fair fishing for rainbow-cutthroat hybrids that range from 6 inches to maybe 3 pounds. About 2/3s of the way up from Lower to Upper Camas are a couple of adjoined shallow (unnamed) ponds that freeze out each winter but are usually restocked annually by migrants from above and below.

Cameron Creek. An open, meadowland stream that flows for 7 miles to the East Fork of the Bitterroot River at Sula. It is fair-to-good fishing for 8 to 10 inch cutthroat, brook and Dolly Varden plus an occasional lunker rainbow trout. Some of it is on posted land, all is accessible by county and private roads.

Camp Creek. A small tributary of the East Fork of the Bitterroot River paralleled by U.S. 93 for 7 miles from Sula to Gallogly Hot Springs. It is excellent fishing for 8 to 10 inch cutthroat and brook trout plus an occasional Dolly Varden.

Canyon Creek. The outlet of Canyon Lake, it flows eastward for about 7 miles to the Bitterroot River at Hamilton. The lower 3 miles are dewatered for irrigation, the upper reaches are followed by a good trail that takes off from the end of a USFS road and are fair fishing for 6 to 8 inch cutthroat trout.

Canyon Lakes. Three lakes in high (elevation around 7300 feet) glaciated country. Only the middle lake contains fish. It is reached by a poor, steep, unmaintained trail 6 miles from the end of the Canyon Creek road (elevation 4000 feet); is about 45 acres, deep and dammed with as much as 20 feet of water level fluctuation. Nevertheless, it's good fishing for rainbow that range from 8 to 12 inches in length.

Capri Lake. Take the Warm Springs Creek trail for 11 miles to the mouth of Fault Creek; then 1½ miles by a Boy Scout trail and finally a quarter of a mile through undergrowth (no trail here) to the lake. Capri Lake is in heavily timbered country, 4.1 acres, mostly very shallow but deep at the lower end and has lots of lily pads around the margins. The spawning grounds are inadequate here but there are a few good cutthroat from an old ('67) plant and is fair fishing for 8 to 10 inchers.

Carlton Creek. Crossed near the mouth by U.S. 93, three miles north of Florence, no trail. The outlet of Carlton Lakes, it flows for 5 miles down a steep, timbered, brushy canyon and then for 3 miles along the Bitterroot valley where it is dewatered for irrigation. The lower reaches have been planted with rainbow and cutthroat trout which have ascended as far as the falls, about 3 miles above the mouth of the canyon.

Carlton Lake. You can drive right to this one with a 4 wheel drive. Take U.S. 93 six miles south from Lolo, then turn right (west) up the rough, crooked, switchback Carlton Lake road 7 steep miles to the lake ½ mile east of Lolo Peak at 7700 feet above sea level. It's 40 acres by 30 feet deep in a rocky, sparsely timbered glacial cirque. It was stocked with 3 inch rainbow in the summer of 1967 and they're fat and frisky now.

Cedar Creek. A very small and steep tributary of the South Fork of Lolo Creek about 3 miles upstream. There is no trail but the lower ½ mile is fair fishing for small (6 to 8 inch) cutthroat.

Chaffin Creek. Crossed near the mouth by U.S. 93, four miles south of Darby and followed by gravel road and trail for 10 miles to Chaffin Lake, west of Sugarloaf Peak. A clear mountain stream flowing through timbered, brushy country. It is fair fishing for 6 inch cutthroat.

Chaffin Lake. The middle (and largest, 26 acres) of 3 glacial lakes beneath Sugarloaf Peak, reached by a good trail 6 miles upstream from the end of the Chaffin Creek road 4 miles above U.S. 93. This lake is dammed at the outlet for irrigation,

CHAFFIN CREEK LAKES
Courtesy Ernst Peterson

CARLTON LAKE
Courtesy USFS

with a resultant 10 feet of water level fluctuation. It is deep in the center, but with shallow dropoffs, and used to be fished heavily for emaciated rainbow in the 7 to 10 inch class.

Charity Lake. See Spud Lake.

Chicken Creek.
A small, clear stream flowing eastward for about 8 miles from the crest of Razorback Mountain to the West Fork of the Bitterroot River, crossed at the mouth by the West Fork highway 2½ miles above Alta and followed by a poor hunters' trail for a few miles upstream. There are lots of little cutthroat here, about the right size for camp fare.

Chrandal Creek.
A very small and steep tributary of Hughes Creek, reached at the mouth by a gravel road 3½ miles above Alta. It's a brushy stream but good fishing for 6 to 7 inch cutthroat and rainbow trout.

Christisen Creek.
Short, small and steep, Christisen Creek flows to the West Fork of the Bitterroot River 8 miles above Conner, and is fair fishing near the mouth for pan-size cutthroat and rainbow trout.

Claremont Creek.
A small tributary of the Burnt Fork, reached at the mouth by road 8 miles east from Stevensville and followed for about a mile upstream by a poor dirt road. It is somewhat polluted by overgrazing but is still fair fishing near the mouth for small rainbow and cutthroat trout.

Clifford Creek.
Take the East Fork road 16 miles from Sula to the guard station, then the Little East Fork road for 2 miles and finally 3½ miles of good trail to the creek's mouth and thence upstream for another couple of miles by trail. This small tributary of the East Fork flows through a heavily timbered, brushy valley and is good fishing for 6 to 8 inch cutthroat trout.

Coal Creek.
Take the West Fork highway from Conner past Painted Rocks Lake for 1 mile to Coal Creek. A small, rapid snow-fed stream in steep, timbered mountains, it is good fishing along the lower reaches for 8 inch rainbow and cutthroat.

Como Lake.
Reached by county road at the mouth of Rock Creek canyon 4 miles west of U.S. 93 and 15 miles southwest of Hamilton. Como is a large (1000 acres) irrigation reservoir with about 50 feet of draw down and a maximum depth of 100 feet at the lowest water level. It lies in a beautiful glaciated canyon and is flanked to the south and east by remnants of great lateral and terminal moraines. It was poisoned in 1960, stocked with rainbow, and is fair fishing now)for 12 inch to 2 pound trout) for vacationers at the public campground north of the dam. A hikers' trail takes you along the north side to an unimproved campground above some falls at the head of the lake. This is undoubtedly one of the most beautiful lakes in western Montana when it is full, prior to the irrigation season, and the ugliest when it is drawn down during the height of the recreation season. It is also heavily used for water-skiing when there's water in it.

Crystal Lake.
A small (8 acres) glaciated lake reached by trail ½ mile above Boulder Lake in high, sparsely wooded talus slope country. It drops off steeply on one side to a maximum depth of about 30 feet, but slopes gradually down the other side from mossy, grassy, rubble-covered banks. It was planted with cutthroat in 1960 and is (reportedly) still good fishing now for 14 to 16 inchers.

COMO LAKE
Courtesy Ernst Peterson

Courtesy USFS

Big Creek. A clear, wadeable stream that drains Big Creek Lake and flows eastward for 16 miles to join the Bitterroot River about 5 miles south of Stevensville. The lower 3 miles are dewatered for irrigation, the upper reaches are followed by trail and are good fishing for 8 to 10 inch cutthroat, brook and Dolly Varden.

Daly Creek. Take the Skalkaho Highway 17 miles east from Hamilton to the mouth of Daly Creek at the Black Bear Guard Station near Camp Cornish, and thence 16 miles along the north side (and generally a few hundred feet above) the creek to its headwaters at Skalkaho Pass. It is a small, clear stream with many rapids, that flows through a deep, heavily timbered canyon and is good fishing for 8 to 10 inch rainbow and an occasional cutthroat trout and Dolly Varden.

Deer Creek. Flows eastward to the West Fork of the Bitterroot River across from the West Fork highway 2½ miles above Alta and is followed by a good trail for 14 miles to headwaters. A high, mountain stream in timbered country, it is good fishing for 7 to 9 inch rainbow and cutthroat in numerous beaver dams along its length.

Dick Creek. Followed by a gravel road from its junction with the South Fork of Lolo Creek 5 miles above that stream's mouth for 9 miles past headwaters. The lower reaches (for about ½ mile above the mouth) are fishable for small 6 inch cutthroat.

Divide Creek. Take the Sleeping Child Creek trail for 5 miles past the Hot Springs to the mouth of Divide Creek and then turn left upstream to headwaters. A small, clear mountain stream that rises at about 7000 feet and descends to less than 5400 feet in 7 miles, it contains fair numbers of pan-size cutthroat trout and is easily fished, a good kids' creek.

Dollar Lake. Take the Boulder Creek trail 12 miles to Turbid Lake, then climb down (way down) for ¼ mile to Dollar Lake, which is about 5 acres, in steep, rocky country with a few trees around the shore. It is only about 20 feet deep with gradual dropoffs, and, in view of its extreme elevation (9300 feet), it seems that it should freeze out but it provides the best fishing of the entire Boulder Lake group for 10 to 12 inch cutthroat. It is seldom fished — just too hard to get there.

BITTERROOT

Duffy Lake. In a high (elevation 7300 feet) cirque at the head of Sweeny Creek. Take a logging road for 6 miles up the north ridge from near the end of the county road at the mouth of Sweeney canyon, then 8 miles by foot trail past Peterson Lake to Duffy Lake. It is a small (18 acres) lake in timbered, steep, and rocky country just east of the crest of the Bitterroot Range; is mostly under 30 feet deep, mud bottomed, and the remains of an old dam are still present at its outlet. It contains a few 8 to 16 inch rainbow but is seldom fished.

East Fork Bitterroot River. Crossed by U.S. 93 at Sula and followed by a good gravel road for about 37 miles to headwaters. The upper reaches flow through conifer and brush-covered canyons, the lower reaches are in pastureland. The East Fork is a clear, easily fished stream that is fair in summer for 10 to 12 inch cutthroat, rainbow, and lesser numbers of brook, brown and Dolly Varden trout; and in winter months, for whitefish ranging to 15 inches. Permission to trespass is required along the lower reaches.

East Fork Camp Creek. Joins the West Fork near Gallogly Hot Springs 7 miles south of Sula on U.S. 93, and is paralleled by the highway, and switchbacked logging roads up through an old burn and then cross-country for 6 miles to headwaters. A small stream with lots of nice holes, it's good fishing for small cutthroat and brook trout.

East Fork Graves Creek. A short, small stream that flows into Graves Creek 2 miles above its mouth, and is good early-season fishing in the lower reaches for pansize cutthroat.

East Fork Lolo Creek. Followed by a gravel road for 6 miles above its mouth about a mile above Lolo Hot Springs on the Lewis and Clark Highway, and then by fair trail for 3 miles to headwaters. A clear, mountain stream that contains a few 8 to 10 inch cutthroat and Dolly Varden. It is generally only poor-to-fair fishing.

AST FORK OF THE BITTERROOT
ourtesy USFS

EAST FORK OF THE BITTERROOT
Courtesy Ernst Peterson

BITTERROOT

DOLLAR LAKE
Courtesy Ernst Peterson

A STUDY IN REFLECTIONS
Courtesy USFS

East Fork Piquett Creek. Take the West Fork highway for 8 miles above Conner, then turn south across the West Fork for 1½ miles to the mouth of this stream, and thence on upstream on a logging road for another 3 miles and finally fight your way up the bottom for another and final 4 miles to headwaters. A clear, snow-fed stream in timber and brush, the lower reaches contain a good population of small rainbow and cutthroat trout.

Eastman Creek. Is crossed at the mouth and followed upstream for 1½ miles by the Butterfly Creek road (1 mile above the Butterfly Guard Station, 8 miles east of Corvallis). It is a very small creek but fair fishing for about a mile above the mouth for pan-size rainbow and cutthroat trout that have migrated from Willow Creek.

Eight Mile Creek. Take the East Valley road from Florence for 2 miles across the Bitterroot River, then turn north and east again on the Eight Mile road for 12 miles to headwaters. This clear little mountain stream flows through timber and brush and is impounded by numerous beaver dams. The lower reaches go dry due to irrigation. The ponds are good early-season fishing for small cutthroat.

El Capitan Lake. Fifteen acres, deep, around 6500 feet elevation in a deeply glaciated but heavily timbered mountain region, El Capitan is reached by the Rock Creek trail 8 miles above Lake Como, then a hunters' trail for 5 miles up El Capitan Creek to the shore. It used to be barren but has plenty of feed, was stocked with golden trout in 1963 but few people (if anybody) have been there since to check them out. It's good elk country too, there's lots of grass for the horses, and a beautiful campsite.

Elk Lake. Reached by a USFS trail up Rock Creek 12 miles above Como Lake, Elk Lake occupies part of a steep walled, glaciated canyon in heavily wooded country. It is very deep, covers about 20 acres, and has a moderate amount of aquatic vegetation around the shore. Because it has no spawning grounds, it must be restocked periodically but is fair fishing at present for 1 to 2 pound cutthroat.vc

Faith Lake. See Legend Lake.

Fish Lake.
A high (elevation 7500 feet) mountain lake in heavily timbered country reached by a good USFS trail 3 miles from the Martin Creek logging road. This is a small (4 acres) shallow lake with boggy shores that is good fishing for 6 to 10 inch cutthroat.

Fish Lake.
A high lake (about 6500 feet elevation), near the crest of the Bitterroot Range, which is reached either by a good horse trail 12 miles up the South Fork from the Lost Horse Creek road 6 miles west of U.S. 93, or by an equally good trail 6 miles from the end of the Lost Horse road over Bear Lake Pass. It's a fair size lake (about 50 acres), deep at the upper end and shallow below in heavily timbered mountains. The fishing is fair for 6 to 12 inch rainbow.

Flat Creek.
A very small, clear, mountain stream reached at the mouth by the Burnt Fork road 16 miles east from Stevensville, and followed by trail for 2 miles to the forks. It is good fishing for small cutthroat and rainbow trout migrating from the Burnt Fork.

Fool Hen Lake.
See Willow Lake.

Fred Burr Creek.
Go 3 miles south of Victor on U.S. 93, turn west on a county road for 5 miles to the mouth of Fred Burr canyon and continue (if you can get permission) up a private access road for 8 miles to the Fred Burr Reservoir, and then for another 10 miles by trail to the head of the creek at Fred Burr Lake. A rapidly flowing stream in a timber and brush covered canyon bounded by sheer rock walls that in some prices rise almost vertically 200 feet above the creek, it is excellent fishing for 7 to 9 inch cutthroat, and a few rainbow and brook trout.

Gash Creek.
The lower 3 miles of this creek flow through open pastureland, the upper 6 miles through a heavily wooded, glacial canyon. It is reached by county road 5 miles southwest from Victor and followed by a USFS road for 4½ miles and finally by trail for 3 miles upstream to headwaters. This is a small, steep, snow-and-rain-fed stream that was planted years ago and is reported to be fair fishing along the lower reachefs for small cutthroat and rainbow trout.

HEAD OF THE FRED BURR DRAINAGE
Courtesy USFS

FRED BURR LAKE (poisoned out)
Courtesy USFS

Gird Creek.
About 16 miles long, the lower 2 miles are dewatered for irrigation, the upper 13 miles are followed by road through timber and park land of the Sapphire Mountains east of Hamilton. A small stream that is generally only poor fishing, it contains a few 10 to 20 inch rainbow, brook and brown trout.

Gleason Lake (or Jones Reservoir).
Take a good county road for 15 miles from the mouth of Willow Creek canyon to within a mile (but what a mile) of the lakeshore, one mile west of Palisade Mountain. Gleason is a small (16 acre) lake, mostly less than 30 feet deep with a 16 foot irrigation dam and as much as 15 feet of water level fluctuation. The bottom is mud, there's lots of aquatic vegetation and the fishing used to be good for nice fat rainbow but they've been done-in by hordes of shiners.

Goat Lake.
A cirque lake, about 5 acres in timber and park country a few miles south of Como Peaks. It is reached by 2½ miles of foot trail from the Tin Cup Creek trail about 6 miles above the canyon mouth. This is a fairly shallow lake with boggy margins and the remains of an abandoned dam at the outlet. It was planted with cutthroat years ago and is excellent fishing now for small (6 to 8 inch) trout.

Gold Creek.
Take the Burnt Fork road 14 miles east from Stevensville to the mouth of this stream, then a good trail for 7 miles to its headwaters on Willow Mountain. It's in good game country and is fair fishing for camp-size rainbow and cutthroat trout. Planted.

Granite Creek.
Take the Lewis and Clark Highway for 1½ miles above Lolo Hot Springs to the creek's mouth, and then a logging road and trail upstream 8 miles to headwaters. Numerous beaver ponds in the lower reaches provide fair fishing for small, pan-size cutthroat.

Graves Creek.
From Lolo take the Lewis and Clark Highway for 16 miles to the Petty Creek road, then 3 miles up Graves Creek and finally a USFS trail for another 3 miles to headwaters. A slow flowing, marshy stream in heavily timbered mountains, it contains a few 6 to 8 inch cutthroat, brook and Dolly Varden but is not much of a fishing stream.

Haccke Creek.
A small tributary of the Burnt Fork, it is reached at the mouth by road 10 miles east from Stevensville. There is fair fishing for a few miles upstream for small rainbow and a few cutthroat trout.

Hackett's Pond.
A 1½ acre, 15 foot deep, PRIVATE pickup-water irrigation reservoir on the Willoughby Creek Drainage. Take the East Side highway 4 miles south from Stevensville, and then the Willoughby Creek road 2½ miles east to the Joe Hackett place. The pond has been planted periodically, with brook trout, rainbow trout, and large mouth bass. Its ongoing status as a fishery is strictly a function of the owner's wishes but this I can say. The water here is about as fertile as any I've seen and whatever is in it — is no doubt doing exceedingly well.

Hauf Lake.
A small (12 acre) lake situated high (elevation 7100 feet) above the deeply glaciated canyon of Mill Creek in timbered, rocky country. The lake is reached by a very poor foot trail 1 mile south and 2200 feet up from the bottom of the canyon about 5 miles above its mouth. It's fairly deep, dammed at the outlet and has as much as 18 feet of water level fluctuation in accordance to irrigation demand. It has been stocked a number of times during the past several years and is now fair fishing for cutthroat that reportedly average about 1 pound but are seldom shown a hook. This is good mountain goat country. Note: The dam is now being raised — so the fishing potential will no doubt be lowered.

HIDDEN LAKE
Courtesy Ernst Peterson

Hayes Creek. A very small stream flowing eastward for 6 miles from the flanks of the Bitterroot Mountains to the Bitterroot River near Charlos Heights. The lower reaches, which are mostly dewatered for irrigation, are followed by county roads and heavily fished by children for small cutthroat and brook trout.

Hayes Creek. Crossed near the mouth by U.S. 93, six miles south of Missoula and paralleled by a poor gravel road to headwaters, this small stream provides some fair kid's fishing in the middle reaches for small cutthroat trout.

Hidden Lake. A small (14 acre) lake about ½ mile southwest of West Pintlar Peak in heavily-timbered country reached by the Little East Fork road for 2 miles and then a good USFS trail up the East Fork of the Bitterroot 10 miles above the guard station. It's quite deep with lots of boulders around the margins, no spawning beds and contains a few large rainbow, remnants of an early plant.

Hidden Lake. In fact as well as name, it's 10 acres at about 6900 feet, and occupies a small cirque on the south side of and 2200 feet above Big Creek Canyon, 6 miles above its mouth. To reach it, follow the Big Creek-Sweathouse Divide cross-country for 2 miles northwest from barren Glen Lake (you can get to within about 4 miles of here by a logging road 11 miles from the Big Creek road, and then there's a trail) to a vantage point about 700 feet above, and a steep scramble down to the shore. It's seldom fished, and then only by a stray angler (in the know) for a basketfull of half-pound rainbow when you hit it right. At other times you can go away skunked.

High Lake. Thirty nine acres, 85 feet maximum depth at low water, a 25 foot dam at the outlet and consequent 20 foot water level fluctuation, elevation 7100 feet. Take the Blodgett Creek trail 4 miles above the mouth of the canyon, then a poor goat trail south up the rocky canyon wall for 1400 feet of elevation and 2 miles travel distance to the lake. It was stocked with rainbow years ago that now average about 1 pound and provide good fishing for the few who will go and get 'em.

Holloman Creek. About 3 miles of fishing water in this small tributary of Miller Creek. It is crossed at the mouth by the Miller Creek road and reportedly produced fair catches of small cutthroat in the past, but is on private, posted land.

Hope Lake. Ten acres, maximum depth 50 feet, with steep dropoffs all around, some large boulders, and a few lilies along the edges, elevation 7500 feet. Hope Lake is reached via 8 miles of fair USFS trail from the campground at Mussigbrod Lake (see in Beaverhead drainage in *Montanans' Fishing Guide, East*) to the Continental Divide, and then 2½ miles east along the Divide and finally another mile down to the shore. Note well; this last mile is closed to all but foot travel. The surrounding country is very steep and rocky, but heavily wooded with conifers growing right down to the water. The fisherman's report, cited here, was to the effect that the fishing is good now for 18 to 20 inchers and he really enjoyed the solitude.

Howard Creek. A small stream flowing to Lolo Creek opposite the Lolo Ranger Station and followed by a USFS dirt road for 6 miles to headwaters. The first mile is poor-to-fair fishing for 6 to 8 inch brook, cutthroat and a few Dolly Varden trout.

Howard's Lake (or Mary's Frog Pong). Two acres, maximum depth 35 feet, immediate shore boggy and marshy, set in heavily timbered mountains. Take the Dick Creek road 9 miles from its junction with the Lewis and Clark highway, then a poor, unmarked trail ¼ mile west to the lake. Howard's was 'discovered' and stocked in 1960 (it is not shown on most maps). It now provides good fishing for 10 to 18 inch Rainbow — lots of mosquitoes and a few cutthroat, but is too small to stand much fishing pressure — and too hard to find to get much.

Hughes Creek. Take the West Fork of the Bitterroot River road for 4 miles above Painted Rocks Lake to the mouth of Hughes Creek, then a poor gravel road for 12 miles upstream through steep, timbered mountains, and finally a USFS trail for 6 miles past the headwaters. A rapidly flowing, snow-fed creek, it's easily fished for good catches of 8 to 10 inch cutthroat, some brook and dolly varden in summer, plus whitefish in winter (when accessible).

Jack The Ripper Creek. Take a good trail from the end of the Blue Joint Creek road for 6 miles to the mouth of Jack the Ripper, and thence 3 miles to headwaters. This is a real pretty little snow-fed stream with good fishing in its lower reaches for 8 inch cutthroat and some rainbow trout.

Johnson Creek. From Painted Rocks Lake take the West Fork road for 12 miles to the mouth of Johnson Creek, and then a foot trail upstream for about 2 miles or 1/3 of the way to headwaters. Johnson Creek is a very small, clear stream that is good fishing for about 1½ miles above the mouth for small cutthroat and rainbow.

Jones Reservoir. See Gleason Lake.

Kenck Lake. See Sears Lake.

Kerlee Lake. Reached by the Tin Cup trail 5 miles above the end of the USFS road in the mouth of Tin Cup canyon, and then a poor foot trail due north for 1 mile to the shore in a small cirque on the southern, alpine slopes of Como Peaks. A small lake (37 acres) and quite shallow although it doesn't freeze out, it is excellent fishing for 6 to 8 inch cutthroat trout.

Kidney Lake. Go to Lower Camas Lake, hoof it around to the south side and hike? — climb? up the little inlet 440 yards south and 500 feet up to Kidney Lake which is almost two lakes, 12.2 acres by 16 feet maximum depth and fair fishing for small rainbow.

KOOTENAI FALLS
Courtesy USFS

KERLEE LAKE
Courtesy Ernst Peterson

Kootenai Creek.
The clear, snow and lake-fed outlet of the 3 Kootenai lakes flows due east through a deep, rock-walled, timber and brush covered glacial canyon for 14 miles to the Bitterroot Valley, and thence for 3 miles across farmland to the river. It is crossed near the mouth by U.S. 93 just west of Stevensville and is followed by a good trail from the mouth of the canyon to headwaters. The lower reaches (below the canyon) are completely dewatered for irrigation. Within the canyon it is a beautiful stream with many holes and riffles containing a fair number of 6 to 8 inch cutthroat, and some brook (from South Kootenai Lake) and rainbow trout.

Kootenai Lakes.
See Middle, North, South.

Lappi Lake.
A 10 acre lake with maximum depth over 40 feet, elevation 6900 feet. Lappi occupies a small, subalpine cirque on the south side and 1100 feet above Bass Creek canyon. It is reached by way of the Bass Creek trail for 5 miles above the end of a USFS road at the Charles Waters Memorial campground at the mouth of the canyon and then a very poor foot trail around ledges and over rocks up the steep, cliff-like side of the canyon for 1 mile to the lake shore. It was planted 20 years ago and used to be good fishing but is now frozen out.

Larry Creek.
A small, steep stream about 4 miles long and completely dewatered for irrigation at the lower end. It's crossed by a county road 3 miles south of Florence and produces a few small rainbow and cutthroat for the local kids.

Lavene Creek.
A very small, (about 3 miles long) tributary of the West Fork of the Bitterroot River, crossed at the mouth by the West Fork highway about 8 miles above Conner, no trail but a logging road 4.3 miles long from the West Fork road gets you into the head of it except there's no fishing here. There *is* some fair fishing in the lower reaches near the mouth though for small cutthroat and rainbow trout.

Legend (or Faith) Lake.
A real small lake, only a couple of acres, deep but boggy along the east side, in heavily-timbered country at the head of Moose Creek. Reached by good trail 5 miles from the Martin Creek logging road. This lake was planted with cutthroat years ago, and is fished a lot for 6 to 12 inchers. A popular spot.

Little Blue Joint Creek. A clear, snow-fed tributary of Blue Joint Creek paralleled by a USFS road taking off from the Blue Joint road along a rocky, timbered canyon for 4.5 miles and then a trail for 2 miles to headwaters. This small creek is good fishing for a couple of miles above the mouth for 8 inch cutthroat and rainbow trout.

Little Boulder Creek. A clear, dashing inlet of Painted Rocks Lake, crossed at the mouth by the West Fork highway and followed along the north side by a USFS logging road for 1½ miles and then a trail for 3½ miles to headwaters. The lower reaches are fair fishing for small (pan-size) cutthroat and rainbow trout.

Little Burnt Fork Creek. A wee small tributary of the Burnt Fork, crossed at the mouth by the Burnt Fork trail about 2½ miles below the lake. It's not much of a stream but does provide fair fishing along the lower reaches for "camp fare" cutthroat.

Little Rock Creek. Drains 3 unnamed cirque lakes between El Capitan and Como peaks, for 2 miles to Little Rock Creek Lake and thence 4½ miles to Como Lake. Take the Como Lake road to a point one mile below the dam, then cross Rock Creek (to the south) which is the outlet of Como Lake, and proceed to the southeast corner of the dam and follow the Bunkhouse logging road (which is closed to motor traffic) and trail for 11 miles upstream to Little Rock Creek Lake. A pretty little stream and good fishing for 10 to 12 inch cutthroat that migrate up from the lake.

Little Rock Creek Lake. Reached as described above. It's about 20 acres, shallow (although it doesn't freeze out) boggy around the margins and contains lots of feed. It was planted years ago with cutthroat. The fish are all about 10 to 12 inches long now and provide excellent fishing for the very few stray hunters (it's almost never hit for fishing alone) who find their way into the lake. This is real good elk country.

Little West Fork Creek. A dashing, snow-fed tributary of the West Fork of the Bitterroot River, crossed at the mouth by the West Fork highway 4½ miles above the West Fork Ranger Station. The lower reaches are excellent fishing for small cutthroat and rainbow. There's no trail.

LOOKING FOR A SPOT TO "SET"
Courtesy Ernst Peterson

LOST HORSE CREEK IN WINTER
Courtesy Ernst Peterson

BAILEY LAKE
Courtesy USFS

Lockwood Lake.
A 10 acre, 15 to 20 foot deep, cirque lake at the headwaters of Mill Creek, reached by a fair trail from Mill Creek Lake. It's barren now but has been suggested as a likely golden trout habitat.

Lolo Creek.
A good size mountain stream flowing for 30 miles from Lolo Pass to the Bitterroot River at the town of Lolo and followed the whole way by the Lewis and Clark highway. The upper reaches are timbered, the lower reaches are pasture and farmland. An easily and heavily fished stream, it is planted and produces good summer catches of 8 to 14 inch rainbow, cutthroat and brook, plus a few Dolly Varden and brown trout, and is a popular creek for whitefish in the winter.

Lost Horse Creek.
The outlet of Twin Lakes, and Ten and Twelve Mile Lakes, Lost Horse flows for 20 miles down a deeply glaciated, brush and timber covered canyon to the Bitterroot River 10 miles south of Hamilton. It's crossed by U.S. 93 near the mouth and followed by a gravel road to headwaters and is heavily fished for fair-to-good catches of 7 to 9 inch cutthroat, and a few brook and Dolly Varden trout.

Lost Horse Lake.
Sixty seven acres, deep, with boggy shores in timbered mountains at the headwaters of the South Fork of Lost Horse Creek. Lost Horse Lake is reached by a 'hunters' trail 3 miles southeast from Fish Lake. It was planted years ago with cutthroat which are now about 12 inches and provide excellent fishing for the few anglers who make the trip.

Lyman Creek.
A very small tributary of Cameron Creek in timbered mountains with no trail. Lyman Creek is seldom fished but fair here and there for pan-size cutthroat and brook trout.

Martin Creek.
Flows through heavily timbered mountains for 15 miles to the East Fork of the Bitterroot River where it is crossed by the East Fork road ½ mile above the Guard Station. It is followed by a logging road upstream for about 3 miles and from here-on-in cross-country for 5 miles to an upper logging road that takes you to headwaters. There are about 11 miles of fishable water that produces good catches of 6 to 10 inch cutthroat.

Mary's Frog Pond.
See Howard's Lake.

McClain Creek.
A very small, steep stream flowing to the Bitterroot River 5 miles south of Lclo. It's fair fishing for 2½ miles above the mouth (to the forks) for 6 to 8 inch rainbow and cutthroat, a kids' creek.

Meadow Creek. Take a good gravel road for 7 miles south from the East Fork of the Bitterroot road (13 miles above Sula), then a trail for 3 miles to headwaters. A brushy, beaver-dammed stream in timbered mountains, it's seldom fished but does contain some small cutthroat and Dolly Varden.

Meadow Creek. Only a couple of miles long and not really a fishing stream although it does contain a few small cutthroat for ½ mile above its junction with the South Fork of Lolo Creek, where it's crossed by the South Fork trail.

Middle Kootenai Lake. Fifteen acres, mostly 15 to 30 feet deep, in heavily timbered mountains at the head of Kootenai Creek, reached by 12 miles of good horse trail from the end of the county road at the mouth of the canyon. This lake was planted with golden trout in the late '50s but only a few have survived — poor fishing.

Mill Creek. Take the Lewis and Clark highway for 8 miles above Lolo to the mouth of Mill Creek, then a logging road and trail upstream for about 2½ miles or as far as you can fish for fair catches of 6 to 8 inch cutthroat and brook trout. A small stream but fished quite a bit.

Mill Creek. The outlet of Mill, Lockwood, Sears, and Hauf Lakes, Mill Creek flows through a steep, rocky canyon in heavy timber and brush for 15 miles and thence for another 4 miles through pasture and farmland to the Bitterroot River near Woodside. The mouth of the canyon is reached by county road, and the stream is followed by road and trail to headwaters. Its upper reaches are fair fishing for small rainbow, the lower reaches for 10 to 12 inch brook, browns and (unfortunately) for trash fish.

Miller Creek. A small stream about 14 miles long, it flows to the Bitterroot River 5 miles south of Missoula. The lower reaches are completely dewatered for irrigation; the upper reaches flow through timbered mountains and pastureland where they were heavily fished for sunfish, largemouth bass, cutthroat and brook trout but are now closed to fishing.

TRAIL ABOVE MAUDE LAKE (Barren)
Courtesy USFS

CAMPING ON LOST HORSE LAKE
Courtesy Ernst Peterson

MIDDLE KOOTENAI LAKE
Courtesy USFS

NORTH KOOTENAI LAKE
Courtesy USFS

Miller Creek Gravel Pit.
Take the Lower Miller Creek road 1 mile south from U.S. 93, climb over a fence and take a short hike across a private (unposted) field to the pond. It is fished by a few kids and fewer adults for largemouth bass, pumpkinseed and rough fish that come in during high water from the Bitterroot River.

Mine Creek.
A small stream, about 5 miles long, that flows to Hughes Creek on the opposite side from the road 8 miles upstream from Alto. It's followed by a logging trail for 3.2 miles above the mouth and is generally fair-to-good fishing for 8 to 10 inch cutthroat and rainbow trout and excellent fishing in some large beaver ponds a mile above the mouth.

Moose Creek.
A high, mountain stream, the outlet of Shadow, Legend, Spud and Fish Lakes, it flows for 9 miles through heavy timber to the East Fork of the Bitterroot River at the East Fork Guard Station and is followed by a good USFS trail to headwaters. A few people fish it for 6 to 10 inch cutthroat, and some Dolly Varden.

Mud Creek.
A real small tributary of the West Fork of the Bitterroot River, crossed at the mouth by the West Fork highway a mile below Painted Rocks Lake. There is no trail here but the creek is clear, fast and fair fishing near the mouth for pan-size rainbow and cutthroat trout.

Nelson Lake.
Thirth five acres, maximum depth about 30 feet, dammed — with 10 foot fluctuations of water level. Nelson Lake occupies a barren alpine cirque 2 miles southeast of Bare Peak. It is almost reached by trail 4 miles from the end of the Gemmel Creek road, (takes off from the north side of the Nez Perce highway, 4 miles above the West Fork Ranger Station) then one mile down (steep down) to the lake. It used to be good fishing and was very popular for skinny cutthroat trout, but they don't reproduce well here and now the fishing is only fair to 8 to 12 inches.

Nez Perce Fork Granite Creek.
A large, clear tributary of the West Fork of the Bitterroot River followed by a good road for 20 miles from mouth to headwaters. An easily fished stream with many beaver dams in the upper reaches, it contains mostly 8 to 11 inch cutthroat, and a few brook, Dolly Varden and browns.

North Fork Granite Creek.
A very small stream crossed at the mouth by the Granite Creek logging road and followed by a jeep road for about ½ mile upstream past occasional pools that contain a few 6 to 8 inch cutthroat.

North Fork Lost Horse Creek. A very small stream flowing southeast to Lost Horse Creek where it's crossed by the Lost Horse road, and is followed upstream by a poor trail to headwaters. It's not much of a creek, but is fair "pan fish" fishing for ½ mile above the mouth.

North Fork Sweeney Creek. About 5 miles long, the outlet of Mills, Holloway, Duffy and Peterson Lakes, the lower couple of miles are good fishing for 7 to 9 inch cutthroat and rainbow trout.

North Kootenai Lake. About 15 acres, and deep, in a rock-bound cirque at the headwaters of Kootenai Creek, reached by trail 12 miles up from the mouth of the canyon. This is a beautiful lake and fair fishing (when they bite) for beautiful silvery rainbow that'll run from 1 to 4 pounds.

O'Brien Creek. A small stream flowing for 12 miles through timber above and hayfields below to the Bitterroot River just across the bridge from Target Range (West Missoula). It's followed upstream by a gravel road for about 7 miles. The lower reaches are generally dewatered for irrigation but there is a limited amount of fair pan-size cutthroat and rainbow fishing in the middle and upper reaches. A good creek for kids.

One Horse Creek. Flows for 7 miles from One Horse Lakes through a brush floored, rock walled canyon to the Bitterroot Valley and thence for 3 miles to the river near Florence. It used to be accessible all the way but it's now washed out. The lower reaches are mostly dewatered for irrigation; the upper reaches are fair fishing for small cutthroat.

One Horse Lakes. Three of them, but only the southernmost contains fish. It is about 15 acres with a mud bottom and a moderate amount of aquatic vegetation around the shore and is not more than 20 to 25 feet deep anywhere. It is reached by trail about 2 miles south from Carlton Lakes. The fishing is excellent for 8 to 10 inch rainbow trout.

Overwhich Creek. This one is better known for the big game hunting in the area than for its fishing. It flows through conifer covered mountains for 20 miles to the West Fork of the Bitterroot River 1 mile above Painted Rocks Lake, and is followed by road and trail to headwaters. The fishing is excellent for small cutthroat from the mouth to some high (about 75 feet) falls 14 miles upstream.

Painted Rocks Lake (or West Fork Bitterroot Reservoir). About 4 miles long by ½ mile wide, and mostly good and deep, in a heavily wooded canyon followed clear around on the south side by the West Fork highway. It was formerly poor fishing due to excessive drawdown and no doubt still is after being drained almost dry during the summer and fall of 1973, for a few hardy rainbow, cutthroat, Dolly Varden, whitefish and most everything else. There is a good campground at the upper end and despite the scarcity of fish its a very pretty reservoir when full, which (unfortunately is mostly before the recreation season. Also its a nice drive and there are lots of good creeks and small mountain lakes in the vicinity.

Pearl Lake. Take the Big Creek trail for 9 miles from the mouth of the canyon, then the South Fork trail for 6 miles, and lastly an unmaintained elk and mountain goat trail for ½ mile to the lake shore in a beautiful alpine setting. It's about 10 acres, fairly deep, and used to be fair cutthroat fishing but is reportedly barren now.

EARL LAKE
Courtesy USFS

WATER SKIING ON PAINTED ROCKS
Courtesy Ernst Peterson

Peterson Lake. Take a logging road from the north side of the Sweeney Creek canyon for 6 miles of switchbacks, then 6 miles of mostly good trail, to the lake shore. Peterson covers about 30 acres, is mostly shallow (up to about 20 feet deep) with a muddy bottom and is overpopulated with 6 to 11 inch rainbow (mostly 6) that are reportedly caught 2 at a time. This one ought to be hit harder so some of these little ones could grow up.

Pierce Creek. A very small stream flowing to the West Fork of the Bitterroot River about 5 miles above Conner. It supports a few small cutthroat and rainbow in the first mile above the mouth and is fair early-season fishing.

Piquett Creek. Take the West Fork highway for 7 miles above Conner, and then a logging road for 5 miles and finally a trail for 10 miles upstream to headwaters. A clear, easily fished mountain creek in heavily timbered country, it's good for 6 to 8 inch rainbow and cutthroat trout.

Piquett Lake. Six and half acres, about 20 feet deep with steep dropoffs all around, in a sparsely timbered alpine cirque with barren talus slopes above, reached by the Little Boulder Creek trail 6 miles from the lower end of Painted Rocks Lake. It's very spotty (but sometimes fair-to-good) fishing for small cutthroat.

Porcupine Creek. Take the Warm Springs Creek trail for 7½ miles to the mouth of Porcupine Creek, which is a very small mountain stream with no trail but fair kids' fishing for 5 to 7 inch cutthroat trout.

Railroad Creek. From the Black Bear Guard Station (on the Skalkaho highway) drive wouth for 2 miles to the mouth of this creek, which is followed for 1½ miles by a logging road and then by trail to headwaters. A small stream, the lower reaches of which provide poor fishing for small (6 to 7 inch) cutthroat.

Reimel Creek. This little stream flows to the East Fork of the Bitterroot River a couple of miles above Sula, and is followed by a good (but closed to motorized traffic) gravel road and trail for 9 miles to headwaters. The lower 2 miles are poor fishing for small cutthroat trout.

Ripple Lake. A shallow, 7½ acre lake, with muskeg margins in heavy timber just west of Pintlar Peak in the Anaconda-Pintlar Wilderness Area; reached by 2 miles of 'Little East Fork' road and then a good trail 8 miles up from the East Fork of the Bitterroot Guard Station. It originally contained only cutthroat but was planted with rainbow several years ago. It's now slow fishing for 8 to 12 inchers and a few that'l maybe hit four or five pounds.

Roaring Lion Creek. From the mouth of Roaring Lion Canyon (reached by county roads 5 miles southwest of Hamilton) continue for 3½ miles on a USFS access road and then by trail for 13 miles to headwaters. The lower reaches (below the canyon) are completely dewatered for irrigation but the upper reaches are excellent fishing for eat'n size cutthroat. A good place to take a beginner.

Rock Creek. Crossed at the mouth by U.S. 93 four miles north of Darby and followed by a county road for 3.2 miles to Como Lake thence by 2 miles of USFS logging road and finally by trail for 9 more miles to Elk Lake for 6 to 10 inch cutthroat, and below Como Lake in early season for 8 to 10 inch rainbow and brook trout. The lower reaches are completely dewatered for irrigation.

Rombo Creek. A very small tributary of the West Fork of the Bitterroot River, crossed at the mouth by the West Fork highway 4 miles above the ranger station. It is no better than poor fishing in the lower reaches for small rainbow and cutthroat trout.

Rye Creek. About 16 miles long, flowing through steep, timbered mountains to the Bitterroot River 3 miles north of Conner, is crossed at the mouth by the east side road and is followed by county road and a USFS trail to headwaters. It doesn't run much water and is only poor-to-fair fishing for pan-size cutthroat.

Sage Hen Lake. See Willow Lake.

Salt Creek. A very small tributary of the West Fork of the Bitterroot River; reached at the mouth by the West Fork highway 5½ miles above the Alta Guard Station. Salt Creek is fair fishing for the first couple of miles above the mouth for 7 to 8 inch cutthroat and rainbow trout.

Sawmill Creek. This small tributary of the Burnt Fork is reached at the mouth by the Burnt Fork road and is fair fishing for small (pan-size) rainbow and cutthroat — in the lower reaches only. There's about a mile of fishable water accessible by a Forest Service trail.

Sawtooth Creek. Crossed near its mouth by U.S. 93 a mile and half south of Hamilton. The lower reaches are all fouled up for irrigation, but the upper reaches flow for 12 miles down a beautiful, glaciated canyon and are followed the entire distance by a good USFS trail but now hear this (to get here you've gotta side-hill it around from Roaring Lion Creek because of restricted access). It's worth the trip though just for the scenery, but the fishing is fair too, for small (6 to 7 inch) cutthroat and rainbow.

Sears (or Kenck) Lake. About 20 acres, 20 feet maximum depth, with considerable aquatic growth around the upper end. It was stocked years ago with rainbow, and more recently with cutthroat. However, there is very little inflow during the summer months and it is drawn so very low for irrigation that all have apparently perished. The lake is reached by a very poor foot trail 1½ miles from the main Mill Creek trail at a point about 8 miles above the mouth of the canyon.

ROCK CREEK
Courtesy Ernst Peterson

Courtesy Danny On

Sharott Creek. A small stream crossed at the mouth by U.S. 93 a mile west of Stevensville and followed by road and an unmaintained foot trail for 5 miles to headwaters below St. Mary's Peak. There is fair fishing here for 6 to 7 inch rainbow and cutthroat in the lower reaches only, as it's very steep and rocky above.

Sheafman Creek. This stream is crossed at the mouth by U.S. 93, four and a half miles south of Victor, and is followed by a good road and then trail for 7 miles to Sheafman Lake. The upper reaches flow through a deeply glaciated, timbered canyon; the lower reaches (within the Bitterroot Valley) are mostly dewatered for irrigation. It sometimes goes dry the whole length except for some good pools in the upper reaches wherein a few 8 to 10 inch cutthroat manage to survive.

Sheafman Lakes. Two lakes about ½ mile apart, but only the lower one has fish. It is reached by a poor, steep trail 7 miles from the end of the county road at the mouth of Sheafman Canyon in beautiful scenic country; is about 6 acres, 15 feet deep, and is held up by an old dam that hasn't been used in years. There's not much inflow or outflow during the summer and it's fair fishing for 8 to 12 inch cutthroat.

Sheephead Creek. Sheephead Creek is crossed at its mouth by the Nez Perce highway 11 miles above the West Fork Ranger Station and is followed along the north side by trail for 6 miles to headwaters. It's an easily fished (good for kids) stream for 8 to 9 inch rainbow and cutthroat trout.

Shelf Lake. A high (elevation 7600 feet) alpine lake, about 10½ acres with a maximum depth of 20 feet, in a sparsely timbered talus-slope locale, reached by a steep climb ¼ mile down the talus slope from Piquett Lake. The fishing is fair-to-good for 10 to 14 inch cutthroat.

Signal Creek. Take the Burnt Fork trail to the mouth of this creek in Boulder Basin, thence for 4½ miles to headwaters. The lower 2 miles above the mouth are fair fishing for camp-size cutthroat and rainbow. The upper reaches are too steep for fish.

BITTERROOT

Silverthorn Creek. A wee small tributary of the Bitterroot River, crossed near its mouth by U.S. 93, two miles south of Stevensville. This one is fair kids' fishing in the lower reaches for small cutthroat and rainbow.

Skalkaho Creek. An easily accessible, easily fished stream closely followed by the Skalkaho highway for 14 miles from its headwaters near Camp Cornish to its mouth 3 miles south of Hamilton. The lower 5 miles are almost completely dewatered for irrigation, the upper reaches are stocked annually with rainbow and provide good fishing for 8 to 10 inchers, plus some Dolly Varden and brook, and a few big browns.

Slate Creek. A small, wadeable inlet of Painted Rocks Lake, crossed at the mouth by the West Fork highway. It's now much of a fishing stream but does support a few small cutthroat.

Sleeping Child Creek. Take U.S. 93 for 2½ miles south from Hamilton to the Skalkaho junction, then continue due south on a county road for another 2½ miles to the mouth of Sleeping Child; and thence upstream for 7 miles to the Hot Springs, from which point you must get out and hike up a good USFS trail for 8 miles to headwaters. It's a heavily, and easily fished stream that produces a few Dolly Varden, brook and rainbow trout.

Slocum Creek. A very small tributary of the Burnt Fork of the Bitterroot River crossed at the mouth by a county road 7 miles east of Stevensville. Slocum Creek flows through heavily grazed pastureland and becomes quite warm in summer, but nonetheless is a fair kids' stream for small rainbow and cutthroat trout.

Smith Creek. This tiny tributary of Sweathouse Creek is reached by a good gravel road, 2 miles west from Victor and is fair fishing along the lower 2 miles for small cutthroat and rainbow — a kids' creek.

Soda Springs Creek. Take the Nez Perce road 4 miles west from the West Fork Ranger Station to the Little West Fork Creek, then a USFS logging road upstream for 1½ miles to the mouth of Soda Springs Creek. There are about 5 miles of fair fishing here for small rainbow and cutthroat trout.

SKALKAHO FALLS
Courtesy Ernst Peterson

SOUTH FORK BIG CREEK LAKE
Courtesy USFS

South Fork Big Creek.
A small stream about 6 miles long, the outlet of the South Fork and Pearl Lakes; reached by trail 9 miles up Big Creek to its mouth, and followed by trail to headwaters. It is good fishing for small rainbow, cutthroat and brook trout but would be worth the trip just for the scenery.

South Fork Big Creek Lake.
Forty acres, deep, with considerable aquatic vegetation; this is a cirque lake in timber and park country reached by a poor, unmaintained trail 4 miles from a point on the Big Creek trail about 1 mile below Big Lake. It used to be good fishing for 8 to 12 inch rainbow but escessive fluctuation of the water level (due to leakage from an old, abandoned dam) has destroyed the spawning grounds and the lake is now reported to be barren.

South Fork Lolo Creek.
From Lolo take the Lewis and Clark highway for 10 miles upstream to Woodman, thence 1 mile south on a good gravel road to the mouth of the South Fork, and a good USFS trail on to headwaters. There are about 10 miles of fishing water (to the Meadow Creek junction). The lower reaches are on private land but permission is usually granted on request. It's good for 6 to 14 inch cutthroat and small Dolly Varden.

South Fork Lost Horse Creek.
The outlet of Fish Lake, the South Fork flows for 9 miles down a heavily forested, glaciated canyon to the main stream where it is reached by a good road 5 miles from Charlo Heights. It is followed to the lake by a USFS trail and is good fishing for 7 to 9 inch cutthroat...but is mostly passed up for the lake.

South Fork Skalkaho Creek.
Flows through timbered mountains from an elevation of about 8000 feet at headwaters for 9 miles to the head of Skalkaho Creek to an elevation of about 4700 feet (near Camp Cornish) on the Sklakaho highway 20 miles east of Hamilton. It's paralleled by road and trail for its entire length and is easily fished for fair catches of cutthroat and Dolly Varden.

South Kootenai Lake.
Fifty acres, mostly less than 20 feet deep, with brushy margins and talus along the northwest side. South Kootenai Lake lies in a heavily wooded cirque reached by a good trail up Kootenai Creek 13 miles above the mouth of the canyon. It is overpopulated with 8 to 10 inch brookies that provide excellent fishing for even the rankest neophyte. A beautiful lake for a weekend trip — and good elk hunting country too.

Spud (or Charity) Lake.
A tiny (2 acre) lake, deep, with boggy shores, in heavily timbered country at the head of Moose Creek, reached by a good trail 5 miles from the Martin Creek logging road. Spud Lake was planted with cutthroat years ago and their progeny (mostly 6 to 11 inchers) do a real good job of supporting moderate-to-heavy fishing pressure now.

Stuart Creek.
This small stream flows through a heavily grazed area to Willow Creek about ½ mile above the Butterfly Guard Station. It's followed by trail and is fair fishing for 1½ miles above the mouth, for pan-size rainbow and cutthroat trout.

Sweathouse Creek.
A clear, fast stream flowing east for 12 miles down its steep, rocky canyon to the Bitterroot Valley and then for another 3 miles through farmland to the Bitterroot River near Victor. The lower reaches go completely dry in summer (due to irrigation diversions), but the upper 'canyon' reaches are good fishing for 6 to 9 inch cutthroat, and a few rainbow and brook trout. There's a trail for 7 miles above the canyon mouth.

BITTERROOT

Sweeney Creek. About 6 miles long from the junction of its North and South Forks to its confluence with the Bitterroot River 1½ miles south of Florence; the upper reaches flow down a heavily forested canyon, the lower reaches (within the Bitterroot Valley) are completely dewatered for irrigation. It's fair fishing above for small cutthroat and brook trout. There's no trail.

Tag Alder Lake. Eight and seven tenths acres, 32 feet deep, in a small timbered cirque just below Mill Point on the Mill Creek-Blodgett Creek divide. There is no trail — just gird your loins and head due west up the mountain front for a 2½ mile, 3300 foot climb from the county road 1 mile south of the mouth of Mill Creek Canyon. This one is very seldom fished but still contains a few (8 to 12 inch) rainbow.

Taylor Creek. A real small tributary of Hughes Creek; crossed by a gravel road at the mouth (7 miles above the Alto Ranger Station) and followed by an old unmaintained trail for 3 miles past headwaters. There is fair 'camp' fishing here for small cutthroat and rainbow trout.

Teepee Creek. A little bitty stream reached by trail 4 miles up Howard Creek from its mouth near the Lolo Ranger Station. Teepee was washed out in 1952 and is poor fishing now for 6 to 8 inch cutthroat in scattered pools for about 1 mile above the mouth.

Tenmile Lake. A deep, 20 acre lake in high (elevation 6700 feet) barren muskeg-meadow country at the head of the Roaring Lion-Lost Horse Divide; reached by foot (if you are ready, willing and able) 4 miles from the Lost Horse road up Ten Mile Creek. This lake is good fishing for 8 to 10 inch cutthroat — descendants of an old, 1965 plant.

Threemile Creek. A small stream, crossed at the mouth by county roads about 4 miles north from Stevensville and followed by a good gravel road for 20 miles to headwaters. The lower 5 miles are mostly dewatered for irrigation, but still seem to produce some good catches of 6 to 14 inch brook and rainbow — the upper reaches are fair for small cutthroat. A kids' creek.

SOUTH KOOTENAI LAKE
Courtesy Ernst Peterson

Tin Cup Creek.
The outlet of Tin Cup Lake, flowing north down a steep walled, glacial canyon for 18 miles to the Bitterroot River 2 miles north of Darby; followed by a road for 6 miles and then a trail to headwaters. The surrounding country is mostly heavily timbered but the creek is easily fished for plentiful 6 to 12 inch cutthroat and some brook trout. There's a campground near the mouth of the canyon.

Tin Cup Lake.
A real high (elevation 9000 feet) lake in a forested, and talus strewn cirque; reached by a good horse trail 12 miles up Tin Cup Creek from the end of the road. A fair sized (125 acre) lake, shallow at either end but fairly deep in the middle and with steep dropoffs along the sides. There is an irrigation dam that raises the water level as much as 15 feet. In spite of this, the lake is overpopulated with 8 to 16 inch, big headed, snake bodied cutthroat.

Tolan Creek.
A good kids' stream for pan-size cutthroat, it's about 10 miles long and flows to the East Fork of the Bitterroot River 3 miles east of Sula. No trail.

Trapper Creek.
A small, steep stream that flows for 12 miles down a rock-walled, glacial canyon to the West Fork of the Bitterroot River about 4 miles above Conner. It is good kids' fishing for 7 to 9 inch rainbow and cutthroat trout. There's a road for about 3 miles and then a trail for another five upstream.

Twelvemile Lake.
Take a poor foot trail for 4 miles to this lake from a point on the Lost Horse road about 14 miles up from U.S. 93. It's fairly deep, about 10 acres, with some swampland about the margins; set in scattered timbered, rocky country. The fishing is good for 8 to 11 inch cutthroat but they seldom see a hook.

Twin Lakes.
Two (Lower and Upper) lakes a few hundred yards apart, reached by the Lost Horse Creek road 18 miles from U.S. 93. About 65 and 40 acres respectively with a maximum depth of about 25 feet and gradual dropoffs, they are situated in a high cirque in predominantly barren rocky country. Both are good fishing for small (8 to 10 inch) cutthroat and rainbow trout, and Upper Twin has some hybrids that run up to 18 inches or better.

TWIN LAKES
Courtesy USFS

TIN CUP LAKE
Courtesy USFS

TIN CUP LAKE
Courtesy Ernst Peterson

CAMP FARE
Courtesy USFS

Two Bear Creek. A small stream about 6 miles long, flowing to Sleeping Child Creek 4 miles above the guard station. The lower reaches are good early-season fishing for cutthroat up to 12 inches long. There is no trail.

Unnamed Lake on the Headwaters of Hope Creek. A high (elevation 7500 feet) cirque lake just north of the Continental Divide in the Anaconda-Pintlar Wilderness Area. To get there take county roads 15 miles north from Wisdom to the Spannuth ranch at the mouth of Plimpton Creek, then 8 miles of fair trail past its headwaters to the Continental Divide, thence 1 mile east along the divide and finally 1 mile down to the headwaters of Hope Creek and the lake which drains eventually to the East Fork of the Bitterroot River. It is quite small (about 1½ acres) with some blue water and lots of lily pads around the margins. The immediate shore is mostly grassy meadow. All observations were made from a high promontory overlooking the lake in its beautiful setting — wonder if there are any fish?

Warm Springs Creek. Eleven miles long and easily fished, this small tributary of the Bitterroot River is followed by a paved road for 1½ miles from U.S. 93 eight miles above Conner to Medicine Springs Hotel and then a gravel road for 4 miles on upstream to the Crazy Creek campground and finally a USFS horse trail for 10 miles to headwaters. It produces good catches of 8 to 10 inch cutthroat, rainbow and a few brook and Dolly Varden trout.

Watchtower Creek. The outlet of (barren) Watchtower Lake, this little stream flows down its deep, ice-scoured canyon for 12 miles to the Nez Perce Fork of the Bitterroot River where it is crossed by the Nez Perce road 9 miles above the West Fork Ranger Station. It's available to its source by a good trail and is excellent fishing for 7 to 9 inch cutthroat.

West Fork Bitterroot Reservoir. See Painted Rocks Lake.

West Fork Bitterroot River. A beautiful, clear, easily fished stream about 50 miles long from its mouth at Conner to its head at the Montana-Idaho Divide and followed the entire way by paved and gravel roads. There are numerous beaver dams in the area between Buck and Ditch Creeks and in the upper headwaters. The West Fork is not heavily fished but produces good catches of rainbow, cutthroat, brook and whitefish plus a few Dolly Varden.

West Fork Butte Creek. A small tributary of the South Fork of Lolo Creek, there is fair early-season fishing here for small cutthroat for the first 3 miles above its mouth, which is about 2 miles above the mouth of the South Fork. The first 1½ miles are reached by a good gravel road; there is no trail.

West Fork Camp Creek. A 4 mile long tributary of Camp Creek, it joins the main stream a mile north of Gallogly Hot Springs on U.S. 93 and is crossed a couple of times by logging roads with hunters trails in between. The fishing is good for pan-size cutthroat and brook trout.

West Fork Lolo Creek. Paralleled by the Lewis and Clark highway from its mouth at Lolo Hot Springs for 7 miles to its headwaters at Lolo Pass. This is a small stream that is heavily fished for fair early-to-middle season catches of 10 to 12 inch cutthroat and a few Dolly Varden.

Wiles Creek. A small tributary of Warm Springs Creek; the lower reaches are followed by an unmaintained USFS trail and are fair fishing for small rainbow and cutthroat trout.

Willow Creek. A fair-sized, easily wadeable, fishing stream that used to flow from Willow Lake to the Bitterroot River near Corvallis but is now completely dewatered in the lower reaches. It's accessible for about 17 miles by a gravel road, the upper 3 miles by trail and is heavily fished for fair-sized cutthroat and small brook trout.

Willow (or Fool Hen and Sage Hen) Lake. Take the Willow Creek road to its end 4 miles past the guard station, then a good trail for 1 mile, and thence a poor unmaintained trail for the next 2 miles to the lake on the slopes of Skalkaho Mountain (elevation 7300 feet). It's about 10 acres, 30 feet deep, with lots of aquatic vegetation around the shores and is situated in timber and parkland country. Willow Lake was planted with rainbow many years ago and contains a good population of 6 to 12 inchers, plus a few old lunkers reported to 10 pounds. There's an old abandoned dam at the outlet.

UPPER TWIN
Courtesy Ernst Peterson

LOWER TWIN
Courtesy Ernst Peterson

Woodchuck Reservoir. Take the "East Side" highway from Florence across the valley and over the river to a point maybe a half mile beyond where the main drag bends to the south, but you take a little old gravel road to the north for ¾ of a mile and then break off to the east again for 2¾ miles and finally north once more on a farm road for 1¼ miles and you're there — at the Woodchuck Reservoir which is no more than a couple of acres if that, gets pretty well drawn for irrigation in the summer, has some trees and brush along the south side, a road across the lower end and up the north side, and a few stray trout you can pick up now and then if you're the "patient" kind. Not a bad spot to while away a summer afternoon if you like the peaceful bucolic atmosphere and don't wish to be bothered too much with fish.

Woods Creek. A small, eastward flowing tributary of the West Fork of the Bitterroot River, crossed at the mouth by a gravel road 9 miles above Painted Rocks Lake and followed by a logging road for 8 miles to headwaters. It's good fishing for medium size cutthroat and rainbow, but note! — the lower reaches are on private property with limited access.

Wornath Reservoirs. Upper & Lower, 1/3 and 1/2 acres, 1/4 mile apart in a suburban development, bordered by lawns, horse barns etc...1 mile southeast across the railroad tracks from Lolo. These have been planted with catchable (10 to 12 inch) rainbow. The upper road is home to one Lady Mallard. And brother, they're strictly private!

WEST FORK BITTERROOT RIVER
Courtesy Ernst Peterson

The Blackfoot Is Beautiful

Beaver, muskrat, deer and elk watch it pass. Wild trout feed in its shallow riffles and rest in deep pools. Nothing interrupts the progress of this free-flowing stream as it leaves the wild country near the Continental Divide, travels westerly through scenic mountains and wide valleys and ends its journey at the Clark Fork River over 100 miles away.

The waters of this mountain stream are an attraction all their own – whether you're a fisherman or not. Many are the times I've simply sat and watched the water flow by – its extreme clarity making it sparkle in the sunlight as it dances over a riffle to become suddenly still in a deep pool. These clear waters contain, in spite of man's technology, wild populations of sporty game fish. The only supermarkets these creatures know are the riffle "shelves" which contain the variety of insect life needed to sustain them. And if watching the water itself isn't enough, you can set up your fly rod or bait dunker and try to outwit a lunker brown, rainbow, cutthroat, or Dolly Varden. Knowing that your success in this will simply add to an already successful day means you're in intimate contact with the beautiful Blackfoot.

Whether you prefer sitting on the rocks at Rainbow Bend or wading up to the brush-covered undercuts near Lincoln, you will find fishing the Blackfoot a memorable experience. Those who know the river have their preferred fishing spot. Newcomers have the pleasure of finding one.

Few streams traversed by a busy highway can still impart a feeling of wildness. Perhaps it's those continual meanders which hide unknown waters around each bend, or that next deep hole or log jam that awaits your anxious cast. Perhaps it's the unspoiled shorelines, the beaver-cut cottonwoods, or that mallard you just spooked off a backwater. Perhaps it's because you know the wild source of these waters. Perhaps it's all of these, but whatever the reason for your feeling, you know you're on the Blackfoot.

There is plenty of elbow room here even during the summer season if you are willing to exercise those little-used muscles. There are a number of access points to the river, but you'll find some stretches difficult to reach without a boat. Nearly all of the river is floatable, but special precautions are needed in some areas. Check with someone familiar before setting out into the unknown.

If you really want the stream to yourself, try winter fishing. The clear summer water assumes a special brilliance between untrampled winter snowbanks. And you'll likely have it all to yourself. Winter whitefishing is popular and productive and a good share of the stream is open to fishing year-round for all species. And don't forget those balmy days in April when the sun is warm and the water still low – a great time to limber up that rod again if you've been cooped up all winter.

Species distribution in the Blackfoot finds cutthroat, brook trout, whitefish and Dolly Varden above Landers Fork. Brown trout take over rather abruptly at Lincoln and comprise most of the trout population down through the Lincoln canyon. After it leaves the canyon, the river becomes more difficult to get to as it flows many miles through private ranchland toward its confluence with the North Fork of the Blackfoot. There are very few public access points and floating is about the only way to see this slow-moving stretch. Take plenty of time because you'll be meeting yourself coming back on the winding river bends.

Below the North Fork there is some rough water as the stream straightens out and picks up speed. But if you're cautious you will make it through and find some excellent fishing along the way. Catches in the lower river will consist of browns, rainbows, cutthroats, brookies, whitefish and Dolly Varden. And you'll find some good ones if you're persistent.

Public and private campgrounds are available along the way and some of them are located at river access points. Food and fishing gear can be found in abundance in Missoula, or if you prefer less crowded shopping conditions, you'll likely find the necessary items in Seeley Lake, Ovando or Lincoln. For the non-camper, food and lodging facilities are also available in these communities.

So whether you swear by nightcrawlers or a number 14 Wulff, whether you prefer an outing with the family or one by yourself, you'll find that fishing adds to the pleasure of just being on – the beautiful Blackfoot.

LITER SPENCE
Fisheries Biologist
Montana State Fish & Game Department

Blackfoot Drainage

Alice Creek. Crossed at the mouth by the Lincoln highway 12 miles east of Lincoln and followed upstream for 10 miles to headwaters. A popular stream in timber above and meadow below, that is good fishing for 9 to 10 inch brook, rainbow, Dolly Varden and a few cutthroat trout.

Alva Lake. See Lake Alva.

Archibald Creek. A real small creek, flowing to the Clearwater River at the outlet of Seeley Lake, and crossed at the mouth by the West Lake road. There are only a few fingerling in this creek but in early spring it is good largemouth bass fishing (up to 4 pounds), in the slough at its mouth.

Arrastra Creek. Flows 11 miles south from Arrastra Mountain to the Blackfoot River where it is crossed by the Lincoln highway about 2½ miles east of the Nevada Creek road. Arrastra is followed upstream for 4 miles by road across mostly posted land and thence for 7 miles by abandoned trail through heavily timbered country to headwaters. The best access is via the Beaver Creek road which crosses Arrastra Creek about 3 miles below headwaters. A popular stream that is fair fishing for 7 to 9 inch cutthroat; the lower 1½ miles sometimes go completely dry, the upper reaches are therefor the best fishing.

Baking Powder Creek. A short, steep, inaccessible tributary of Falls Creek, reached at the mouth by the Landers Fork trail 2 miles above the end of the road. The lower reaches are fair to good fishing for pan-size cutthroat. It is fished only occasionally by horse or backpackers.

Beartrap Creek Reservoir. Take U.S. 20 sixteen miles east from Lincoln to the point where the main highway bears off to the left (north) up Pass Creek. Here you break off to the right and follow the Blackfoot River (secondary) road for 2 miles right to this 10 acre PRIVATE reservoir. It is long and narrow, followed by a service road along its west side, was planted with brown trout some time ago and is good fishing now (for the owners) for 1 to 2 pounders; or so I am told.

Beaver Creek. A very small tributary of Placid Creek, crossed at the mouth by the Jocko Lake-Placid Lake road and more or less followed by road for 3 miles to headwaters. It's excellent fishing for small (6 to 8 inch) cutthroat and a few rainbow trout.

Beaver Creek. A rough, narrow, steep and twisty stream flowing south to the Blackfoot River a couple of miles west of Lincoln. It is crossed at the mouth by the Lincoln highway and is easily accessible by road to headwaters (about 9 miles in all). Beaver Creek is good early-season fishing near the mouth for cutthroat, brook trout and a few browns but: its mostly all posted.

BLACKFOOT

This map is not intended for navigational purposes. Navigational hazards are not shown. Access areas shown are public. Other places may be open to public use through the consent of individuals.

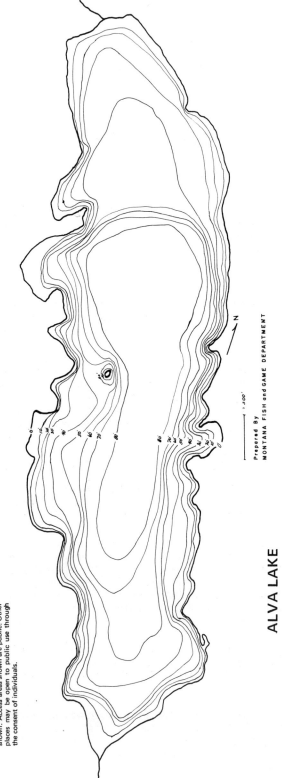

ALVA LAKE

Prepared By
MONTANA FISH and GAME DEPARTMENT

BLACKFOOT

Belmont Creek. Take the north 'river road' 7 miles from the Ninemile Prairie school house (12 miles east of Potomac) to the mouth of this creek, and then upstream for 8 miles. A small creek in timbered country, it's fair fishing for 7 to 10 inch rainbow, brook trout and cutthroat.

Bertha Creek. A small, swampy stream flowing from Summit Lake for 3½ miles to Rainy Lake and paralleled by the Swan Lake-Seeley Lake highway. It's easily accessible and good fishing (especially in the numerous beaver ponds) for 8 to 12 inch cutthroat trout.

Bertha Lake. See Colt Lake.

Bighorn (or Sheep) Lake. A deep, 10 acre, cirque lake in the rough Lincoln Backcountry just below the Continental Divide. It is reached by a poor, unmarked trail about 2 miles west and 'straight up' from the Bighorn Creek trail 12 miles above the end of the road; also from the Continental Divide trail about 9 miles from Wildcat Ridge on Alice Creek. A real hard-to-get-to lake that is very seldom fished but reportedly excellent for large cutthroat and rainbow trout.

Big Sky Lake. See Fish Lake.

Black Lake. See Elbow Lake.

Blanchard Creek. A small tributary of the Clearwater River crossed at the mouth by the Lincoln highway 1 mile west of the Seeley Lake turnoff and followed through steep, timbered mountains for 7 miles by logging road and thence by trail to headwaters. An easily fished stream with moderate brush cover, it is good for 9 to 10 inch brookies, and a few rainbow, cutthroat and whitefish.

Blanchard Lake. Really a shallow, 10 acre, beaver-dammed canyon on the Clearwater River in logged-over land, reached by a logging road 1 mile north of the Seeley Lake turnoff from the Blackfoot highway. It is moderately popular, contains mostly 6 to 10 inch cutthroat and brook trout.

ICE FISHING IN BROWNS LAKE
Courtesy USFS

BIG HORN LAKE.
Courtesy USFS

Blind Canyon Creek. A small stream, partly open and partly brushy that ends in swampy country 3 miles east of Seeley Lake. It is available by an old logging road ½ mile north of Seeley and then 3 miles east on the Cottonwood Lake road. The fishing here is fair for 6 to 8 inch cutthroat trout.

Boles Creek. Flows for 12 miles through timbered country to the west side of Placid Lake. Reached at the mouth by a gravel road around the south end of the lake and on to headwaters. However, much of the stream (5 miles or more) is accessible only by trail. Its pretty good fishing for 7 to 8 inch rainbow, brook trout and small largemouth bass.

Boulder Lake. Reached by a 2 mile overland hike from the end of a logging road off the South Fork of the Jocko road (Flathead drainage). A 20 acre lake, timbered on three sides but with talus on the south end, it's moderately popular and real good fishing for good size (10 to 13 inch) cutthroat.

Brown's Lake. An open rangeland, 500 acre lake, maximum depth about 25 feet with moderate dropoffs, a popular camping area, mostly fished through the ice in the winter. You can drive right to it on county roads 10 miles southwest from Ovando. It has been excellent fishing (especially through the ice) for large rainbow, and is still good for 7 to 14 inchers, along with an occasional 10 to 15 inch coho salmon, a very few brook trout, and a gunny sack full of suckers.

Bull (or Deer Park) Creek. Sometimes called Deer Park Creek in the upper, timberland reaches; Bull Creek flows through Tupper Lakes to Deadman Lake and eventually across the Lincoln highway (11 miles east of Ovando) to Browns Lake. It is a small stream on private land and is seldom fished but good for 8 to 10 inch cutthroat and a few rainbow trout. Now here this, it's posted and THEY MEAN IT!

Cabin Creek. Take a USFS trail from the North Fork (of the Blackfoot River) Guard Station to the mouth of this stream, then another trail for 6 miles upstream through timbered country to headwaters. The lower reaches (about 3 miles) are good fishing for 6 to 8 inch cutthroat trout.

THROUGH THE ICE AT LAKE INEZ
Courtesy USFS

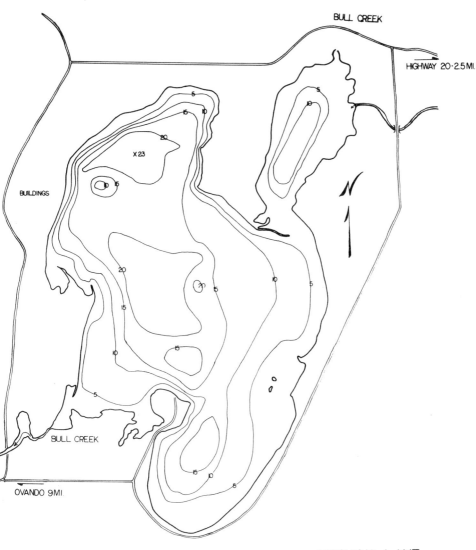

BROWNS LAKE

T14N-R11W-S17,20,29
POWELL COUNTY
TOTAL SURFACE ACRES - 500
CONTOUR INTERVAL - 5 FT

SCALE

MONTANA FISH AND GAME DEPT - 1966

This map is not intended for navigational purposes. Navigational hazards are not shown. Access areas shown are public. Other places may be open to public use through the consent of individuals.

Camas Creek. A small, open meadowland stream that flows to Union Creek ¼ of a mile west of Potomac and is followed by a road for 6 miles to headwaters. It's real good fishing for 8 to 10 inch brook trout.

Camp Creek. A small tributary of the East Fork of the North Fork of the Blackfoot River; reached by trail 1½ miles above the confluence of the East and North Forks 4 miles east of the North Fork Guard Station. Camp Creek flows through heavy timber with an open meadow here and there and is good fishing for the first 3 miles above the mouth for 6 to 8 inch cutthroat trout.

Camp Lake. Thirty acres, in heavily timbered country north of East Spread Mountain; reached by the Spread Creek trail 8 miles above the Monture Guard Station or the Lake Creek trail 10 miles above Coopers Lake. There is excellent fishing here for nice fat rainbow up to 14 inches, and long, snaky lunkers from 16 to 20 inches.

Canyon Creek. A fair size, wadeable stream in timbered country, the outlet of Canyon Lake, reached at the mouth by a USFS horse trail 7 miles up the Dry Fork of the North Fork of the Blackfoot River and followed by a good trail for 5 miles to Canyon Lake. This one is good fishing for 8 to 10 inch cutthroat with an occasional lunker.

Chamberlin Creek. A small stream in steep, timbered, mountain country; reached near the mouth by a gravel road over Scotty Brown's bridge across the Blackfoot River 5 miles west of Ovando and followed by a logging road for 7 miles to headwaters. There is fair fishing here for 6 to 7 inch cutthroat trout.

Chimney Creek. A little stream flowing through steep, timbered country above, and open benchland below to Douglas Creek 5 miles west of Helmville. Chimney Creek is seldom fished but contains a few pan-size brook and cutthroat trout in the lower reaches.

Clearwater Lake. Reached by trail 4 miles from the Summit Springs Campground on the Swan-Seely Lake road. A mountain lake in timbered country, about 125 acres, 40 feet deep with steep dropoffs and a beaver dam at the outlet. Clearwater was poisoned in 1958 and planted with cutthroat in 1960, has been excellent and still is fair-to-good fishing for trout that range in size from 7 or 8 inches to 3 or 4 pounds.

Clearwater River. A heavily fished creek in timber and mountain meadows, it flows for 35 miles from Clearwater Lake through Rainy, Alva, Inez, Seeley, Salmon and Elbow Lakes to the Blackfoot River at Clearwater. It is easily accessible for its full length by the Seeley Lake-Swan Lake highway. A fair-to-good summer fishing stream for 7 to 11 inch brook trout, rainbow, cutthroat, whitefish and yellow perch in that order, and excellent whitefishing in winter. There is also a small fall run of Dolly Varden between Seeley and Salmon Lakes.

Colt Creek. A tiny stream, the outlet of Bertha Lake, flowing 5 miles through timbered mountains to Rainy Lake and followed to headwaters by a logging road. It is fair fishing for pan-size cutthroat near the mouth and in the middle (swampy) reaches.

Colt (or Bertha) Lake. Is reached by the Colt Creek logging road 7 miles above the Clearwater River, in logged-over country (looks like the wrath of God), and was stocked with cutthroat in '66. They're around 12 inches now if they've not frozen out.

IT'S MY LAKE
Courtesy USFS

Colt Creek Reservoir. Is located on Colt Creek about 2 miles up from the mouth by the Colt Creek logging road. This reservoir exists only during the high spring runoff when the creek overflows a 3 acre meadow partly dammed by a log jam. It's in a logged-over area and is excellent fishing for 8 to 12 inch cutthroat (it looks like the water is boiling) but you need a rubber boat or waders to get out to them. Lots of folks fish this one.

Conger Creek. A real small stream flowing through steep, timbered mountains to Canyon Creek ½ mile below Canyon Lake. It is followed by a trail for 5 miles to headwaters. The lower 2 miles are fair-to-good fishing for 6 to 8 inch cutthroat trout.

Coopers Lake. A beautiful, clear mountain lake in heavily timbered country, reached by road above Copenhaver's ranch, 18 miles northeast from Ovando. Coopers is about 180 acres, deep with extremely steep dropoffs, and is (unfortunately) infertile. It was rehabilitated in 1967 and was replanted with cutthroat trout in 1968. They're just doing so-so now.

Coper Creek. A tributary of Landers Fork, followed by a modern sophisticated road from its mouth (2½ miles above the Lincoln highway) for 14 miles through a narrow, timbered canyon below and then on up a timbered, flat-bottomed valley to the end of the fishing water. The lower reaches (below the campground) are generally poor fishing except for a few 3 to 4 pound Dolly Varden that run up in the fall. The upper reaches are fair fishing in numerous beaver dams, for 10 to 15 inch cutthroat and a few rainbow and Dolly Varden. See also Snowbank Lake.

Copper Lake. A small (3½ acres) fairly deep lake, in high (elevation 7000 feet), timbered country reached by a poor horse trail ¼ mile east from the Stonewall Creek trail 2½ miles above the end of the jeep road. Do not confuse Copper Lake with three barren lakes from 1 to 2 miles distant and west of the Stonewall Creek trail. Copper Lake contains lots of good size Cutthroat (?) trout that can be seen rising and feeding but (reportedly) very few people can catch 'em, fact of the matter is there aren't many who can find their way in and fewer yet who find it worthwhile to return.

BLACKFOOT

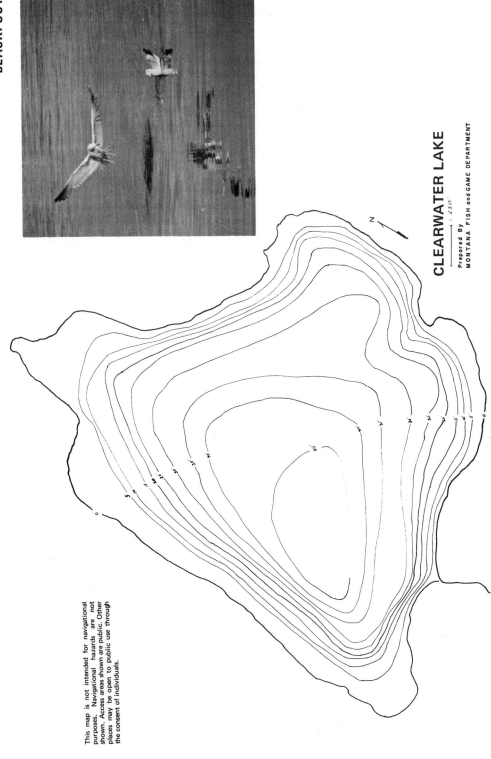

CLEARWATER LAKE

Prepared By
MONTANA FISH and GAME DEPARTMENT

This map is not intended for navigational purposes. Navigational hazards are not shown. Access areas shown are public. Other places may be open to public use through the consent of individuals.

Cottonwood Creek. A small tributary of Douglas Creek that flows from timbered mountains above to meadowland below. The lower reaches are mostly used for irrigation but are fair fishing for 6 to 10 inch brook and brown trout; the upper reaches are good fishing for small cutthroat. The lower reaches are easily accessible by country roads.

Cottonwood Creek. The outlet of Cottonwood Lake; flows for 15 miles through hilly, meadow-and-pothole country (ground moraine) to the Blackfoot River about 1 mile above Sperry Grade, and is followed by a good gravel road for the full length. A brushy creek that is good fishing for 3 to 18 inch Loch Leven and 2 to 15 inch whitefish in season, fair for 3 to 11 inch rainbow and an occasional brook trout, 5 to 22 inch Dolly Varden and 7 to 10 inch cutthroat trout. Lots of everything.

Cottonwood Lakes. One puddle and two 15 acre lakes a few hundred yards apart in steep timbered country; reached by Cottonwood Creek road 9 miles from the Seeley Lake Campground. Both lakes contain big cutthroat (reportedly up to 4 pounds) that are sometimes seen but seldom caught; poor fishing. The lower lake also has a good population of 6 to 9 inch brook trout.

Deadman Lake. A 10 acre, shallow lake with floating logs all around; on the edge of farmland below heavily timbered Markham Mountain. Deadman is 10 miles east of Ovando and can be approached to within a few hundred yards by the Whitetail Ranch road. It was poisoned in 1957 and planted with rainbow in the late '50's and '60's. Although seldom fished now, it is fair for rainbow plus a few brook and cutthroat trout from Bull Creek.

Deer Creek. A small stream flowing for 12 miles through steep, timbered mountains to the northwest end of Seeley Lake, and easily accessible for most of its length by road. It's moderately popular and real good fishing for 6 to 8 inch cutthroat trout and a few brook trout.

Deer Park Creek. See Bull Creek.

Dick Creek. A slow, meadowland stream flowing to McCabe Creek, crossed at the mouth by a county road 1¼ miles north of Ovando and easily accessible for its entire length (about 5 miles) by private road. Dick Creek is good fishing for mostly 12 to 14 inch rainbow, quite a few 10 to 12 inch cutthroat and an occasional large brown trout.

Dinah Lake. A 34 acre lake near timberline at 6700 feet elevation on the east side of the Mission Range and reached by driving up the Marshall Creek road and hiking 2 miles or by a 2 mile hike over the Clearwater-Jocko Divide from Lake Elsina. It can be fished either from shore or from a small island but is generally poor fishing (due to an overabundance of feed) for rainbow up to 5 pounds or better. In truth 'tis said that most of the fish caught here are probably poached from the small inlet during spawning season — or through the ice in winter.

Douglas Creek. A slow, turbid stream flowing to Nevada Creek 3 miles northwest of Helmville. It is followed along the lower reaches by the Helmville-Drummond road, then by poor (private) road to headwaters. It contains mostly small (6 to 8 inch) cutthroat with a few brook, rainbow and Dolly Varden. A marginal stream.

Dry Fork North Fork Blackfoot River. Reached at the mouth by the North Fork trail at the guard station, and followed by a good trail through steep, timbered country for 15 miles to headwaters. A fair sized stream, the lower reaches

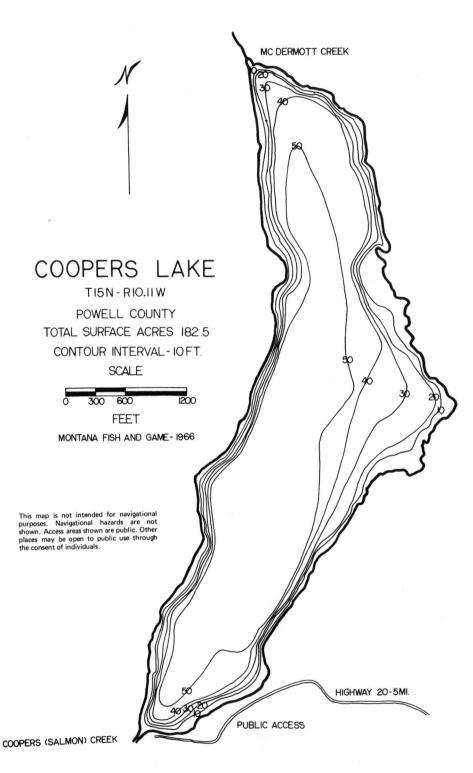

(about 3 miles) go almost dry except for a few holes. There is excellent fishing, especially in the upper reaches, for very nice 10 to 14 inch cutthroat trout.

Dwight Creek. A small tributary of the Dry Fork of the North Fork of the Blackfoot River, crossed at the mouth by the Dry Fork trail 8 miles above the North Fork Guard Station and followed by a good USFS trail for 5 miles through steep, timbered mountains to headwaters. The lower 3 miles are good fishing for small (6 to 8 inch) cutthroat trout.

East Fork of the North Fork of the Blackfoot River.

An excellent fishing stream in mostly timbered country; followed by a good USFS horse trail for 10 miles from the mouth near the North Fork Guard Station to headwaters. The East Fork flows through many open meadows and silted-in-beaver ponds and supports large populations of 6 to 8 inch cutthroat and rainbow trout.

East Twin Creek. A tiny stream in logged-off country, crossed at the mouth by the Blackfoot highway 8¼ miles east of Bonner. It used to be excellent fishing for small cutthroat trout but has been ruined by logging.

Elbow (or Black) Lake. This lake is a long, wide spot (about 40 acres) in the Clearwater River 2½ miles above the Blackfoot highway or a ½ mile hike across open meadowland, west from the Seeley Lake road. It is considered to be fair fishing for largemouth bass and yellow perch, but is poor for trout — cutthroat, rainbow, brook brown and Dolly Varden.

Elk Creek. A small, clear stream that is followed along the east side by a jeep road from its junction with the Blackfoot River below Nine Mile Prairie to headwaters near "Top O' Deep." The upper reaches are in a timbered, brushy valley; the lower reaches are in open meadowland. It is mostly posted but good fishing for 7 to 8 inch cutthroat, some brook, and a very few rainbow trout.

Elsina Lake. Take the Marshall Lake road 2 miles west from the Clearwater highway to the Deer Creek turnoff, then 7½ miles to the lake shore. This road is up for closure in favor of a new Burlington Northern access road now in the finishing stages. The lake is real shallow, about 15 acres, in timbered country right on top of the Clearwater-Jocko divide, quite popular and good fishing for 7 to 9 inch cutthroat and rainbow trout.

Falls Creek. A short (6½ miles), steep stream in timbered country, it flows to Landers Fork 4 miles east of the Indian Meadows Guard Station and is reached at the station by the Landers Fork trail. It is reported to contain plentiful, small native cutthroat trout.

Falls Creek. Heads in Camp Pass and is followed by a good horse trail for 6 miles down a narrow, heavily timbered mountain valley to Monture Creek 1 mile above the guard station. An open stream, it is fair fishing for 6 to 10 inch cutthroat and rainbow trout.

Fawn Creek. A very small tributary of Deer Creek, crossed at headwaters by the Jocko road 4 miles east of Upper Jocko Lake (in the Flathead Drainage) and followed by road for its entire length — about 3 miles. The fishing is excellent for 10 to 12 inch cutthroat but it is too small a stream to stand much pressure.

Finley Creek. A small stream in timbered mountains above, and swampy land below; flows to Placid Creek 1½ miles above the lake and is reached by a logging road around and past the south end of the lake for 1 mile to its mouth, and then 6½ miles on to headwaters. It is excellent fishing for 6 to 8 inch cutthroat, brook and a few rainbow trout.

Fish Creek. A tiny stream, about 3 miles long, flowing northwest from timbered country to the Blackfoot River at the east side of Nine Mile Prairie. It is very good fishing for such a small creek, contains 7 to 10 inch brook, and a few rainbow and brown trout.

Fish Creek. A little stream that drains Fish Lake and Sam Drew's rearing ponds, for 1½ miles through private timberland to Salmon Lake. It is followed all the way by a road and the lower end is fair fishing "the opening day of the season" for small brook trout.

Fish (or Big Sky) Lake. Drive 2½ miles north on State Highway 35 from Salmon Lake to Morrell's Ranch, then double back 5 miles to the southwest and you're there. It's a private hatchery, 1 mile long by ⅛ mile wide, in dense timber right down to the water, and good fishing for 14 to 16 inch rainbow trout.

Frazier Creek. A well small tributary of the Blackfoot River, reached at the mouth by a poor road 8 miles south from Ovando. A brushy stream in timbered country without a trail, it is seldom fished but contains quite a few small cutthroat.

Gold Creek. A clear, mountain stream flowing for 25 miles through heavily timbered country to the Blackfoot River 1 mile above McNamaras bridge. Take the Blackfoot highway 10 miles east from Bonner to the Gold Creek logging road turnoff (between East and West Twin Creeks), then 9 miles north to the Gold Creek logging road Guard Station, and thence 15 miles by a good trail to headwaters at Boulder Lake; also available for 2 miles above the mouth by jeep road. A heavily fished stream that produces good catches of 9 to 10 inch rainbow and brook, and large (15 to 21 inch) Dolly Varden trout.

Green Lakes. See Tupper Lakes.

Hardscrabble Creek. See Krohn Creek.

Harper's Lake. Take the Clearwater (Seeley Lake) highway 1 mile north from the Lincoln highway, then 1/5 mile west through the Boyd Ranch gate to the lake. A "landlocked," 18 acre lake in the Boyd Ranch pasture, about 28 feet maximum depth with a gradual dropoff all around. Harper's suffers heavy winter kill now and again but is good early-season fishing now for 10 to 12 inch rainbow and winter fishing for grayling plus a very few good sized cutthroat trout.

Heart Lake. Take a good USFS horse trail from the public corrals on Copper Creek near Indian Meadows 4½ miles to Heart Lake in steep, timbered mountains. It is 32 acres with a maximum depth of 50 feet in spots-and steep dropoffs; and moderately popular for good July catches of the prettiest 10 inch grayling you ever saw.

Hogum Creek. Take the Lincoln highway 8 miles east from Lincoln to just opposite the mouth of this stream on the south side of the Blackfoot River, and then a logging road for 3 miles upstream which leaves you on a side hill a mile above the creek in timbered, brushy country (there's some private meadowland near the mouth that is being subdivided and so here goes another Montana trout stream). Anyhow, it is at this writing still good early season fishing for 9 to 10 inch brook trout, some rainbow and a very few whitefish.

Horsefly Creek. A real small stream in burned-over, brushy country, flowing for 3 miles to the Blackfoot River on the opposite side from the highway 9 miles east of Lincoln. It is followed by a jeep road but is very seldom fished, mostly by

NEZ BEAUTIES
Courtesy W.E. Thamarus, Jr.

HEART LAKE
Courtesy USFS

sheep-herders, although it produces good catches of 8 to 9 inch cutthroat, some brook and a very few brown trout.

Horseshoe Lake.
A wide spot, 30 acres, in the Clearwater River, in open meadowland 1¼ miles above the Blackfoot highway near the Seeley Lake turnoff. Horseshoe is poor fishing for trout (cutthroat, rainbow, brown, brook and Dolly Varden) and fair fishing for yellow perch and largemouth bass.

Hoyte Creek.
A tiny, meadowland stream that heads near Ovando and flows 3½ miles to McCabe Creek; is crossed near the mouth 2½ miles west of Ovando by the Lincoln highway; and is fished occasionally by local residents for small (8 to 9 inch) brook trout.

Humbug Creek.
A small, meadowland stream flowing for about 6 miles to the Blackfoot River a mile south of Lincoln. It is too small for a real fishing creek but does produce a few good early-season catches of 8 to 10 inch cutthroat and brook trout. Access is with 'permission only' as its all on posted land.

Inez Lake.
See Lake Inez.

Jones Lake.
A very shallow, 40 acre lake in private, sagebrush-and-meadowland, 1¼ miles northwest of Ovando. Jones Lake partly freezes out each year and is fished mostly by local people for 6 to 10 inch yellow perch.

Keep Cool Creek.
The outlet of Keep Cool Lake, flows for about 8 miles to the Blackfoot River near Lincoln and is mostly accessible by a poor road. It's an open meadowland stream that is easily fished for good catches of 8 to 10 inch brook, and some cutthroat and rainbow trout. A popular creek with local residents.

Krohn (or Hardscrabble) Creek.
The outlet of Krohn Lake, drains 3½ miles through mostly slough country to the Blackfoot River where it is crossed by Highway 200 eight miles east of Lincoln. It contains a few pan-size brook trout near the mouth.

BLACKFOOT

INEZ LAKE

1" = 500'

Prepared By
MONTANA FISH and GAME DEPARTMENT

This map is not intended for navigational purposes. Navigational hazards are not shown. Access areas shown are public. Other places may be open to public use through the consent of individuals.

Lake Alva. One mile north of Lake Inez on the Seeley Lake-Swan Lake highway in timbered mountains, Lake Alva has an area of about 300 acres and a maximum depth of 90 feet with lots of aquatic vegetation. There is a real nice public campground near the south end but no resorts. It was rehabilitated in 1967 and stocked with cutthroat. It's good now for 8 to 12 inch cutthroat and 5 to 9 inch yellow perch, fair for 8 to 16 inch Dolly Varden, poor for 9 to 10 inch whitefish, and there are (unfortunately) hordes of pumpkinseeds and suckers. A real popular spot.

Lake Creek. Drains Camp and Otatsy Lakes for 6½ miles through steep, timbered country to the North Fork of the Blackfoot River; is reached near the mouth by a good USFS logging road 11 miles north of the Whitetail Ranch turnoff (North Fork of the Blackfoot River road) from State highway 200 and is followed by trail to headwater. The lower reaches go dry in summer but the upper reaches are fair fishing for small (6 to 8 inch) cutthroat trout.

Lake Dinah. See Dinah Lake.

Lake Inez. Take the Blackfoot highway 38 miles east from Missoula to the Sean Lake-Seeley Lake highway, and then north for 20 miles to Inez. Its statistics are: elevation 4100 feet, 293 acres, maximum depth 70 feet, in timberland, mountainous country, with 2 or 3 dozen private cottages, and a public campground on the north end. It was rehabilitated at the same time as Lake Alva and, if anything, is a little better fishing for cutthroat, kokanee, and yellow perch mostly, with quite a few Dollys, and Pumkinseeds and suckers by the cartload.

Landers Fork of the Blackfoot River. Crossed at the mouth by the Lincoln highway 6 miles east of Lincoln and followed by road and trail for 30 miles through timbered mountains to headwaters. The lower reaches are fair, the upper reaches good fishing for 8 to 10 inch brook trout.

Lincoln Gulch. A tiny stream that flows to the Blackfoot River a few miles west of Lincoln and is fair fishing through a couple of barnyards in the lower reaches only for small brook trout. Incidentally the lower reaches are further enhanced by dewatering for irrigation.

Liverpool Creek. A very small stream that almost dries up in the summertime but contains 8 to 10 inch brook and very few cutthroat trout in the lower reaches for about ¼ mile between the highway and its junction with the Blackfoot River 2 miles west of Lincoln.

Lodgepole Creek. A small tributary of Dunham Creek, reached at the mouth by a logging road 6 miles from the Monture Campground and followed for another 6 miles through real steep, timbered country to headwaters at Youngs Pass. It's a good fishing stream for the little ones (6 to 8 inch) cutthroat trout.

Marshall Creek. A small stream that flows for 4 miles through steep, heavily timbered mountains to Marshall Lake and thence for 2 miles to the West Fork of the Clearwater River. It is reached by a good road 2½ miles west from the Clearwater highway below Lake Alva and is followed by road to headwaters. Marshall is fair-to-good fishing for 8 to 12 inch rainbow, Dolly Varden, cutthroat and whitefish.

Marshall Lake. One hundred and fifty acres, maximum depth 55 feet, with much aquatic vegetation around the margins in heavily timbered country. Marshall is reached by a county road 3 miles from the south end of Lake Alva on the Swan Lake-Seeley Lake highway. It's good fishing in summer for 9 to 10 inch whitefish, rainbow, Dolly Varden and cutthroat but is generally inaccessible in winter.

McCabe Creek. Flows southward for 7 miles from Spread Mountain through a narrow, heavily timbered valley past the "Little Red Hills" to open meadow and pothole country and then for 5 miles to junction with Monture Creek 3 miles west of Ovando. The lower reaches are accessible by county road, the upper reaches by USFS trail. It's excellent fishing for 9 to 10 inch cutthroat and a scattering of brook trout and whitefish.

McDermott Creek. A very small stream flowing for 3½ miles through timbered country to the north end of Coopers Lake. The upper reaches go dry but the lower reaches, which are really a slough, contain 10 to 12 inch cutthroat and rainbow trout from the lake.

McElwain Creek. A small stream flowing from timberland above, through meadowland below to the Blackfoot River 1 mile east of the Blackfoot Bridge school, and crossed near the mouth by the river road. There is no road upstream and it is seldom fished although it does support a few small cutthroat in the upper reaches.

Meadow Creek. A clear mountain stream flowing north from timberland above and through meadowland below to the East Fork of the Blackfoot River 5½ miles up the East Fork trail. Its followed by a good trail for 8 miles to headwaters and beyond to the North Fork road. There is a commercial campsite at the mouth and another 3 miles above. The fishing is excellent for 8 to 10 inch rainbow, cutthroat and rainbow-cutthroat hybrids.

Meadow Creek Lake. Is really an oversized (7 acres) shallow beaver pond with a boggy west shore and a guest ranch on the east shore, in timbered 'way back' country reached by a good horse trail about 15 miles from the end of either the Beaver or Arrastra Creek roads. It is lightly fished but has been very good indeed for light-yellow colored 12 to 13 inch cutthroat and a few lunkers that will go 3 to 4 pounds. But...the busy little beaver who dammed this lake have flown the coop (been trapped out) and the dammed lake is now dammed dry.

Middle Fork Landers Fork Creek. A little strem in high, open park country, crossed at the mouth by the Landers Fork trail and followed to headwaters by a trail along the east bank. It contains small cutthroat trout.

REFLECTIONS IN MARSHALL LAKE
Courtesy USFS

NORTH FORK BLACKFOOT
Courtesy W.E. Thamarus, Jr.

Monture Creek. A good "fish'n" tributary of the Blackfoot River crossed near the mouth by the Blackfoot highway 3 miles west of Ovando and accessible by gravel roads for 12 miles up from the mouth through pothole and swampy meadowland to the Monture Campground, thence by a good USFS horse trail for another 18 miles through steep, timbered mountains to headwaters. A popular stream in good hunting country, heavily fished for 8 to 10 inch cutthroat, brook, rainbow trout and whitefish, and a few 10 to 16 inch Dolly Varden trout.

Moose Creek. This small stream flows north through timbered country to the Blackfoot River opposite the highway about 8 miles west of Lincoln. It contains a few small cutthroat and rainbow trout.

Morrell Creek. Tributary to the Clearwater River, crossed near its mouth by the Seeley-Swan Lake highway about 2 miles south of Seeley. It is followed by logging roads for 7 miles, and you can scramble up along an old unmaintained trail for another 7 to headwaters above Morrell Lake in real steep, heavily timbered mountains. Morell is a popular stream that is fair fishing below the falls (10 miles up) for 6 to 9 inch cutthroat, and brook trout along with a few loch leven spawners and lots of Kokanee spawners in the fall along the lower reaches. It's barren above the falls which is just as well because this is excellent GRIZZLY country; so good in fact that lots of mighty nimrods do their hunting elsewhere!

Morrell Lake. This lake is reached by a good horse trail 3 miles above the end of the Morrell Creek road in Grizzly country. It's about 3 acres, fairly shallow with lots of aquatic vegetation and a mud and gravel bottom. The fishing is poor-to-fair for cutthroat, brook, Dolly Varden and rainbow trout in that order — nothing here but 8 or 9 inchers.

Mountain Creek. A very small stream, only a couple of miles long, flowing north from Morrell Mountain to lose itself in swampland about 4 miles east of Seeley Lake. It's crossed at the lower end by the Cottonwood Creek road and is fair fishing for pan-size cutthroat trout.

Mud (or Penny) Lake. Mud Lake is about 5 acres, 20 to 25 feet deep, and in heavy timber. It is hard to find but reachable by a USFS trail ¼ of a mile north from Coopers Lake. It's mostly an unknown quantity, but was rehabilitated in 1967 and replanted with cutthroat in 1968. A rubber boat will come in handy here.

Nevada Creek. In the lower reaches it is mostly a slow, meandering stream in brushy swamp and meadowland easily reached by the Helmville-Avon road and county roads for 25 miles from Finn to its junction with the Blackfoot River 4½ miles northwest of Helmville (1 mile east of the Blackfoot Bridge schoolhouse). The upper reaches are followed by a logging road and trail for 15 miles through timbered mountains to headwaters below Granite Butte on the Continental Divide. It is mostly a 9 to 11 inch rainbow, brook and cutthroat stream (plus a few whitefish and Dolly Varden) below, and contains small cutthroat and rainbow trout above.

Nevada Creek Lake. This is a narrow, 100 acre reservoir reached by the Nevada Creek road 10 miles south from Helmville. It's fairly popular but is almost completely dewatered in late summer and only poor fishing for 8 to 12 inch rainbow, a very few whitefish, cutthroat, brook and brown trout to 16 inches. Furthermore it went completely dry in the summer of '73 and, unless restocked, likely is completely barren now.

Nirada Pond. See Nevada Creek Lake.

BLACKFOOT

North Fork of the Blackfoot River.
A beautiful, clear stream that flows for 22 miles through steep timbered mountains above and for 14 miles through open meadowland below to the Blackfoot River 2 miles south of Ovando. The North Fork is crossed by the Lincoln highway 4 miles east of Ovando and is followed by road and trail to headwaters. This is a popular stream with a campground 1 mile above the North Fork Guard Station (near the falls), and is good fishing for 10 to 12 inch cutthroat, rainbow, and a few whitefish, brook and Dolly Varden trout.

North Fork Blanchard Creek.
A small stream followed by a road for 7 miles from its mouth to headwaters in timbered, brushy country. The lower reaches are fair fishing for 10 to 12 inch brook and a few cutthroat trout.

North Fork Elk Creek.
A small stream crossed at the mouth by the Elk Creek trail 7 miles above that stream's mouth and followed by a poor jeep road for 2 miles upstream. It is seldom fished but does contain a very few 6 to 8 inch cutthroat trout.

North Fork Placid Creek.
The outlet of Elsina Lake, it flows for 3 miles through timbered mountains to Placid Creek and is accessible all along by road. There is some fair fishing here for 6 to 8 inch cutthroat.

Owl Creek.
Drains Placid Lake for 3 miles to the Clearwater River ½ mile above Salmon Lake and is followed along the north side by a county road. It's an easily fished stream that produces excellent catches of 8 to 10 inch cutthroat, some brook and a few rainbow trout.

Otatsy Lake.
Is ¼ mile east by a USFS trail from Camp Lake, or 1¼ miles south and west from Canyon Lake. There is timber all around it (right down to the water's edge on the south and west sides) but it's swampy to the north and northeast. All in all, this is a very pretty 30 acre lake that's full of 8 to 10 inch cutthroat and fairly popular because of it. Come to think of it the easiest way in is via the Spread Creek trail ¼ of a mile north of the Monture Guard station.

Parker Lake.
On the East Fork of the North Fork of the Blackfoot River in heavily timbered mountains at about 6000 feet elevation, Parker Lake has an area of 12 acres and a maximum depth of about 15 feet (owing, in part, to a beaver dam at the outlet). A mostly mud-bottomed lake, it has a few rocky bars, gravel beds and some nice springs near the inlet. It is good fishing for 8 to 12 inch cutthroat plus a very few to three or four pounds. They're hard to catch and are so lightly colored they are often mistaken for goldens. Parker Lake is moderately popular; there's a USFS campground and it is reached by trail 2 miles from the Webb Lake Guard Station which is 7 miles above Indian Meadows. Because its in the Scapegoat Wilderness area there is no improved campground but there *are* lots of good camp spots nearby.

Penny Lake.
See Mud Lake.

Placid Creek.
About 11 miles long, flowing from steep, timbered mountains above, through swampy flatland below to Placid Lake easily reached by road all the way. A pretty good cutthroat creek, the lower 2 or 3 hundred yards are good kokanee snagging in season (with permission only).

Placid Lake.
Take the Swan Lake-Seeley Lake road ½ mile past (north of) Salmon Lake, and then a good gravel road 3 miles west of Placid Lake; in timbered, mountainous country on three sides but low and swampy to the west. An 1143 acre lake with a maximum depth of about 85 feet, marshy shores, 40 or 50 cottages but no public campground. It's fair fishing for about anything you might want: 10 to 12 inch

kokanee, brook trout, Dolly Varden, whitefish, and lots and lots of suckers. A good place to get run over by water-skiiers.

Poorman Creek. A brushy beaver-dammed stream in timbered mountains above and brushy meadows below; followed by a gravel road for 18 miles from headwaters below Stemple Pass to junction with the Blackfoot River 2 miles west of Lincoln. It's good fishing for 8 to 10 inch cutthroat, brook and a few rainbow trout.

Rainy Lake. Seventy acres, 30 feet maximum depth with steep dropoffs, dammed; in steep, timbered country reached by a dirt road ½ mile off the Swan Lake-Seeley Lake highway 1 mile above Lake Alva. Rainy Lake was poisoned in 1958 and planted with cutthroat that now average between 10 and 11 inches and run up to 14 inches — plus quite a few good-sized Dollys and lots of suckers. Good fishing!

Ringeye Creek. This is a small, steep, hard-to-fish stream in lodgepole and spruce country. It drains Webb Lake for 1½ miles to Landers Fork 6 miles above the end of the road at Indian Meadows. No one fishes above the lake but it's fair fishing below for 8 to 10 inch cutthroat trout.

Salmon Creek. A "private" meadowland stream that flows to the North Fork of the Blackfoot River about 5 miles east of Ovando and is crossed many times by various county roads. It's good early-season fishing for cutthroat and rainbow trout that average 8 to 10 inches and range up to 3 pounds.

Salmon Lake. About 45 miles from Missoula by good oiled highway, in heavily timbered mountains, elevation 3850 feet. Salmon Lake is a popular summer home area with a public campground on the east side. It contains fair numbers of 10 to 11 inch cutthroat, rainbow and kokanee salmon; large populations of small yellow perch and sunfish; and a few Dollys and browns. A popular picnicking, water-skiing and swimming lake, as well as a good place to go fishing.

Sauerkraut Creek. A small, northward flowing stream in timbered country, Sauerkraut joins the Blackfoot River about 4 miles west of Lincoln. The lower reaches contain a few small brook trout.

Seeley Lake. Elevation 400 feet, in heavily timbered mountains, 1025 acres, maximum depth 125 feet with mostly moderate dropoffs and much vegetation around the margins. It is reached by a good paved highway about 55 miles east from Missoula. Seeley is a popular resort lake with many cottages, resorts, public beaches, a great deal of water-skiing, swimming, picnicking, etc., and fair fishing in spite of it all. It contains a fair number of 8 to 12 inch whitefish, some fair-size Dolly Varden and brown trout, hordes of yellow perch and pumpkinseed and too many trash fish.

Seven Up Pete Creek. A partly inaccessible 'private' land creek that flows to the Blackfoot River about 5 miles east of Lincoln and is followed by a logging road for 4 miles to headwaters. Its mostly reached from the 7-Up Dude Ranch 5 miles east of Lincoln by State Highway 200. The lower reaches sink but its good fishing for pan-size cutthroat in beaver ponds above, along the middle reaches.

Sheep Lake. See Bighorn Lake.

Shoup Creek. A small, short stream that drains Shoup Lake to Montour Creek across private timberland and is crossed at the mouth by a logging road. It contains cutthroat, rainbow, and Dolly Varden but is very seldom fished. It just ain't worth it.

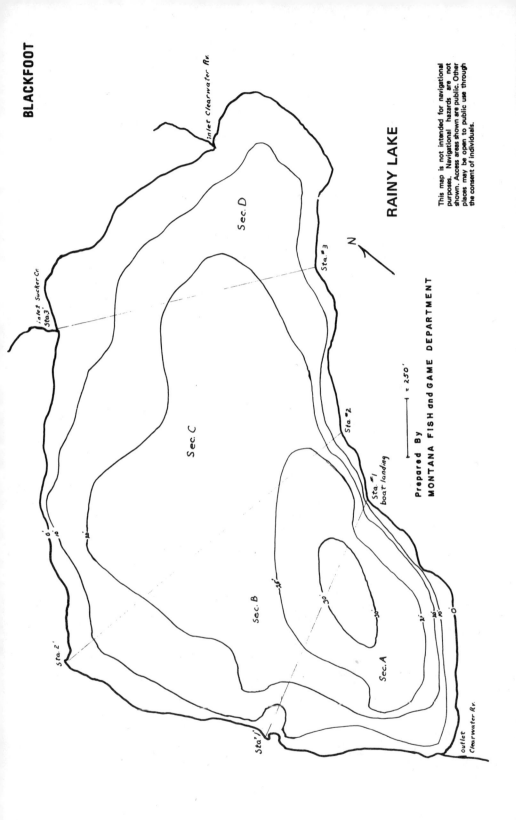

PARKER LAKE
Courtesy USFS

FOOL HEN
Courtesy USFS

Shoup Lake. A private lake in timbered country reached by driving 10 miles northwest from Ovando on gravel roads, and then hiking for maybe a mile through swamps and bogs past two mean dogs and an irate owner. Shoup is about 15 acres, fairly shallow with lots of grass and cattails around its margins and a tremendous population of fresh-water shrimp. It contains good-size (14 to 16 inch) cutthroat and rainbow, plus a few 10 to 12 inch Dolly Varden that are so well fed they're wonderfully hard to catch.

Silver King Lake. A "private," 75 acre lake, real deep with steep dropoffs all around, in heavily timbered mountains; drains to Krohn Lake but is reached by trail 2 miles from Alice Creek about 5 miles above the Lincoln highway. It has some really nice (18 to 26 inch) rainbow trout but they're hard to catch.

Snowbank Lake. A 6½ acre, 35 foot maximum depth glacial "kettle" lake in timbered country right at the junction of Snowbank Creek with Copper Creek. It is reached by a good road right next to the Copper Creek Campground. It has steep enough dropoffs to successfully fish from shore; was originally barren but is periodically stocked with 10 inch rainbow and is good fishing now. Too cold here for swimming.

Spook Lake. Take the Vaughn Creek road (pretty small for fishing) about 6 miles south and then west from the public campground on the east end of Placid Lake to a jeep trail which takes you another mile west to sub-circular Spook Lake at the head of Boles Creek. It's about 30 acres, in timber all around and with relatively shallow dropoffs so that it's best fished from a boat which is okay because you can make it in easily with a pickup. The fishing is good too, for cutthroat ranging from a few inches up to 6 pounds or better — the results of alternate year planting by the State Fish & Game Department. You'll have company at this one.

Spring Creek. A small, spring-fed stream flowing through open meadows for 3 miles to Nevada Creek 3 miles northwest of Helmville. It's easily accessible by county roads and fair fishing for 8 to 10 inch brook and cutthroat trout.

Stonewall Creek. A small tributary of Beaver Creek, crossed at the mouth by a county road 1 mile west of Lincoln and followed to headwaters by road and trail. The upper reaches are mostly in timber, the lower reaches flow through numerous

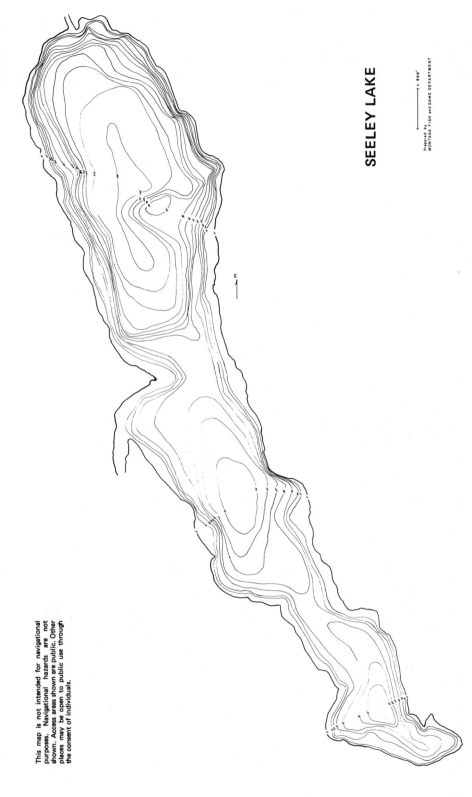

SEELEY LAKE LOOKING WEST
Courtesy USFS

ONE THAT DIDN'T GET AWAY
Courtesy Ernst Peterson

beaver dams in brushy meadows. It's good fishing for 8 to 12 inch brown, cutthroat and brook trout — plus Dolly Vardens to 10 or 12 pounds.

Summit Lake. Three hundred yards west of the Swan Lake-Seeley Lake highway on the very top of the Clearwater-Swan Divide in steep, timbered country, Summit is a 25 acre lake, only about 10 feet deep, with overhanging bog all around the margins and brushy as all get out. It was poisoned out in 1958 to protect Rainy Lake from reinfesting with rough fish, and planted with cutthroat in 1959 — is excellent fishing now for 9 to 12 inch cutthroat. You need a rubber boat.

Swamp Creek. A very small mountain stream that ends in swampland ½ mile east of Trail Creek; reached by a gravel road ½ mile north from the town of Seeley Lake and then 3 miles east on the Cottonwood Lake road. Its fair fishing for pan-size cutthroat trout.

Tobacco Valley Creek. Really the headwaters of the North Fork of the Blackfoot River; reached by the North Fork trail 8 miles above the guard station and followed by trail for 4 miles to headwaters on the Continental Divide. It is almost an alpine stream and reportedly fair-to-good fishing near the mouth for 8 to 10 inch cutthroat trout.

Tupper (or Green) Lakes. Big and Little. Big Tupper is about 14 acres, 20 feet deep with a shallow dropoff all around, on PRIVATE ("stay out, and I mean it," the owner says) timberland. It is reached by the Kleinschmidt Flat road 13 miles east from Ovando and was planted with rainbow in 1958, '59 and '60. However, there was no reproduction and few if any are left. It is fair-to-good fishing though for some 10 to 14 inch brook trout come up from Bull Creek. Little Tupper is just a barren pothole.

Twin Creeks. See East and West Twin Creeks.

Twin Lakes. Upper and Lower. One mile apart in timbered country on the East Fork Drainage (of the North Fork of the Blackfoot River in the Scapegoat Wilderness country); reached to within a few hundred yards by a good USFS horse trail 8 miles from the North Fork Ranger Station or 14 miles from the Beaver Creek-Arrastra Creek road crossing. Upper Twin is 7 acres, shallow, open around the east end and slow fishing for 12 to 16 inch cutthroat (plus an occasional lunker to

5 or 6 pounds). Lower Twin is 9½ acres, fairly deep in the middle but with shallow dropoffs all around. It is fair fishing for smaller (6 to 7 inch) rainbow and rainbow-cutthroat hybrids.

Uhler Creek. A small stream in timbered country, Uhler flows southeastward to the Clearwater River midway between Alva and Inez Lakes and is reached for most of its length (7 miles) by a road around the south end of Lake Alva. It is good fishing for 8 to 10 inch cutthroat.

Union Creek. A small (jumpable), pastureland stream flowing westward for about 10 miles down the Potomac Valley to the Blackfoot River ½ mile above McNamaras bridge. It's fished some (by kids mostly) for lots of 6 to 7 inch fingerlings.

Upsata Lake. Take U.S. 200 two miles west from Ovando, then a Montana State Game Management road northwest for 4½ miles to the south end of the lake in pothole, grass and timber country. It is about 85 acres with shallow dropoffs and a maximum depth of 40 feet. Upsata is very popular and good fishing (especially near the center and deeper parts) for 5 to 17 inch rainbow trout. A good ice fishing lake. There is a State Fish and Game campground here with one antique, malodorous Chick Sales.

Wales Creek. A small stream flowing northeastward to the Blackfoot River, reached by a county road 10 miles south from Ovando and thence for a few miles upstream through open meadows. It contains lots of 4 to 6 inch cutthroat and rainbow trout.

Wales Reservoir. Is on the Wales Ranch on Wales Creek about 1½ miles above the ranch buildings. You take the Ovando road 10 miles northwestward from Helmville, then a ranch road for about 3 miles. It's real good fishing for 10 to 12 inch cutthroat trout *if you can get permission.*

Warren Creek. A slow, meadowland stream flowing southwestward for 15 miles to the Blackfoot River 3 miles southwest of Ovando, and easily accessible all along by county roads. It's good fishing for 9 to 12 inch cutthroat and rainbow trout.

Washington Creek. A little tributary of Nevada Creek, crossed by the Helmville-Avon road a mile south of Finn and followed by 3 miles of poor dirt road north of the Keily Ranch. Washington goes dry in the summer but is poor-to-fair fishing for 8 to 10 inch cutthroat in some dredge ponds about a mile above the Ranch.

A MISERABLE DAY...

...BRIGHTENS
Courtesy Cary Hull

THROUGH THE SUNFLOWERS
Courtesy USFS

Webb Lake.
This is a shallow, 18 acre beaver pond (the site of the Webb Lake Gurad Station) on the Ringeye Creek drainage in steep, timbered mountains. It is reached by a good trail 7 miles above Indian Meadows on the Copper Creek road, or 9 miles up the East Fork (of the North Fork of the Blackfoot River) from the North Fork Guard Station. It is good early-season fishing for 10 to 11 inch cutthroat and (reportedly) a very few rainbow trout.

West Fork Clearwater River.
An easily wadeable stream that flows through steep timbered country above Seeley Lake and is followed upstream for 4 miles by a gravel road which is washed out above. There are about 10 miles of good fishing here for 7 to 8 inch cutthroat and brook trout.

West Twin Creek.
A very small stream crossed at the mouth by the Blackfoot highway 8 miles east of Bonner and followed upstream to headwaters by logging roads. It used to be good fishing for small cutthroat but few, if any, are left now.

Willow Creek.
A brushy stream that heads below Dalton Lookout, and flows for 8 miles to the Blackfoot River a mile upstream from the mouth of Sauerkraut Creek. It is followed most of the way by a county road from 2½ miles west of Lincoln and supports a few small rainbow but isn't much of a fishing stream.

BLACKFOOT

Courtesy USFS

Clark Fork Expose'

The Clark Fork of the Columbia River has it all. Riffles, pools, beautiful scenery, pollution, channel changes, and, surprisingly, even brown trout, rainbows, and record whitefish.

If one were to begin at the headwaters, Silver Bow Creek would catch the eye. This stream has functioned as an industrial and municipal sewer. For many years, the Anaconda Company discharged toxic mine water waste into the stream. The primarily brown trout fishery between Warm Springs and Deer Lodge depends entirely upon the rate of pollution abatement programs currently underway by the Anaconda Company and the Butte Metro sewer. In 1967 following an extensive period of red water, the fishery was nearly eliminated, but now a small trout population exists and the fishery can best be described as marginal.

From Deer Lodge to Drummond, tributary streams enhance the water quality and fishing improves. In this stretch of the river many deep holes with excellent stream bank vegetation provide fishermen good brown trout fishing with many in the two pound class. Most fishermen use lures including Mepps and the "Cyclone." Night crawlers are effective at all times of the year, but especially so during the spring runoff. In the summer, as the algae blooms, fishermen have problems with lures and baits as they entangle in the green mosses caused by the addition of nutrients of agricultural fertilizers and municipal wastes from towns along the river. However, at this time, dry fly fishermen watching the hatch may have good success. While fishing, anglers can view the beautiful Flint Creek Range or the busy Interstate Highway 90, depending upon where they stop.

From Drummond to the mouth of Rock Creek, the Clark Fork parallels the interstate highway. Access may be difficult, but good brown trout fishing is available to those willing to walk a short distance. Float fishermen frequently reach water untouched by other fishermen. Using large wet flies in late spring, they do well in this stretch and commonly take two to three pound browns.

Rock Creek flows into the Clark Fork River about twenty miles above the town of Missoula. It is well known for its trout fishing. Beautifully clear, cool water with an abundance of riffles and runs greets the angler. This same water with its excellent quality and cooling effect enhances fishing in the Clark Fork. Below Rock Creek, rainbows, browns, and an occasional cutthroat may be taken by the experienced angler. Again one has the choice of fishing alongside the interstate highway or walking to more remote stretches of the stream. This stretch of water is becoming more popular and fishing solitude may be short-lived. If one were to take a rock out of a riffle, he would view a rich and varied aquatic insect community. In this reach, the "Salmon Fly," or Hellgrammite, is abundant and a common bait. The past effects of highway channelization are also evident as one sees the Turah area and the shifting channel trying to regain length it once had.

During the spring runoff, fishermen should be prepared for a change in the Clark Fork River. During this time, the stream runs deep brown as sediment is carried in many instances from over-grazed banks and logging areas. Baits are most effective, including sucker and squawfish meat and night crawlers. Near the town of Bonner, a small hydroelectric dam slows the water. The reservoir from this dam is very small, although over the years sediments, including some from previous "red water" periods, have collected and fishing above the dam is poor. Past flushing of the sediments has had detrimental impact on the aquatic communities and fishery below.

Below Milltown Dam through the city of Missoula, the river can offer excellent fishing for trout and whitefish, although previous flushing have caused some lean years. Fish concentrate below Milltown Dam and bait fishermen commonly congregate here. In the stretch of the river below Milltown Dam, fishermen use the spectrum of trout enticers including bait, lures, and flies. Rainbow is the most abundant fish and the spring run of rainbow trout up Rattlesnake Creek draws many fishermen to its mouth during the months of April and May. Squawfish, suckers, and peamouth can and do plague the bait fisherman in the main river, and yet this stretch of the Clark Fork River has and is capable of producing trout in the five pound plus class.

A greenway protecting the stream through the city of Missoula would be a valuable asset and would provide a more pleasing aesthetic experience for fishermen. The river below Missoula flows through a series of well-vegetated channels with deep holes and under-cut banks. Unfortunately it flows past a sewage treatment plant which has a measurable effect on the taste of the fish below. However, anglers have been promised an upgraded sewage treatment facility and hopefully this will improve the water conditions. Fishermen commonly take rainbows and browns in this area.

The Clark Fork River below its junction with the Bitterroot River is slower moving, with long holes and fast riffle areas. Rainbow trout are common and many winter fishermen frequent this area to try for whitefish using the whitefish flies with a maggot or two on the tip of the hook. This water provides a challenge for any angler. West of Missoula, the fisherman views the Hoerner-Waldorf pulp mill, which annually discharges mill wastes into the Clark Fork River. Fishermen may smell both the unmistakable odors that ensue from the plant and the effluent wastes which seep continuously into the Clark Fork.

Ninemile Creek, Fish Creek and Petty Creek flow into the Clark Fork near Alberton. The mouths of these streams are favorite fishing holes for many bait fishermen. Occasionally brightly red-slashed "Cut" are caught. From Alberton on down to the mouth of the Flathead River the Clark Fork roars through deep canyon areas accessible only by foot, or "lazes along in slow meanders." Float fishermen are cautioned about hazardous stretches – many a trip has ended up with overturned rafts and lost gear in the white waters of the steep canyons. Fishing along these stretches is good, and beautiful one to two pound rainbow and cutthroat are taken. Fishermen are known to concentrate at the mouth of the St. Regis River for good brown trout and rainbow fishing. Waters are well suited for dry flies, baits and lures, but again bait fishermen may be hampered by abundant squawfish.

The Clark Fork has the potential to be a great stream, but the question is whether or not the rate of man-caused sedimentation, industrial or municipal pollutants, and other water quality problems will be decreased. The decision on the fate of the Clark Fork lies with the people who use it and abuse it. Think of what it can be and let's strive to achieve it.

RON MARCU
Fisheries Biologist
Montana State Fish and Game Department

CLARK FORK OF COLUMBIA NEAR PLAINS
Courtesy Ernst Peterson

Clark Fork of the Columbia
Lower Drainage

Acorn Lake. This one is only a couple of acres but maybe up to 15 feet deep in spots, deep enough to keep from freezing out; in real rocky cliff country reached cross-country up the drainage a couple of miles from the end of the Eddy Creek road. It's real good fishing for 9 to 12 inch cutthroat trout and not fished a heck-of-a-lot either.

Albert Creek. A small stream flowing for 12 miles through logged-off country from headwaters on the northern slopes of Petty Mountain to the Clark Fork across the river from Frenchtown. Albert Creek is crossed at the mouth by the old "River road" about 18 miles west of Missoula and is followed for about 5 miles upstream by logging roads. It was formerly excellent fishing for (6 to 9 inch) native cutthroat but has been ruined by logging and very few fish are left at this writing.

Arrowhead Lake. This 12½ acre lake has a maximum depth of about 40 feet, is in heavily timbered country at the head of Big Spruce Creek, and is reached by a USFS trail 13 miles up from the West Fork Thompson River road, — or you can also get there by a one mile cross-country hike from the end of the Four Lakes Creek road. It's not fished a whole lot but does support a good population of 8 to 14 inch cutthroat trout. You can watch 'em while you catch 'em.

Ashley Creek. A tiny stream flowing for about 3 miles through a steep, timbered canyon above, to flat ranchland below where the lower reaches are entirely diverted for irrigation. Ashley is part of the Thompson Falls municipal water supply and is reached by road 3 miles northwest from town. It's seldom fished except by kids for small (6 to 7 inch) cutthroat trout.

Baldy Lake. In subalpine country ½ mile southwest of Mount Baldy, 7 miles west as the eagle flies from Hot Springs, it is reached by a USFS trail 2 miles above the end of the McGinnis ditch road. Baldy is about 15 acres, deep in the upper end and shallow near the outlet, with steep cliffs along the southwest side. A lightly fished, spotty lake, it is only poor-to-fair fishing at best for 8 to 12 inch cutthroat trout.

Beatrice Creek. A very brushy little stream that is paralleled by a logging road for 3 miles through timbered country to its mouth on Fishtrap Creek. It is lightly fished and only poor-to-fair for 6 to 8 inch cutthroat and Dolly Varden.

Beaver Creek. A fair sized, easily fished stream formed by the confluence of Big and Little Beaver Creeks, it flows for 8 miles across ranchland on the Clark Fork floodplain to empty into the Noxon Reservoir a few miles below Trout Creek. It is fairly popular and good fishing for mostly 10 to 12 inch rainbow plus an occasional cutthroat, brook trout, whitefish and loads of trash fish.

Big Beaver Creek. Flows eastward through a heavily timbered, flat bottomed canyon for about 11 miles and then onto the Clark Fork floodplain for another 4 miles to junction with Little Beaver Creek; paralleled by a good gravel road from mouth to headwaters. An easily, and heavily fished, stream for mostly 8 to 10 inch cutthroat in the upper reaches, 6 to 14 inch brookies in beaver ponds along the middle reaches, and 10 to 12 inch rainbow below.

BIG LAKE
Courtesy H. Newman

BULL RIVER FLOATING
Courtesy Lance Schelvan

Big Creek. About 3 miles of fishing water here from the junction of the East, West and Middle Forks to its mouth on the St. Regis River a couple of miles above De Borgia; followed all the way (through timber and brushy bottomland) by a logging road. Big Creek, and its several forks, is very popular locally and fair fishing for mostly 6 to 9 inch cutthroat, plus an occasional brook and rainbow trout.

Big Lake. Thirty-eight acres, in open subalpine country, bare to the north but in timber on around; drained by Lake Creek (too small for fish) to Rattlesnake Creek. Big Lake is reached by a trail of sorts about ¾ of a mile north of Worden Lake. It is fair fishing for skinny 8 to 12 inch rainbow trout. P.S.: You will leave most of the motor scooter crowd down below on this one; a mile hike and 800 feet of climb is too danged much for their atrophied legs.

Big Rock Creek. Flows west from Bassoo Peak to the Thompson River 3 miles below the Bend Guard Station. Big Rock Creek flows over a falls 3 miles above its mouth and then through a steep, rock and timbered gorge below. There's good fishing all along for 8 to 12 inch cutthroat and brook trout. It's paralleled by a logging road for the first mile upstream, and a USFS trail off and on past the falls.

Big Terrace Lake. See Terrace Lakes.

Blossom Lakes. Two lakes that are almost one 31 acre lake with some swampy areas about two-thirds of the way toward the upper end. The upper one has a maximum depth of 17 feet, the lower one about 75 feet. They are in timbered country ½ mile east of the Idaho line, and are reached by a steep USFS trail 2½ miles from the Prospect Creek road, 30 miles west from Thompson Falls. Both are overpopulated with skinny (and ravenous) 6 to 8 inch brookies that provide fair-to-good fishing for not very many fishermen.

Blue Creek. Flows for about 1 mile through private ranchland from the junction of its East and West Forks to its mouth on the Clark Fork 1 mile east of the Idaho line. It's very poor fishing (reportedly) due to 2 big culverts under U.S. 10A, through which fish from the Clark Fork refuse to swim. The West Fork flows through a small, steep walled gorge where it forms a long narrow lake that is equally poor fishing. Neither are recommended.

Boiling Spring (or Dupont) Creek. A small, brushy, marshy stream in timbered mountains; crossed at the mouth by a logging road around the south side of Lower Thompson Lake and about 2½ miles upstream by the Powers Peak road from Twin Lakes. This creek is seldom fished and usually poor for 7 to 9 inch cutthroat trout.

Bonanza Lakes. Two lakes a few hundred yards apart in open alpine country (a sheepherder's paradise), about a third of a mile east of the Idaho line and 15 miles by crow-flight from Superior. Take the Cedar Creek road past headwaters to the state line trail, then north on the trail for 2½ miles, and finally east on a jeep trail to the lakes. Right Bonanza is about 19 acres; Left Bonanza 10 acres. Both are deep, easily accessible all around and good fishing for 6 to 9 inch brook trout.

Buck Lake. A beautiful, alpine lake, fairly deep, 3 acres, in a high, glaciated valley ½ mile below Wanless Lake. Buck Lake is reached by a USFS trail 9½ miles above the end of the McKay Creek road (around and beyond Goat Peak); or you can get in by trail 7 miles from the end of the Swamp Creek road. It's lightly fished (too far to hike) but good for them that makes it in for 7 to 9 inch cutthroat trout.

Buford Pond. See Two Mile Pond.

Bull River. A fair sized, easily fished, moderately popular stream that flows from Bull Lake for about 22 miles through an open, flat bottomed, steep walled mountain valley to the Clark Fork 7 miles below Noxon. It is followed by a good road all the way. It's generally good fishing for mostly 10 to 14 inch rainbow, plus a large variety of other species including cutthroat, brown, brook trout, Dolly Varden, whitefish, kokanee salmon, yellow perch and trash fish.

Burdette Creek. Take the South Fork Fish Creek road 9 miles west from Lolo Hot Springs, then a new logging road for 2 miles up Burdette Creek and a USFS trail for another 3 miles to headwaters in high, mountain meadowland. Burdette is a real swampy creek with lots of beaver ponds and lots of fish, 6 to 10 inch cutthroat.

Butler Creek. From Nine Mile House on U.S. 10 (25 miles west of Missoula) take the Nine Mile road for 8 miles to the mouth of Butler Creek, and then a county road for 4 miles (or about half way to headwaters) in timbered mountains. This is a very small stream but fair fishing for small (6 to 8 inch) cutthroat and rainbow trout.

Cabin Lake. A 13½ acre by 40 foot deep-with shallow-dropoffs lake, in high (elevation 7050 feet), timbered country reached by a USFS trail 1 mile from the end of the West Fork Thompson River road. It was planted with cutthroat in 1971, is heavily fished now — and good, too — for trout that'll average around 9 inches and range up to 2 feet or more in length.

Cabinet Gorge Reservoir. Extends from across the line in Idaho, eastward to about 5 miles past Thompson Falls. Cabinet Gorge is about 12 miles long and has a maximum width of about 1 mile, is accessible by U.S. 10A, and is a popular boating and fishing area. It contains a small population of 10 to 14 inch rainbow, as well as brown trout, Dolly Varden, largemouth bass, sunfish and yellow perch, but is (unfortunately) being "taken" by hordes of trash fish.

Cache Creek. A moderately popular, good fishing stream in timbered mountains, reached at the mouth by the South Fork Fish Creek road and followed by a good horse trail to headwaters. It produces good catches of 10 to 13 inch cutthroat and rainbow trout from all along its 9 miles of fishing water.

CARTER LAKE
Courtesy Phyllis Marsh

CLIFF LAKE
Courtesy H. Newman

CLARK FORK of the Columbia (Lower)

Canyon Creek. A tiny creek in timbered country, crossed at the mouth by the Vermilion Creek road and followed upstream for about a mile by road, and then another three miles by trail. Only the lower mile is fair fishing for 6 to 8 inch cutthroat trout.

Carters Lake. Lies ¼ mile east (down through the trees) from the old cabin near the middle of the Rattlesnake lake group — see directions to Worden Lake. It's in timber all around but with mostly open shores, 10 acres in size, about 20 feet deep at the most with shallow dropoffs and water level fluctuations of as much as 12 feet. It's not fished much, and usually with poor results, for skinny 8 to 12 inch rainbow trout.

Cataract Creek. Take the Vermilion Creek road for 5 miles to its mouth, then a USFS trail for 5 miles around a gorge below and up a narrow, flat bottomed canyon above to its headwaters on Seven Point Mountain. This moderately popular stream is good fishing for 6 to 9 inch cutthroat trout.

Cedar Creek. A nice little stream about 15 miles long (in country that was heavily mined in the early 1900's), followed by a good gravel road for its entire length down a steep, timbered canyon to the Clark Fork 1½ miles east of Superior. Cedar Creek is mostly open, easily and heavily fished for fair catches of 8 to 10 inch cutthroat, and an occasional rainbow, brook, brown or Dolly Varden trout. It's mostly on private land and fished by locals.

Cedar Log Creek. Flows from Cedar Log Lakes for 7 miles through real rough, timbered country to the West Fork of Fish Creek; reached at the mouth by the West Fork trail and paralleled for a few miles upstream by a trail along the ridge (¼ mile to the northwest). There's some pretty steep water below but it flattens above with quite a few nice holes. Cedar Log is seldom fished but fair for 6 to 8 inch cutthroat trout.

Cedar Log Lakes. Two lakes, a mile apart, in alpine timber just east of the Idaho line, 22 miles southwest (as the crow flies) from Alberton. The best way in is up the Montana Creek jeep road for 15 miles from its junction with the Fish Creek road to its upper end near Kid's Lake, Idaho (the last couple of miles are closed), and a steep ¼ mile hike by USFS trail to East Cedar, or a 2½ mile hike to West Cedar.

CLARK FORK of the Columbia (Lower)

They are 44 and 14 acres in size, deep enough to stand up and drink in except that West Cedar has mostly shallow dropoffs, and both are moderately popular. East Cedar is very good fishing as a rule for 6 to 12 inch cutthroat trout, West Cedar is fair for 7 to 15 inches.

Cement Gulch. A real small tributary of Trout Creek; reached at the mouth by the Trout Creek road and followed upstream by a (closed) logging road for 4 miles to headwaters. It is lightly fished but good in the lower reaches for pan-size cutthroat trout.

Cherry Creek. Flows for about 10 miles through steep, timbered mountains to the Clark Fork, 4 miles east of Thompson Falls. It is easily accessible by road along 5 miles of fishing water and is fairly popular for nice catches of 6 to 8 inch cutthroat and brook trout.

Chippewa Creek. A small, brushy stream in a timbered canyon, tributary to the South Fork Bull River. Chippewa is crossed at the mouth by road and followed upstream for a mile or so by an old (real gone) logging road. The lower mile is spotty but sometimes good fishing for 6 to 10 inch cutthroat trout.

Chippy Creek. Crossed at the mouth by the Thompson River road 10 miles below the Bend Guard Station and followed upstream for 1½ miles by a logging road, then by trail for 7 miles to headwaters. It is a small, brushy, hard-to-fish stream that is seldom bothered but fair fishing in the lower reaches for small brook trout, and cutthroat above.

Cirque Lakes 1, 2, 3, 4. See Upper Wanless Lakes.

Clear Creek. Flows east through steep, timbered mountains for 14 miles from Clear Peak to Prospect Creek near Thompson Falls. It is followed by a logging road and is heavily fished along the lower reaches for fair catches of 6 to 8 inch brook and cutthroat trout.

Clear Lake. At the head of Deer Creek a couple of miles west of Ward Peak in a little timbered glacial cirque with some open camping spots lies Clear Lake. It is about 9 acres by 40 feet deep and good fishing for 6 to 10 inch brookies and is reached (by a few) by taking the Little Joe Creek road to the Montana State line, then turning west on the State Line road for about 8 miles and finally dropping off by shanks mares for a final half mile to the shore.

Cliff Lake. Fifty acres, real deep, with aquatic vegetation around the lower end and rock cliffs above the upper end; in subalpine country 1 mile above Diamond Lake, and reached by 1½ miles of good trail from the end of the Dry Creek road about 20 miles from Superior. This is a beautiful lake that is only lightly fished because it supports only a small population of 7 to 16 inch cutthroat.

Combpest Creek. This is a small stream, mostly on river bottom ranchland across the Clark Fork from Plains. The lower 3 miles are readily accessible by road and are poor fishing for 6 to 8 inch rainbow trout.

Cooper Gulch. A small, clear, mountain stream reached at the mouth by the Prospect Creek road and followed by a logging road to headwaters. The lower 2 miles are heavily fished for pan-size cutthroat trout.

Copper (or Silvex) Lake. An old mine reservoir about 3 miles south of Lookout Pass; reached by a jeep trail 1½ miles from Interstate 90. Copper Lake covers 6 acres and is fairly shallow, with some aquatic vegetation, logs, etc., and a real muddy bottom. It was planted years ago and again in 1970 with cutthroat fingerling, and is fair-to-good fishing now for 7 to 13 inch trout.

Crater Lake. Was planted with cutthroat in the early '70's, isn't fished a heck of a lot but is good at times; is 16 acres in steep timbered mountains and is (seldom) reached by taking the North Fork Fish Creek trail for 5 miles to the Crater Creek (too small for fish) drainage and then cross-country for another 1½ miles southwest to the shore.

Crystal Lake. A 118 foot deep, 10 acre cirque lake in timbered mountains, reached by a doggon steep trail ¾s of a mile above the Deer Creek jeep road (for those who are too lazy to hike, it should be noted that scooters are not allowed here) bridge which is due to be removed as of 1975 or '76 or so. Anyhow, once you get there, if you get there, it's fair-to-good fishing for 4 to 9 inch brook trout.

Dalton Lake. A little old 2½ acre cirque lake a third of a mile east of the Idaho line in the old 1910 burn; reached (by the young and hardy) most easily by a ¾-of-a-mile cross-country hike due south across the intervening saddle from Pearl Lake. It was planted with cutthroat in 1969 and they're pretty good fishing now, fat and sassy and not scared of anything. They never see anyone.

Deep Creek. Not much of a fishing stream, it is small, and in a narrow timbered canyon. Deep Creek is crossed at the mouth by a county road 15 miles west of Thompson Falls and followed to headwaters by a USFS trail. There are a few small cutthroat here and there along the lower 3 miles but they seldom see a hook.

Deer Creek. A small stream flowing north through a narrow timbered canyon to the St. Regis River opposite DeBorgia; followed by a logging road for 6 miles upstream. The lower reaches are sometimes fished by local residents for 8 to 12 inch brookies, a few cutthroat and rainbow trout.

Deer Lake. This Deer Lake (there is at least one in most every major drainage) is only about 3½ acres, in rocky bluff country, open around the shores, maybe 15 feet deep beneath cliffs at the west end and shallower towards the outlet at the east end. It was planted with cutthroat in 1972 and is good fishing now for 10 to 12 inchers. To get there take a USFS trail 3½ miles west from the Liver Ridge road about 3 miles in from the West Fork Thompson River road.

Denna Mora Creek. Is reached at the mouth by Interstate 90 right by the rest stop 4 miles east of Lookout Pass, now how's that for service? A real small stream with only about a mile of fishable water in the lower, flat reaches, it is seldom fished but fair for a few messes each spring of 7 to 9 inch brook trout.

Devil Gap Creek. A very small; seldom if ever fished but fair for 6 to 9 inch cutthroat in the flat, upper reaches. The lower 3 miles in the canyon go dry. There is no trail but it's crossed at the mouth by the Martin Creek road.

Diamond Lake. Seventeen acres, in heavily timbered country reached by a USFS trail 1 mile from the end of the Dry Creek road (also about 1 mile below Cliff Lake). Diamond Lake is deep in the upper end, generally logged-up around the margins and mostly fished from raft or boat. It is quite popular and good fishing for 6 to 11 inch brook trout.

CLARK FORK of the Columbia (Lower)

Dominion Creek. A short (about 2 miles of fishable water) brushy little stream, tributary to the St. Regis River, 3 miles west of Saltese. To get there, leave Interstate 90 at interchange no. 1 and follow the river road for a mile by an access road to the mouths of Dominion and Randolph Creeks. It's not fished much except by local people but does produce fair catches of 8 to 10 inch brook and cutthroat trout.

Dry Creek. About 8 miles of fair-to-good fishing here for small brookies and cutthroat along the Dry Creek road and another 4 miles of paralleling trail in open country above and heavily timbered country below — west (and south of the Clark Fork River) from Superior. It's a fairly popular stream with local anglers.

Duckhead Lake. Good and deep, about 500 yards long by 150 yards wide, surrounded by steep grassy slopes and scattered timber with some talus along the east side. It is drained by (barren) Honeymoon Creek and is reached by a trail up that stream 3½ miles from the West Fork Thompson River road. There's some fair fishing here for small cutthroat and some 12 to 14 inch rainbow-cutthroat hybrids but it doesn't get fished much. It's uncommon good Grizzly country too.

DuPont Creek. See Boiling Spring Creek.

East Fork Big Creek. A small stream in a narrow, steep walled, heavily timbered canyon; easily accessible throughout its length of 5 miles by a logging road and lightly fished for 8 to 9 inch cutthroat, Dolly Varden and brook trout.

East Fork Blue Creek. Flows south past the old Scotchman Mine for 3 miles through timbered mountains to Blue Creek just above the Clark Fork floodplain. A real brushy stream that is hard to fish despite easy accessibility by old logging roads, it used to be good for 8 to 10 inch cutthroat and (reportedly) still is.

East Fork Bull River. Drains St. Paul Lake for 9 miles through a flat bottomed, timbered valley to the main river, readily accessible by road for 7 miles and a trail on in to the lake. It's lightly fished (pretty brushy) but good for small (4 to 8 inch) cutthroat, and a very few BIG (to 15 pounds) Dolly Varden.

DRY CREEK
Courtesy USFS

RATTLESNAKE WINTER
Courtesy R. Konizeski

East Cedar Lakes. See Cedar Log Lakes.

East Fork Dry Creek. A very small stream reached by a logging road 5 miles south from Thompson Falls. The lower 2 miles are fair fishing for small (6 to 8 inch) cutthroat. Note: The main stream goes dry BELOW the junction of the East and West Forks.

East Fork Elk Creek. A small stream in logged over, open hillside and brush country, seldom fished (except by a couple of loners) but if you know the holes the lower three miles are excellent at times for 9 to 11 inch cutthroat trout.

East Fork Petty Creek. Crossed at the mouth by the Petty Creek road (either from Alberton or the Lolo drainage) and followed by an old logging road for 4 miles to headwaters. The East Fork is easy to miss and not many people hit it — so, it's pretty fair fishing for pan-size cutthroat trout.

East Fork Rock Creek. Flows from Rock Lake for 1 mile down a barren, glaciated valley, then 1½ miles through Rock Meadows, and thence for another 2½ miles through a narrow, timbered valley to the main stream. The East Fork is followed by a jeep (the Heidelberg Mining Co.) road to the meadows where it widens and meanders about to form many nice pools that are heavily fished for some of the best 6 to 10 inch cutthroat fishing in this part of the state.

East Fork Twelve Mile Creek. A tiny stream crossed at the mouth by the Twelve Mile road and followed by the old U.S. 10 to headwaters. It is not often fished but the lower 3 miles contain a few pan-size cutthroat trout.

Eddy Creek. A small stream in a narrow, timbered valley; crossed at the mouth by U.S. 10, seven and a half miles east of Superior and followed by a logging road to headwaters. The lower reaches go almost dry in late summer and most people don't realize that there is enough water above to float the few trout that are in it. Some of the locals fish it for pan-size cutthroat.

Eddy Creek. A small stream flowing down a real steep, rocky gorge to the Clark Fork about 6 miles west of Weeksville and followed by a logging road for 6 miles to headwaters. The lower 2 miles are fair fishing for 6 to 8 inch cutthroat trout.

Elk Creek. A moderately popular stream flowing north from the junction of its East and West Forks for 6 miles through mostly bottom meadowland to the Clark Fork 1 mile east of Heron. Elk Creek is easily accessible by road and is fair-to-good fishing throughout its length for 6 to 10 inch cutthroat, brook trout, and some fair size Dolly Varden.

Elk (or Sims) Lake. You can drive to within 1½ miles of this one, up the Sims Creek jeep trail. Take off from the Vermilion Creek road about 2½ miles east above the West Fork, then 2½ miles up the Creek and finally a pack trail into the shore at 4100 feet above sea level. It's 7 acres in a steep, little, timbered pocket (brushy around the shore with an old cabin in a cottonwood grove on the southwest corner) in the lower Cabinet Range foothills. There's lots of fish food here and some real red hot fishing for 1 to 2 pound cutthroat — summer and winter. Not too many folks make it though, mostly a few locals.

Farmer's Lakes 5 & 6. Lake No. 5 is about 25 acres, No. 6 is 18 acres, a third of a mile apart (No. 5 is 120 feet higher and to the south) in a couple of small glacial cirques east of and 1300 feet or so straight down from Stuart Peak. No. 5 is in timber all around; No. 6 is open towards Stuart Peak (below rock and talus). Both are

CLARK FORK of the Columbia (Lower)

Courtesy Cary Hull

Courtesy Carry Hull

drained by the headwaters of High Falls Creek (which is too small and much too steep to support fish). The best way to get there is via shanks mares 3 miles right up the creek and west from its mouth on Rattlesnake Creek at a point about 15 miles above the Missoula city limits. When you arrive you will be alone — in a beautiful glacial cirque with no place to go but back out again. The fishing (?): well, No. 5 is fair at times for rainbow trout in the 12 to 14 inch category; No. 6 is sometimes good for 6 to 10 inch cutthroat.

First Creek. A small, brushy stream that occupies a narrow, 4 mile long valley in steep, timbered mountains. It is crossed at the mouth by U.S. 10, seven miles east of Superior and at headwaters by a logging road. The lower reaches go dry, but the upper reaches contain a few small cutthroat trout.

Fish Creek. A good sized stream with lots of nice big holes, in a flat bottomed, timbered canyon; flowing to the Clark Fork across the river and 4 miles east from Tarkio. The lower 8 miles are accessible by trail and the upper 7 miles (to the junction of North and West Forks above Clearwater Crossing) by a logging road from Rivulet. This is a heavily fished, open stream that is almost as good as the famous Rock Creek on the upper Clark Fork Drainage. It contains mostly 9 to 10 inch cutthroat and rainbow (a few will run up to 3 or 4 pounds) and lesser numbers of whitefish, brook and a very few brown trout and BIG (recorded to 12 pounds) Dolly Varden.

Fishtrap Creek. A fair size, but wadeable, stream about 18 miles long, in timbered mountains; crossed at the mouth by the Thompson River road and followed upstream by a logging road. Fishtrap is heavily fished and a consistent producer of 7 to 8 inch cutthroat, plus a few brook and rainbow trout.

Fishtrap Lake. A long, narrow, boomerang-shaped lake, about 50 acres, shallow at the upper end but 60 feet deep or better at the outlet, in timbered country but open around the shore. Fishtrap is accessible by the Lazier Creek road 11 miles above its mouth and is excellent fishing (although lots of folks skip it) for 6 to 10 inch cutthroat trout, a very few brook trout and too many suckers.

Flat Creek. A "kids' only" creek, too small for them to get into trouble but still fair fishing for 6 to 10 inch rainbow and cutthroat trout. There's a dam of sorts at the headwaters and a logging road below to its mouth at Superior, for about 2 miles of fishing.

CANADAS
Courtesy Cary Hull

Flat Rock Creek. A small stream, about 3 miles of fishing, in timbered country; crossed at the mouth by the Twelve Mile Creek road and followed by a good trail upstream to headwaters. It's fair-to-good for 6 to 10 inch rainbow and brook trout plus a few cutthroat in some nice beaver dams scattered along the upper reaches.

Four Lakes Creek. A very small stream (less than 2 miles of fishing) in high timbered mountains; reached by the West Fork of the Thompson River road and followed to headwaters (i.e., Frog, Grass, Porcupine, and Knowles Lakes — all barren) by a logging road. It's moderately popular (with disappointed lake fishermen?) for camp eat'n cutthroat trout.

French Lake. A deep, circular cirque lake, 18 acres, in timbered mountains. French Lake is reached by a USFS trail up the North Fork of Fish Creek 10 miles above Clearwater Crossing. It is seldom fished but contains large numbers of good-sized (7 to 15 inch), hard-to-catch cutthroat trout.

Frenchtown Gravel Pit. Lies between (north of) the old U.S. 10 and the new freeway a mile west of Frenchtown. It is a nice little recreation pond for locals and passers-by; about 5 acres with a maximum depth of perhaps 10 feet or so and some springs that keep it from freezing out. A good place to swim in the summer, iceboat in the winter, and fish for 10 to 15 inch brook trout, 10 to 14 inch large mouth bass, 6 to 7 inch bullheads and little pumpkinseeds the year around.

Glacier Lake. A deep, 25 acre, cirque lake in barren rocky country reached by a 1½ mile hike from the end of the Wrangle Creek (off Rattlesnake Creek) road. Glacier is reported to turn literally red at times with immense populations of fresh water shrimp. It's supposed to be barren (ho-ho).

Gold Lake. Take the Ward Creek road 3 miles above the mouth to the Up-Up Lookout road, then follow that road for about 8 miles to within a couple of miles south and finally hike a last ¾s of a mile like a wild man, west across country to the lake which is only 1½ acres by perhaps ten or twelve feet deep at most but fair fishing for 7-14 inch rainbow trout.

CLARK FORK of the Columbia (Lower)

Grant Creek. A pretty, little stream in heavily timbered mountains above, it is mostly dewatered for irrigation below. Grant Creek is crossed by U.S. 10 near the mouth a few miles west of Missoula and is followed by a county road and trail for 12 miles to headwaters. Grant Creek is popular early-season fishing for small cutthroat trout.

Graves Creek. A small stream, followed by road from its headwaters for 10 miles through a timbered canyon to the Clark Fork 7 miles west of Thompson Falls. It is readily accessible, heavily fished, and fair for 10 to 12 inch rainbow, and a very few cutthroat, brook trout, Dolly Varden and whitefish.

Hazel Lake. Seven acres, deep but with considerable aquatic vegetation around the margins, Hazel lies between Eagle and Ward Peaks in alpine timber and is approached by a good USFS horse trail a couple of miles above the Ward Creek stream crossing. It was planted with cutthroat in 1971, was good fishing in the middle 70s and could still be if it hasn't frozen out.

Heart Lake. A deep, jellybean-shaped, 60 acre alpine lake whose upper end lies below a big cliff and the lower portions in the open country of the old 1910 burn. It is reached by a good trail 3 miles up the South Fork from the Trout Creek road, and is lightly fished for good sized (7 to 16 inches) brook trout.

Heart Lake. Kind of looks like one from the air. Heart Lake lies at the base of a steep cliff in a little timbered cirque maybe an eighth of a mile northeast of Idaho. It is mostly open around the shore, is drained by the East Fork of Big Creek and is reached to within ¾s of a mile by the East Fork logging road and then a tough steep climb on in. It is 6½ acres with a maximum depth of 60 feet and is no better than fair fishing for 6 to 15 inch rainbow trout, but is moderately popular for all of that.

Henderson (or Little Ninemile) Creek. Real small, short (4 miles), in a narrow, timbered canyon across (south of) the river from Interstate 90, twelve miles above St. Regis. It's seldom fished but the lower mile contains 6 to 8 inch rainbow, cutthroat and brook trout that migrate up from the main river. There's an unmaintained trail, of sorts.

Henry Creek. A tiny (2 miles of fishing water) stream in logged-over grazing land, it flows to the Clark Fork 4 miles west of Paradise and is followed by a logging road to headwaters. Henry Creek is seldom bothered and just as well because it contains only a few small (6 to 8 inch) cutthroat trout.

High Falls Creek. A very small, steep tributary of Rattlesnake Creek; closed for the Missoula municipal water supply.

Hoodoo Creek. A little, open meadow creek, about 2 miles of fair 'kids' fishing at the head of Trout Creek a half mile below Hoodoo Pass.

Hoodoo Lake. Ten-and-a-half acres by 35 feet maximum depth, in high alpine meadows at the upper end and timber below; reached by a (closed) jeep road a mile from Hoodoo Pass on the Idaho State line at the head of the Trout Creek road and then a half mile hike on in to the shore. It is only lightly fished but excellent for eating size (6 to 9 inch) brookies.

Hub Lake. Five acres, 18 feet maximum depth, in subalpine timber ½ mile above Hazel Lake by a USFS trail. This one is too far back to be popular and it is just as well because there's not much here anyway, a very few 10 to 12 inch cutthroat trout. You could starve yet!

Illinois Creek.
Real short (about 2 miles), and real small (goes dry in spots), but has numerous beaver dams from ½ mile above its mouth on up. It is followed by trail from its mouth on Cedar Creek, and is fair fishing for 8 to 10 inch rainbow trout below the dams and 6 to 8 inchers in the ponds above them.

Imagine Lake.
Imagine that!

Irish Creek.
This little stream in timbered country is reached at its mouth by the Cache Creek trail and followed by trail for 3 miles to its headwaters. It is good fishing for 6 to 8 inch native cutthroat trout.

Johnson Creek.
About 5 miles long and followed all the way by a new logging road down its narrow canyon to the Clark Fork, 2 miles east of Superior. The lower reaches go dry but the upper reaches are fair fishing for small (6 to 8 inches) cutthroat trout.

Jungle Creek.
Flows through timbered country to Fishtrap Creek 1 mile above the mouth and is followed upstream by a new logging road. Jungle Creek is only occasionally fished but contains fair numbers of small cutthroat trout.

Kreises Pond.
A 9½ acre, fairly deep timbered pocket that was dammed up years ago and water ditched in. It can be reached by a jeep road from either Stony or Butler Creek (in the Ninemile Drainage) a couple of miles above the USFS Remount depot. It is stocked with cutthroat and was formerly excellent, but now only poor fishing. Good deer country.

Lake Creek.
A tiny tributary (2½ miles) of Rattlesnake Creek; part of the Missoula municipal water supply. It's closed to fishing but is (reportedly) fair for small cutthroat, brook and Dolly Varden trout.

Lang Creek.
Flows through open grazing land to the Thompson River 1½ miles east of Lower Thompson Lake. The lower reaches are readily accessible by the Thompson River road but are seldom fished, though they contain a few trout and yellow perch that have migrated up from the main stream.

ARGRASS
Courtesy USFS

LITTLE LAKE
Courtesy H. Newman

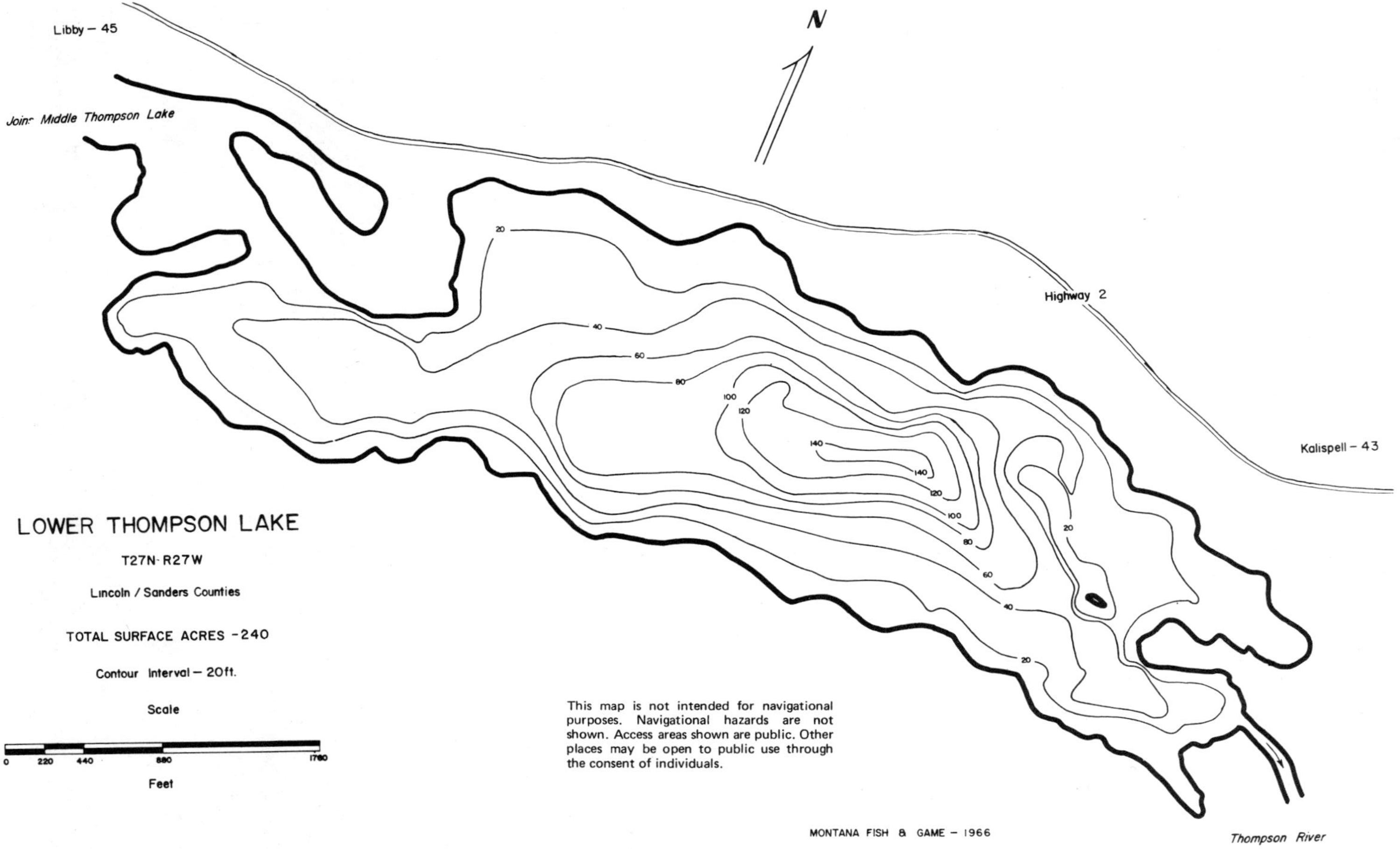

CLARK FORK of the Columbia (Lower)

LOWER THOMPSON LAKE

T27N-R27W

Lincoln / Sanders Counties

TOTAL SURFACE ACRES - 240

Contour Interval — 20 ft.

Scale 0 220 440 880 1760 Feet

This map is not intended for navigational purposes. Navigational hazards are not shown. Access areas shown are public. Other places may be open to public use through the consent of individuals.

MONTANA FISH & GAME – 1966

Libby – 45

Join- Middle Thompson Lake

Kalispell – 43

Highway 2

Thompson River

Lenore Lake. Can't be more than an acre or two in size, in a small cirque in logged over country at the head of a little mile-long intermittent drainage to (barren, too small for fish) Coyle Creek a mile west of Two Mile Creek 7½ miles upstream. There is no trail but the Coyle Creek logging road takes you down 1½ miles of switchbacks to Two Mile Creek. I've never gotten to Lenore but a friend who has tells me he saw something rising there the summer of '73.

Little Beaver Creek. A small, easily accessible by road stream flowing through a timbered canyon above and hay and pastureland below, to Big Beaver Creek 2 miles south of Whitepine. It is posted but the lower 7 miles are good fishing for 10 to 14 inch brook and some rainbow trout.

Little Joe Creek. Plenty of water here, but the creek is only about 2 miles long from the junction of its North and South Forks to its mouth on the St. Regis River a mile above town. Little Joe is followed by a logging road through its very narrow canyon and is fair fishing at times (it often goes dry in summer) for 7 to 9 inch cutthroat and a few rainbow and Dolly Varden trout. It does get fished a fair amount at that.

Little Lake. This little lake is one of the Rattlesnakes, a half mile north and a 500 foot climb up-and-back-down-again from Big Lake. It's open to the southwest, in timber to the northeast, 16 acres by 63 feet deep with mostly shallow dropoffs except along the southeast shore, and pretty good fishing for 6 to 8 inch cutthroat trout. It is one of the best of the rattlesnakes.

Little Ninemile Creek. See Henderson Creek.

Little Rock Creek. Flows through logged-over country for 9 miles to the Little Thompson River, and is accessible for the lower 2 or 3 miles by a logging road. It's moderately popular and fair fishing (in the lower reaches) for 6 to 8 inch rainbow trout and a few Dollys.

Little Terrace Lake. See Terrace Lakes.

Little Thompson River. A good-size stream, crossed at the mouth by the Thompson River road and followed by a logging road upstream for about 10 miles. The lower reaches flow through a rocky gorge, the middle reaches in open rangeland and the upper reaches in timbered mountains. It is heavily fished for the lower 11 miles and is good for 7 to 8 inch rainbow, brook, a few cutthroat, and a very few 10 to 14 inch Dolly Varden and whitefish.

Little Trout Creek. A small stream flowing for 4 miles through steep, timbered mountains to Trout Creek, ¼ mile below the guard station, and followed by a logging road for about 2½ miles upstream. It goes dry in the upper reaches but is fair below for 6 to 8 inch brook trout and a few cutthroat. It is seldom fished, is mostly on private land.

Loder Pond. One quarter of an acre in an excavated bog, bordered by grassy lands and encircled by barbed wire — 3 miles north up the Lynch Creek road from U.S. 10A just west of Plains. It was built in '67 and planted with brook trout for family fishing. Not available to the public but is shared with "the happiest pair of geese in the country."

Lost Creek. Small, in a moderately brushy, timbered canyon, about 6 miles long with 4 miles of fairly popular fishing water. Lost Creek is followed most of its length by a logging road. It is good fishing for 6 to 10 inch rainbow and cutthroat trout. There is a nice meadow with lots of meanders above.

CLARK FORK of the Columbia (Lower)

Lost Lake. A high lake in the old 1910 burn; about 5 acres, fairly deep, with cliffs at the upper end. It's drained by Lost Creek but is reached via the Cedar Creek road to within 3 miles, then by shanks mares along the old road to the shore. Lost Lake is not fished much but is good for 6 to 10 inch brook trout.

Lowell Lake. See Section 32 Lake.

Lower Thompson Lake. Just east of Middle Thompson Lake; 240 acres, over 150 feet deep in places, and easily accessible by U.S. 2 along the north side a few miles west of Yakt. It's fished by boat (if and when, but seldom) for poor catches of 10 to 13 inch Kokanee salmon, smallish yellow perch, good catches of 10 to 14 inch largemouth bass, cutthroat and brook trout.

Lower Trio Lakes. See Trio Lakes.

Lynch Creek. An easily fished, pastureland, 'kid's' creek, with about 3 miles of fair fishing water, available by county roads 3 miles west of Plains. It contains 6 to 8 inch cutthroat and brook trout.

Martin Creek. Flows through a timbered canyon for 10 miles to the Noxon Reservoir; is accessible at the mouth by the south side road 5 miles above the dam site, and is paralleled by a logging road to headwaters. It is generally good fishing for 6 to 10 inch brook and rainbow trout below, and cutthroat and rainbow trout above. The middle reaches usually contain fish also, but sometimes go dry in middle to late summer.

McCormick Creek. Small, about 7 miles long in timbered country, crossed at the mouth by the Ninemile Creek road 1 mile above Stark and followed upstream by a logging road for 2 miles beyond the gold dredgings. It's fair fishing above the dredgings for 6 to 8 inch cutthroat and rainbow trout.

McGREGOR LAKE
Courtesy Ernst Peterson

DDLE THOMPSON LOOKING WEST
urtesy USFS

MCKINLEY LAKE
Courtesy H. Newman

McGinnis Creek.
A small "headwaters" tributary of Little Thompson River, followed by a logging road for 2½ miles of fishing. It used to be good before the road was built, but is now poor at best for small (pansized) cutthroat trout.

McGregor Creek.
The outlet of McGregor Lake, it flows for 5 miles through timbered country to Lang Creek, and is followed by U.S. 10 all the way. It is lightly fished but fairly good in early season for 10 to 16 inch cutthroat trout.

McGregor Lake.
A large (1328 acres) lake in timbered country 30 miles west from Kalispell by U.S. 2. It is a fairly popular recreational area with a couple of dozen summer homes, two resorts, and 2 boat liveries. The fishing is good (mostly from boats) for 16 to 18 inch (a few to 16 pounds) Mackinaw and 10 inch rainbow trout. It was planted with Kokanee in 1964 and they're providing nothing but sport now.

McKinley Lake.
Twenty acres, dammed with 12 foot water level fluctuations and only 20 feet of maximum depth, in timber all around but with mostly open shores in a nice little glacial cirque; reached by a good USFS trail ½ mile southeast from Carter Lake. It is moderately popular but pretty slow fishing for 8 to 12 inch heavy-bodied rainbow trout.

McManus Creek.
About 4 miles long in timbered country; crossed by the old U.S. 10, two miles above the mouth and by the new U.S. 10 at the mouth, 3 miles east of Saltese. This little stream is seldom fished but the lower reaches are fair for 6 to 8 inch cutthroat and brook trout.

Middle Fork Big Creek.
Six miles long in a beaver-dammed, timbered canyon; followed by a logging road for about 4 miles of fishing. The Middle Fork isn't fished much but is fair for small cutthroat, Dolly Varden and brook trout.

Middle Fork Bull River.
Crossed at the mouth by a road 2½ miles below Bull Lake, and followed by a poor trail to headwaters. There are about 4 miles of fishing water in a timbered canyon with lots of down cedar, log jams and the like. It's not bothered much (too tough), but good for 6 to 10 inch cutthroat trout.

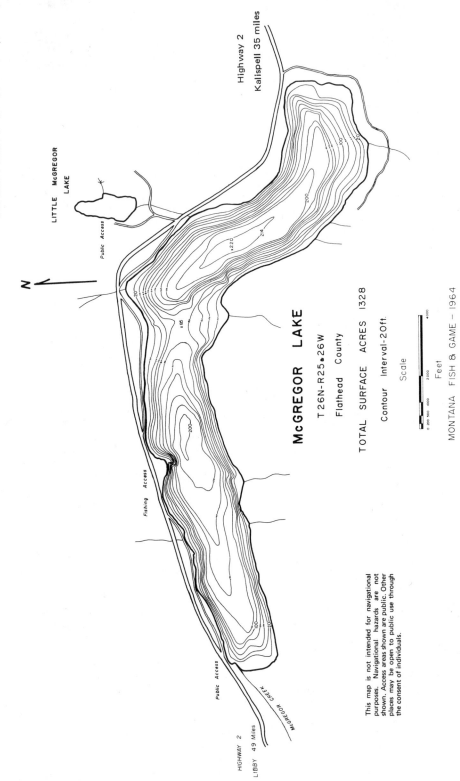

ourtesy USFS

GLACIER LAKE PRODUCES
Courtesy H. Newman

Middle Fork Indian Creek.
At the headwaters of the West Fork Fish Creek in logged-over country accessible by trail for 4 miles from end to end. The Middle Fork has a 2 acre beaver pond about 2½ miles above its mouth that is good fishing for small cutthroat trout. Otherwise, this stream is a flop.

Middle Thompson Lake.
A big (602 acre), deep lake in timbered country 8 miles west of McGregor Lake on U.S. 2. It is a fairly popular recreational area with about 20 summer homes and 1 boat livery. The fishing is good (mostly from boats) for 10 to 14 inch Kokanee Salmon and largemouth bass, smallish yellow perch and sunfish, and fair for large (to 8 pounds) cutthroat trout.

Middle Trio Lakes.
See Trio Lakes.

Mill Creek.
Ten miles of fishing water in open meadow above and in a steep, rock and timbered gorge below. The lower reaches are almost completely dewatered for irrigation and what's left is crossed by U.S. 10 at Frenchtown; the upper reaches are followed by a logging road for about 8 miles to headwaters. It is real good fishing in the gorge for 4 to 10 inch cutthroat. Up in the meadows it's mostly brook trout. This is an excellent kids' creek but the newly improved logging road may well finish it.

Missoula Gulch.
Small and brushy, it drains Missoula Lake for 2 miles to Bonanza Gulch and thence to Oregon Gulch. It is crossed by the lake road ½ mile above the mouth. The lower ¼ mile is in a big flat and is fair fishing for 8 to 12 inch cutthroat trout.

Missoula's Kids' Pond.
On 39th Street about 1 mile below the mouth of Pattee canyon. A shallow, man-made, gravel-bottomed pond of about 2 acres. It's stocked annually with 7 to 12 inch rainbow trout (about 2000 per year) and is at times excellent fishing "for kids only."

Missoula Lake.
Reached by the Cedar Creek road, 25 miles southwest from Superior in the old 1910 burn. It's about 12 acres, over 50 feet deep, with mostly steep dropoffs, partly brushy and partly open around the shore. It used to have lots of small brook trout that were poisoned out in 1962. Since then it has been planted with rainbow trout and has provided fair fishing. There's an unimproved campground and rafts.

CLARK FORK of the Columbia (Lower)

This map is not intended for navigational purposes. Navigational hazards are not shown. Access areas shown are public. Other places may be open to public use through the consent of individuals.

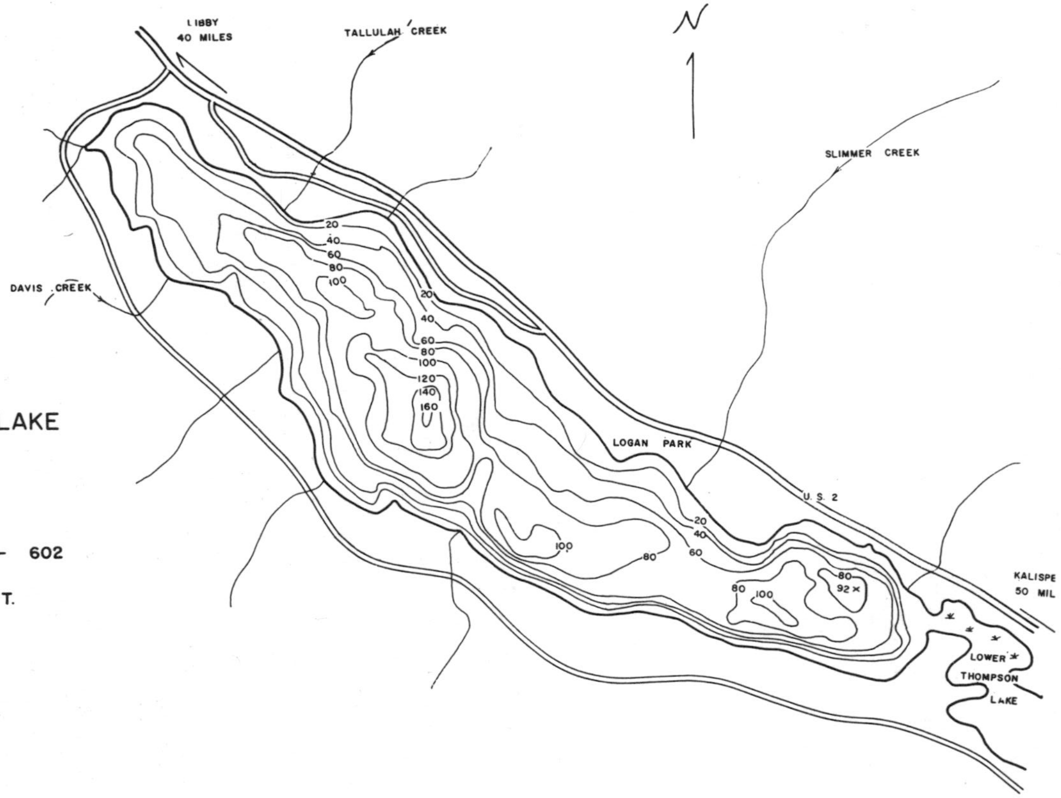

MIDDLE THOMPSON LAKE

T26N R27W

LINCOLN COUNTY

TOTAL SURFACE ACRES — 602

CONTOUR INTERVAL — 20 FT.

MONTANA FISH & GAME — 1965

OUT OF SEASON
Courtesy Meir's Studio

Montana Creek. A small tributary of Cache Creek in a little, flat bottomed, timbered valley. There are about 3½ miles of meandering fishing water in the lower reaches, the first mile of which is accessible by trail. It is good for 6 to 8 inch cutthroat trout.

Moore Lake. A deep, 13 acre sub-alpine lake in timbered country, accessible by the South Fork Little Joe Creek road, 14 miles from St. Regis. Moore Lake lies below talus slopes at the upper end and a small cliff on the east side. It has quite a few aquatic plants around the margins and is heavily fished for good catches of 6 to 13 inch brook trout.

Moran Basin Lakes. Two lakes, but the lower one is a barren pothole. Upper Moran is a deep, 14 acre lake in high, timbered 'Wilderness Area' reached by a USFS horse trail 2 miles from the end of the East Fork of the Bull River road. It isn't fished much but has been excellent for skinny, 10 to 11 inch cutthroat and surely does support a burdgening population of ravenous mosquitoes. It's worth your manhood to heed nature's call.

Mosquito Creek. A small "kids'" creek flowing from timbered mountains above, across the open floodplain below, to the Clark Fork river at Belknap. It is followed by a road for about 3 miles of fair fishing for 6 to 8 inch brook trout.

Murr Creek. This little stream flows through rocky, sparsely timbered country. It is crossed at the mouth by the Thompson River road and followed upstream by a logging road and trail for about 2 miles of good fishing for 6 to 8 inch cutthroat trout. It's only lightly fished.

Nemote Creek. A brushy little stream with about 2 miles of fishing in hay and pastureland, crossed by U.S. 10 three miles west of the Tarkio turnoff. This seldom fished creek has some good holes which are fair in early season for 6 to 10 inch brook and some rainbow trout.

Ninemile Creek. A fair-sized, easily wadeable stream in open hay and pastureland, followed by county roads from its mouth near U.S. 10, (25 miles west of Missoula) upstream for about 25 miles of fishing water. Ninemile used to be an excellent cutthroat stream but logging and dredging operations have hurt it badly. It is now mostly an early-season stream that is good for nice size rainbow, Dolly Varden, cutthroat trout and whitefish, that migrate up from the Clark Fork.

North Fork Bull River.
About 5 miles of fishing water in a beautiful, timbered canyon, accessible by a USFS horse trail all the way. The North Fork is only lightly fished but good for 6 to 10 inch cutthroat, and a few BIG Dollys.

North Fork East Fork Bull River.
A real small stream in a steep, timbered canyon; crossed at the mouth by the East Fork road and accessible by trail for about 2½ miles of fishing water. It's only lightly fished but good for small (pan size) cutthroat, and believe it or not, a very few BIG (recorded to 15 pounds) Dolly Varden trout.

North Fork Fish Creek.
A good fishing stream in a narrow, timbered- and brushed-up canyon; followed to headwaters (7 miles) by a good trail. This creek is lightly fished for small rainbow and cutthroat trout.

North Fork Little Joe Creek.
A good fishing stream followed by a two lane logging road for 7 miles through a heavily timbered canyon from its mouth to Broadaxe junction. It contains 6 to 8 inch cutthroat and Dolly Varden trout.

North Fork Little Thompson River.
About 3 miles of fair fishing here for 6 to 8 inch rainbow and brook trout; crossed at the mouth by the Thompson River road and followed upstream by trail. A moderately popular stream.

North Fork Second Creek.
This is a very small, brushy stream in a steep, timbered canyon. It is hard to fish and not bothered much for the few small (6 to 7 inch) cutthroat it contains. If you're still interested, there's a logging road for the first 2 miles.

North Fork Trout Creek.
Here is a small, seldom fished stream in heavily timbered country at the headwaters of Trout Creek. It is paralleled by a jeep road for 2 miles of fair fishing for pan-size cutthroat trout, but you gotta walk cause there're no vehicles allowed here.

CAMPERS
Courtesy USDI,BR

NOXON RAPIDS RESERVOIR
Courtesy Ernst Peterson

TROUT CREEK SWIMMING AREA(?) ON NOXON RESERVOIR
Courtesy Danny On

Noxon Rapids Reservoir.
A large reservoir on the Clark Fork, about ½ mile wide at the maximum by 15 miles long; accessible along the north side by U.S. 10A, and on the south side by county roads. Noxon is fished (mostly by boat) and used to be excellent for 10 to 14 inch rainbow along with a few small cutthroat and brook trout, and some lunker Dolly Varden, brown trout and largemouth bass. Unfortunately, it is being taken over by trash fish but is still fair-to-good in places if *you know the water.* Fished mostly by local folks.

Oregon Gulch.
A brushy gulch where the creeks come in, in timbered country through which flows this good size stream for 3 miles to Lost Creek. The upper and lower reaches are reached by the Cedar Creek and Lost Creek roads, the middle reaches by foot trail only. It is moderately popular and fair fishing for 7 to 8 inch cutthroat plus a very few rainbow trout.

Oregon Lakes.
Three moderately popular, deep cirque lakes in high (but heavily timbered) country at the head of Oregon Gulch. Lower Oregon Lake is 3 acres in size and accessible by three-quarters of a mile of steep trail from the Cedar Creek road. Middle Oregon is 27 acres and accessible by trail ½ mile beyond the lower lake. Upper Oregon Lake is 11 acres by about 40 feet deep and lies ½ mile up the drainage from Middle Oregon. However, it is most easily reached by a three hundred yard hike from the Idaho State Line trail. All three lakes are spotty, but often good, fishing for 6 to 10 inch brook trout.

Outlaw Lake.
Is reached by a cross-country hike about 2½ miles up the (barren, too small for fish) drainage south from the Clark Fork at a point about ten miles east of Thompson Falls. It was planted with cutthroat in 1970 and is good fishing now for 10 to 12 inchers but not many people know about it.

Packer Creek.
Flows right through Saltese and is followed by logging roads for 2½ miles of fair fishing in brushy alder choked country for pan size cutthroat. It used to be closed for the municipal water supply.

CLARK FORK of the Columbia (Lower)

Pearl Lake. A 14 acre alpine lake in the old 1910 burn; reached by a rough cross-country hike ½ mile up the drainage (east) from Heart Lake (the one on Trout Creek). It is seldom fished but was planted with cutthroat in 1969. They have done well through the summer of '74 and should still provide good fishing now.

Petty Creek. An open ranchland stream reached via the Graves Creek road (in the Bitterroot Drainage) and followed by a county road for 12 miles from headwaters to its mouth on the Clark Fork 1 mile east and across the river from Alberton. The upper reaches have lots of nice water; the lower end sinks in summer. It's heavily fished in the early season and good for 6 to 9 inch cutthroat, brook and whitefish, and a few rainbow trout.

Pilcher Creek. This small (3 miles), steep tributary of Rattlesnake Creek has been closed to fishing for many years as it is utilized for the Missoula municipal water supply.

Pilgrim Creek. Followed upstream from its mouth at the Noxon Guard Station by a good gravel road for 6 miles through hay and pastureland to the junction of its South and West Forks. A moderately popular creek and good fishing for 8 to 10 inch rainbow, cutthroat and brook trout; and fair for 16 to 24 inch brown trout.

Poacher Lake. You don't have to be one to fish it but you do have to hike cross-country a ways from the Quartz Creek logging road to get to it at about 5600 feet above sea level in the headwaters. Its 8½ acres by 34 feet maximum depth, was planted with cutthroat in the early 70's and is good fishing indeed for 10 to 12 inch trout.

Prospect Creek. A fair-sized stream, 21 miles long, mostly in timbered country and followed by a good gravel road from headwaters to the Clark Fork opposite Thompson Falls. The lower 5 miles are heavily fished by locals for a few 10 to 14 inch rainbow, brook, Dolly Varden, and once in awhile a brown trout. The upper reaches support a fair population of 6 to 10 inch cutthroat.

Quartz Creek. A small stream crossed at the mouth by a county road along the south side of the Clark Fork 1 mile west of Tarkio and followed upstream by a logging road through about 5 miles of steep, timbered country. Quartz Creek has been polluted by mining but is fair fishing along the lower 3½ miles for small "resident" rainbow and a few large spawners from the main river.

Rainy Creek. Quite a small stream (about 2 miles of fishing) in fairly open country south of Lookout Pass; reached at the mouth by Interstate 90 four and a half miles west of Saltese and followed past headwaters by a logging road. It is lightly fished and only poor to fair in early season for small rainbow, brook trout and an occasional whitefish.

Randolph Creek. A small, brushy stream, crossed at the mouth by Interstate 90 four miles west of Saltese (see Dominion Creek); is seldom fished but fair for a few miles above the road for pan-sized cutthroat and brook trout, and a few good-sized whitefish.

Rattlesnake Creek. The City of Missoula municipal water supply; Rattlesnake Creek is a clear, mountain stream that flows to the Clark Fork at Missoula and is followed by road for 20 miles up a narrow, steep canyon to headwaters. It heads in as many as 13 subalpine, cirque lakes, some of which contain fish and are open to fishing. The creek itself is closed except for the lowermost reaches, unfortunately so because it is potentially excellent fishing for 8 to 16 inch cutthroat, Dolly Varden and brook trout.

RATTLESNAKE CREEK IN WINTER
Courtesy USFS

SMOKEJUMPER
Courtesy USFS

Rattlesnake Lakes. See Big, Carters, Farmer's, Little, McKinley, Sheridan, and Worden Lakes.

Rock Creek. This little stream is crossed at the mouth by a road 2 miles east and across the Clark Fork from Tarkio and is followed upstream by a logging road and trail through steep, timbered mountains. It's seldom fished but fair for 6 to 8 inch cutthroat trout in early season for about 1½ miles; the lower reaches go dry in summer.

Rattlesnake Lake No. 3. Twenty acres in a small glacial cirque with bare open headwalls to the west and timbered on around. If you want to get there, and I don't believe you *really* do, about the best way in short of a chopper is to hike 2½ miles up Rattlesnake Creek above the mouth of (barren) High Falls Creek, which empties into the Rattlesnake from the west across the creek from the road, or about ½ mile above the mouth of (barren) Porcupine Creek which comes in from the east. Now ford the Rattlesnake (in truth 'tis tough but you will have to leave your smoke-belching motor bikes here) and make like a mountain goat for 1¼ miles up the series of little intermittant waterfalls that are looking you in the eye. Should you make it you will find the fishing generally good for 8 to 10 inch cutthroat trout. Good luck — you'll need it!

Rock Creek. Crossed at the mouth by U.S. 10, one mile north of the Noxon Dam and followed upstream by a logging road to the junction of its East and West Forks. It is heavily fished for about 6 miles and is good for 6 to 12 inch cutthroat trout.

Rock Lake. Forty-two acres, deep, in a barren, glaciated valley below Rock Peak; reached by trail 3 miles above the end of the East Fork of Rock Creek (jeep) road. Rock Lake sustains a moderate amount of fishing pressure for BIG (to 5 or 6 pounds) cutthroat that can be seen but are seldom caught.

Rock Meadow Ponds. Considered to be a part of Rock Meadows on the East Fork of Rock Creek. There is no maintained trail. From the lower meadows (about 2 miles below Rock Lake) take the small tributary due south for 2 miles to the first of a string of 5 ponds, each about ⅛ of a mile apart. They're all reported to be very good fishing for 6 to 10 inch cutthroat trout but you will pay dearly for them (with your blood — once the hordes of ravenous mosquitoes find you).

CLARK FORK of the Columbia (Lower)

Rudie Lake. Three-quarters of a mile south (on the contour through timbered country) from Crystal Lake; reached by a poor trail 2 miles from the end of the Deer Creek logging road. Rudie is about 12 feet deep at its deepest, 7 acres, and seldom fished but excellent indeed for 6 to 9 inch brook trout.

St. Louis Creek. A tiny tributary of Ninemile Creek; crossed at the mouth by the Ninemile road at Old Town, and followed upstream through heavily timbered country for about 1½ miles by old logging and mining roads. This is a moderately popular, early-season stream that is good fishing for 6 to 8 inch rainbow and cutthroat trout.

St. Paul Lake. A deep, 14 acre lake in barren, glaciated country (elevation about 7200 feet) reached by a USFS horse trail 4 miles from the end of the East Fork of the Bull River road. It has been planted with cutthroat but is only poor fishing at best and is scheduled for a replant.

St. Regis Lakes. Two (East and West) lakes in high, barren, alpine country accessible by trail 2 miles west from Sildex about a mile south of Lookout Pass and ½ mile west on Interstate 90. West St. Regis is only 2 acres and barren. East St. Regis is 7.2 acres, 44 feet deep in at least one spot, with some scattered timber around it and fair-to-good fishing, though seldom bothered, for 6 to 9 inch brook trout.

St. Regis River. A fair-sized but wadeable, heavily fished stream followed by Interstate 90 for 40 miles through timbered mountains from its headwaters below Lookout Pass to the Clark Fork at St. Regis. It is fair fishing for mostly 4 to 10 inch cutthroat and some brook trout.

Savenac Creek. A brushy little stream that flows through timbered mountains to the St. Regis River 2½ miles west of DeBorgia; is crossed at the mouth by Interstate 90 and followed upstream by a (private) logging road for about 2 miles of good fishing. It contains mostly 8 to 10 inch cutthroat and is a good kids' creek, they can nock-em-over in the headwaters.

PACKING IN
Courtesy USFS

SHERIDAN LAKE AT MAXIMUM DRAWDOWN
Courtesy H. Newman

Savenac Nursery Pond. A 1 acre rearing pond on the north side of U.S. 10, five miles east of Saltese. It has been periodically stocked with mostly Rainbow trout that are fairly small in the spring and are usually turned into the St. Regis River in the fall. There is limited fishing here for "kids only."

Schley Lake. Two acres, deep, just below, southeast, of the summit of Schley mountain at the head of Montana creek. Reached (if you can find it) by an unmarked "trail" ¼ mile above the end of a logging road from Cache creek past the mouth and up the ridge west of Montana creek. It is surrepititiously planted now and again with cutthroat fingerling and is reportedly red-hot fishing for a year or two — by those in the know. There's no reproduction here.

Second Creek. A tiny stream that dries up below and is seldom fished, though fair for about 1 mile along the middle reaches for 6 to 8 inch cutthroat. It is crossed at the mouth by U.S. 10, six miles east of Superior and followed upstream by a logging road.

Section 32 (or Lowell) Lake. A beautiful, 28 acre by 70 foot deep with steep dropoffs along the southwest side, lake in a northeast facing Cabinet Mountain cirque below talus slopes to the south and west and some stunted alpine fir on the east. To get there you hike 1½ miles cross-country up the right hand drainage from the end of the North Fork Bull River trail in steep Grizzly country. This is truly a superb trip for the scenery as well as the fishing, which is excellent when (and if) you get there for 8 to 12 inch cutthroat. 'Tis overpopulated. You'll like it.

Semen Creek. Not much of a creek, very brushy and seldom fished although it reportedly contains a fair population of 6 to 8 inch native cutthroat trout. It is crossed at the mouth by the Thompson River road 10 miles below the Bend Guard Station.

Sheridan Lake. A shallow, 20 acre, dammed Rattlesnake lake with about 10 feet of water-level fluctuation, in timber all around but with mostly open shoreline; a quarter of a mile east and 250 feet below, down the outlet from Big Lake. It has lots of little 8 to 10 inch rainbow trout in it just right for the skillet.

Shroder Creek. A little tributary of the Thompson River 3 miles above the Bend Guard Station. Shroder is accessible by trail for about 2 miles of fair fishing in a narrow, rocky gorge, and is lightly fished, mostly by local people, for 6 to 8 inch cutthroat trout.

Siamese Lakes. Two; Lower Siamese is real deep, 34 acres, with cliffs along its upper side, in timbered alpine country. It is a steep ½ mile hike from the upper lake. Upper Siamese is 28 acres, deep, and is reached by a ½ mile hike from the State Line trail, or 10 miles by a USFS horse trail from the Clearwater Crossing up the West Fork of Fish Creek, and then 3 miles by a hunters' trail. Both lakes are popular with the dudes, and fair to excellent at times for 9 to 15 inch rainbow and 7 to 14 inch cutthroat trout.

Silver Creek. Closed to fishing above the Saltese water supply dam but reportedly fair fishing for 8 to 10 inch cutthroat and brook trout below the dam.

Silver Lake. A high alpine lake in the old 1910 burn just east of the Idaho line; reached by a very steep jeep road 7 miles south from Saltese. It is 17½ acres, deep but with mostly shallow dropoffs, bordered by rocks at the upper end, timber and willow elsewhere and with a campground at the lower end. It is moderately popular and good fishing for mostly 5 to 10 inch brook trout and some 10-19 inch browns.

Silvex Lake. See Copper Lake.

Sim's Lake. See Elk Lake.

Six Mile Creek. A small (open ranchland) creek crossed at the mouth by U.S. 10, six miles west of Frenchtown and followed upstream by county roads. It's fair "kid's" fishing for mostly small brook trout, and a few cutthroat and rainbow trout up from the Clark Fork.

Sleman Reservoir. A once popular, 1 acre, ice-fishing pond ⅛ of a mile below the Ninemile Remount Station on a blacktop road. It has contained goodly numbers of very nice (14 to 20 inch) cutthroat but was drained dry in 1962. However more trout have found their way in from Ninemile Creek and repopulated this pond.

Slimmer Creek. Four miles long with two lakes and some beaver ponds on the upper reaches; crossed at the mouth by the Thompson Lakes highway (U.S. 2), and followed to headwaters by a USFS trail. It is fair fishing for small brook and cutthroat trout.

South Branch Big Beaver Creek. Four miles long in a steep, timbered canyon; reached at the mouth by the Big Beaver road and followed upstream by a good logging road. There is fair early season fishing here for small cutthroat trout.

South Branch West Fork Trout Creek. A tiny stream (about 2½ miles of fishing water) in a steep, timbered canyon just this side of Idaho; accessible at the mouth by a good USFS trail 4 miles from the Trout Creek road, and followed upstream to headwaters. It is seldom fished as people just don't get this far back, but has some nice pools and is fair for 6 to 10 inch cutthroat trout.

South Fork Bull River. This small stream is reached at the mouth by a road a few miles south of Bull Lake and is followed upstream for several miles through a beautiful, timbered canyon. It is terribly brushy, hard to fish, and seldom bothered but contains quite a few 6 to 12 inch cutthroat trout.

STAHL POND
Courtesy USDA, SCS

SNOW SCENE ON THE THOMPSON RIVER
Courtesy USFS

South Fork Fish Creek.
About 15 miles of fair fishing for 6 to 10 inch rainbow trout, and a few cutthroat and brook trout, plus an occasional large Dolly Varden. The South Fork is available at the mouth by a logging road 3 miles east from Tarkio, and along the headwaters from the Lolo (Bitterroot) Drainage. The middle reaches are followed by an ancient USFS horse trail.

South Fork Little Joe Creek.
Seven miles of good fishing for 3 to 10 inch (this is good fishing?) cutthroat and a few small Dolly Varden in a narrow, timbered canyon west of St. Regis; easily accessible by a logging road all the way.

South Fork Martin Creek.
A small stream in a timbered canyon followed by a jeep road (after fording Martin Creek) for 3 miles of fair fishing for 8 to 10 inch cutthroat trout. It's moderately popular with local residents.

South Fork Nemote Creek.
Three miles of fishing water in meadowland below and timber above; accessible by a logging road. The South Fork is a good early-season creek for 6 to 10 inch brook and rainbow trout.

South Fork Trout Creek.
Reached at the mouth by the Trout Creek road 25 miles from Superior, and followed by a logging road and 2 miles of good USFS trail upstream for about 4 miles of fishing water. It is a nice open stream in high, timbered country that is moderately popular for fair catches of small native cutthroat trout.

Square Lake.
A 12 acre alpine lake in timber a mile southeast of Hazel Lake below Ward Peak. To reach this one see the directions for Clear Lake but take off from the State Line road about 5 miles from the end of the Little Joe road (west). There's a parking spot at the takeoff point. You won't have much company once you've arrived but the fishing is fair for small brook and cutthroat trout.

Squaw Creek.
Six miles long; crossed at the mouth by U.S. 10 and county roads 5 miles west from Thompson Falls, and followed upstream by a USFS trail. The lower 1½ miles go dry in summer, the upper reaches in a narrow, timbered canyon, are fair fishing for 6 to 8 inch cutthroat and brook trout.

Courtesy Danny On

Stahl Pond. Drive 3 miles west from Missoula on the old Mullan road. Park your car for a quick getaway and strike out south over the fence and through the mud for another mile to this very brushy, spring-fed, dammed slough. It was planted with rainbow trout fingerling in '67 that the owner figures on catching himself between duck hunting forays. It's private!

Stony Lake. Eight and a half acres by no more than 7 feet deep (but it doesn't freeze out) in timbered, windfall country; reached by jeep road about 5 miles up the West Fork of Fishtrap Creek from the main stem. Stony is mostly open around the shore, moderately popular, and good fishing for 8 to 10 inch cutthroat trout.

Straight Creek. A really nice stream to fish, with about 5 miles of good water in the old 1910 burn. It is followed by a trail all the way from its mouth on the North Fork of Fish Creek just above the Clearwater Crossing, to some big falls at the head of the fishing water. Not many people come here, but the fishing is good for 7 to 9 inch rainbow, cutthroat, and maybe a few brook trout. There are no lunkers 'tis true, but this one you ought to hit.

Surveyor Creek. Drains Surveyor Lake for 4 miles through timbered country to the South Fork of Fish Creek. The entire creek is accessible by a logging road but is only poor fishing for small (pan size) cutthroat and rainbow trout. Not much of a stream.

Surveyor Lake. Reached by the Surveyor Creek logging road to within ½ mile from the second crossing over the creek, and then a short climb in up a good trail. It is 18 acres, over 50 feet deep at the upper end and shallow at the outlet. The fishing — is no better than fair for 6 to 11 inch rainbow trout.

Swamp Creek. About 7 miles of fishing water in pasture and scattered timberland; easily accessible all the way by a road about 10 miles west and across the river from Plains. A heavily fished, early-season creek for most anything that swims in the Clark Fork; mostly 6 to 10 inch cutthroat, rainbow and a few 12 to 16 inch brown trout.

FISHING THE THOMPSON RIVER
Courtesy Ernst Peterson

Sylvan Lake. Take the Vermilion River road for 20 miles to Happy Gulch, then a logging road up the gulch to within ½ mile of this very shallow 6 acre lake which occupies a cirque high on the southern slopes of Sliderock Mountain and, incidentally, drains to Miller Creek (barren, too small for trout). It is reported to be fair fishing for brook and cutthroat trout. There's a nice public campground at the north end with picnic tables and fireplaces.

Tamarack Creek. Is about 3 miles (at the mouth) north of St. Regis by paved road (State 461) and then you can drive on up it for 7 miles through some open but mostly timbered land. It pretty well dries up in the lower (open) reaches but is fair fishing above for pan-size cutthroat and rainbow trout.

Terrace Lakes. Big and Little, about a quarter of a mile apart in alpine (barren rock and cliff but with some scattered patches of timber) country, reached by 1 mile of steep cross-country hiking from the end of the (abandoned) USFS West Fork Fishtrap Creek logging road. Big Terrace is only 16 acres but has a 160 foot maximum depth with mostly shallow dropoffs and lots of drowned logs, and cliffs at the upper end. Little Terrace is 7½ acres by 24 feet deep at most. Both were planted with cutthroat in 1971 and are good fishing (for the few that hike in) for 10 to 12 inch trout.

Thompson Creek. A small stream, reached at the mouth 1½ miles west of Superior and for 3 miles upstream by road. There are about 4 miles of fishing which is fairly good for the first mile above the end of the road (brushy here) but poor below, for small cutthroat. A kids' creek.

Thompson Lakes. See Upper, Middle, Lower.

Thompson River. A good-sized stream flowing south from Lower Thompson Lake for 50 miles through a wide range of open meadow and timber, private and National Forest land to the Clark Fork 7 miles east of Thompson Falls; easily accessible all the way by public and private roads. This is a moderately popular river (really just a creek above) that is generally good fishing for mostly 8 to 12 inch brook and cutthroat trout in the upper reaches, and rainbow, whitefish, largemouth bass, yellow perch and a few Dollys below — especially right near where the river leaves Lower Thompson Lake.

FANLESS LAKE
Courtesy USFS

TROUT CREEK IN WINTER
Courtesy USFS

CLARK FORK of the Columbia (Lower)

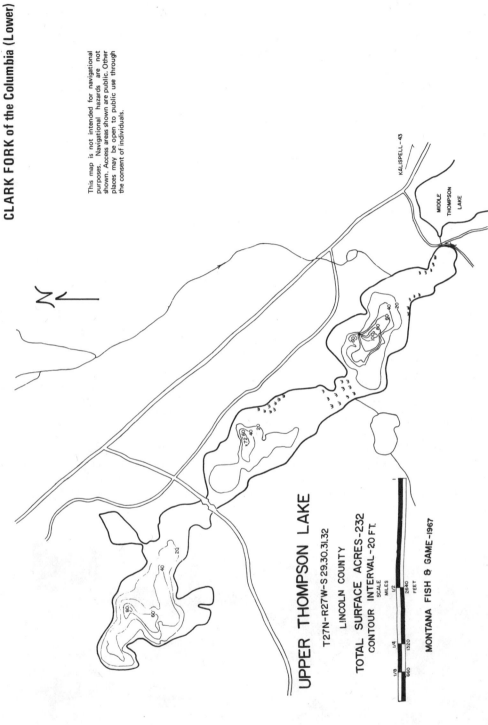

Torino Creek. A small headwater tributary of Dry Creek; accessible by road for 1 mile of poor-to-fair fishing (from the mouth to the big falls). Torino contains mostly 6 to 8 inch brook, and a few cutthroat and rainbow trout.

Trail Lake. Nineteen feet deep at the most, 11½ acres, in a timbered pocket reached by trail 2 miles from the end of the Trail Creek road (on the headwaters of Trout Creek). Trail Lake was planted years ago and is now fair fishing, but spotty, for 6 to 13 inch brook trout.

Triangle Pond. Drive 3 miles west from Noxon along the south side of Noxon Rapids Reservoir to this 7½ acre by 32 foot maximum depth, triangular shaped, flooded borrow pit just south of the road. It is planted with cutthroat trout and provides some good fishing for 12 to 14 inchers, mostly by locals who know it's there.

Trio Lakes. Three seldom-visited alpine lakes about ¼ mile apart in the old 1910 burn. Easiest way in is by way of the Trout Creek road (from the Clark Fork 5 miles east of Superior) to the South Fork, then a trail for 3 miles to the State Line and thence south for 3 miles, and finally a short, steep cross-country hike to the lakes. Upper Trio is about 5 acres in size and fair fishing for 6 to 10 inch brook trout. Middle Trio is only 3 acres, but fairly deep and fair-to-good fishing for 7 to 10 inch brook trout and 6 to 14 inch rainbow. Lower Trio is about 14 acres and fair fishing for 9 to 18 inch rainbow and 12 to 14 inch cutthroat trout.

Trout Creek. A heavily fished stream flowing through a brushy, timbered flatbottomed canyon to the Noxon Reservoir a mile west of Trout Creek and accessible by a washed out logging road for 6 miles to the junction of its East and West forks. Trout Creek is easily (but not heavily) fished for fair catches of 10 to 12 inch rainbow and cutthroat trout, and brook trout in some beaver ponds 3 miles or so above the mouth.

Trout Creek. A fair-sized stream with 15 miles of heavily fished water in a timbered canyon followed by a good road from its mouth on the Clark Fork River 5 miles east of Superior to some nice falls 9 miles upstream, and a trail beyond. There's a steep canyon directly below the falls that is always good fishing, though difficult to work, for 9 to 12 inch cutthroat, Dolly Varden and whitefish. Above the falls it is mostly good fishing too for 8 to 10 inch cutthroat. Below the canyon its only fair fishing due to contamination from old mine workings dating from the 1930s.

Tuscor Creek. About 5 miles of fishing water in steep, timbered mountains above, and ranchland below; all easily accessible by a road from the mouth at Tuscor (or where Tuscor used to be before it was flooded out by the dammed reservoir). This is moderately popular water with local people and fair fishing for 6 to 9 inch cutthroat trout.

Twelvemile Creek. About 12 miles long with 11 miles of fishing water in a narrow, timbered canyon that is easily accessible all the way by a logging road, and is crossed at the mouth by Interstate 90 a dozen miles west of St. Regis. Twelvemile is moderately popular and fair-to-good fishing for 8 to 10 inch brook trout and cutthroat too, plus an occasional rainbow, brown trout or whitefish. The lower reaches are perhaps over fished, the upper reaches are better.

Twentyfour Mile Creek. A tiny, headwater tributary of Prospect Creek in steep, timbered country; crossed at the mouth by the Prospect road and followed upstream by a USFS trail for 2 miles of fair fishing. Contains 6 to 8 inch cutthroat trout, not often fished.

Twin Lakes. Two 4 acre potholes (the upper one deep, the lower fairly shallow) about 200 yards apart in timbered country reached by the Twin Lakes Creek road 6 miles above the Bend Guard Station. Both are poor fishing for 10 to 16 inch cutthroat trout but for some reason lots of people hit them. They were planted with cutthroat trout recently, but there is quite a bit of water level fluctuation and they may not do very well.

Two Mile Creek. A small, fair-to-good fishing stream in mostly timbered country; accessible by road from the mouth (across the river from Interstate 90, four miles west of St. Regis) upstream for 6 miles of fishing water. There are some open meadow beaver ponds a couple of miles above the mouth. It contains mostly 8 to 10 inch cutthroat, brook trout and a few Dolly Varden.

Two Mile (or Buford) Pond. An old cut-off oxbow of the St. Regis River, 2 miles west of town on U.S. 10; has been planted annually with spawners for kid fishing and at present is home to one mamma and her covey of 6 little ducklings.

Unnamed Lake. A tiny (maybe an acre) pothole in timber at the head of (barren) Cataract Gulch which drains for 1 mile to Missoula Gulch ½ mile above the mouth of Bonanza Gulch. I've never been there either, but a fellow I know said he saw baby submarines erupting from the surface. Couldn't have been too big or the lake wouldn't have floated 'em.

Unnamed Lake. A one acre pothole in timber 1½ miles north and a little east of Oregon Peak. To get there you take the Cedar Creek road as though you were going to Missoula Lake, until you hit Oregon Gulch. Now follow along the gulch (still on the road) north for about a mile until you cross a little tributary coming in from the southwest. This is it. Follow it up for maybe 500 yards and you are there. Someone wrote a year or so ago and told me he saw fish rising from the surface, while flying over in a Cessna 180. Why don't you try it — and let me know? R.L.K.

Upper Thompson Lake. Almost a continuation of the Middle and Lower Thompson Lakes, in flat, timbered country on the divide between the Thompson and Fisher River Drainages. Upper Thompson is followed along the north shore by U.S. 2 about 35 miles west of Kalispell, is about 232 acres, and has a maximum depth of around 75 feet. It is full of Yellow Perch, quite a few Largemouth Bass and lots of rough fish.

Upper Wanless Lakes. Take the McKay Creek road from its mouth on U.S. 10A near the upper end of Noxon Reservoir, north for 5 miles to the end, my friend, then the Wanless Lake trail 9 miles on to Upper Wanless (or Cirque Lake No. 1). No. 2 is about 1 mile below, and Nos. 3 and 4 are about ½ mile apart another mile or so down the trail in a steep, partly timbered glacial canyon. No. 1 is about 3 acres. All are deep enough to support fish. Nos. 1, 2 and 3 are very good fishing, indeed, for 8 to 14 inch cutthroat and No. 4 is fair fishing for 6 to 14 inch cutthroat. Quite a few folks stop at No. 3 on their way in to Wanless.

Vermilion River. A fair sized stream that heads in (barren) Imagine Lake anf flows for 25 miles through mostly timbered canyons, and two gorges 12 to 14 miles above the mouth, in a semicircle to the Noxon Reservoir 2 miles east and across the water from Trout Creek. This is an easily accessible stream all the way along by road and is good cutthroat fishing for about a mile above the mouth of Miller Creek, but is generally poor fishing in the lower reaches for small (6 to 8 inch) whitefish, cutthroat, Dolly Varden and rainbow trout, but fair above for cutthroat, brook trout and a few large Dollys.

Wanless Lake. Deep, 120 acres and 165 feet deep, in a big cirque at the head of a barren glaciated valley that is drained by Swamp Creek. You get there by one mile of good trail below Upper Wanless. Wanless is heavily fished for 6 to 16 inch cutthroat. It's also being turned into a garbage dump by careless sportsmen.

Ward Creek. A small tributary of the St. Regis River 6 miles west of town; followed upstream for 5 miles through a narrow, timbered canyon by a logging road, and then for 4 miles by a beautiful trail through towering White Pine and huge Cedar trees, (if they haven't been chopped down), to some falls at the head of the fishing water. Ward Creek is moderately popular and good fishing for 8 to 9 inch cutthroat, brook trout and a few Dolly Varden.

Ward Lake. See Hazel Lake.

Weeksville Creek. A small stream that flows south down its timbered canyon to the Clark Fork at Weeksville; and is crossed at the mouth by U.S. 10, eight miles west of Plains. There are about 1½ miles of fishing water, which are accessible by road, seldom fished, and poor at best for 6 to 10 inch cutthroat trout.

West Cedar Lake. See Cedar Log Lakes.

West Fork Big Creek. This is a small stream with only about 3½ miles of fishable water, accessible by road and trail. There are some beaver ponds, at the mouth and about ¾ of a mile above. On the whole it is moderately popular and mostly fair fishing for 6 to 8 inch cutthroat and brook trout.

West Fork Blue Creek. A dangerous stream to fish with a game warden around, it's half in Montana and half in Idaho. The Montana section includes the mouth and lower 2½ miles which are in rolling, timbered country a few miles east of Cabinet Gorge Reservoir. In this area it is reported to be fair-to-good for 6 to 10 inch cutthroat. There are fishermen trails only.

West Fork Dry Creek. A tiny stream, (the main creek goes dry, below the junction of the East and West Forks), that is readily accessible by a logging road 5 miles south from Thompson Falls. It is fair fishing near the mouth for small cutthroat trout.

WORDEN LAKE AT MAXIMUM DRAWDOWN
Courtesy H. Newman

TWIN LAKES
Courtesy H. Newman

West Fork Fish Creek. Accessible by a good USFS trail for 5 miles of fishing above the mouth, in timbered country at the end of the Fish Creek road at Clearwater Crossing, to about a mile above its junction with Indian Creek. The West Fork is a small, lightly fished stream but has some good water and is fair-to-good for 8 to 10 inch cutthroat trout plus occasional Dolly Varden reported to run up to 9 or 10 pounds.

West Fork Fishtrap Creek. A very small stream that drains Stony, Terrace, and 3 unnamed alpine lakes through a steep, narrow canyon for about 6 miles of fishing. It is not very popular but readily accessible by a jeep road and fair fishing for 6 to 8 inch cutthroat trout.

West Fork Swamp Creek. Just a trickle in a heavily timbered canyon; followed by a jeep road for about ½ mile of fair fishing for pan-size cutthroat, seldom bothered.

West Fork Thompson River. A nice size, "open" stream; easily accessible by road for about 6 miles of fishing. It is heavily fished and fair-to-good for small cutthroat, and a few brook trout.

West Fork Trout Creek. A pretty fishing stream in a brushy, timbered canyon; accessible by a fair trail from the end of the Trout Creek road for about 5 miles of fishing. There are some nice camping spots and it is excellent fishing for 8 to 13 inch brook trout, cutthroat and Dolly Varden.

White Creek. A tiny tributary of Cache Creek, no trail, fair fishing in the lower reaches (1½ miles) for pan-sized cutthroat, too steep for fish above.

White Pine Creek. A small tributary of Beaver Creek; reached at the mouth by a county road a couple of miles west of White Pine and followed by a logging road (past some beaver ponds about 4 miles upstream) for a total of 12 miles to headwaters in logged over country. This one is real good fishing and is hit hard for 7 to 8 inch cutthroat above the ponds and 8 to 10 inch brook trout below. The best fishing is on private land.

Willow Creek. Another small stream with only about 2½ miles of fishing water in a big beaver dammed flat on the upper reaches of the Vermilion River. It is easily accessible by road and heavily fished for plentiful small cutthroat. It is also good moose hunting country, in season.

Worden Lake. One of the Rattlesnake lakes, sort of oblongish-squarish in outline, with a good 750-foot high, bare rocky cliff at the upper end, a dam at the lower end and trees all around. It is 10½ acres with mostly shallow dropoffs and a maximum depth of 19 feet. To get there just amble a quarter of a mile west and a little north from the old cabin, 2½ miles up the Lake Creek (too small for fish) logging road from Rattlesnake Creek about 18 miles up from the Missoula City limits — and watch out you don't get run over by the motor bikes. Worden is a fair producer of pan-size cutthroat trout, planted in 1969.

Wrangle Creek. A short (3 miles) headwater tributary of Rattlesnake Creek; part of the City of Missoula municipal water supply and closed to fishing, but reported to be good for 8 to 10 inch cutthroat, brook and Dolly Varden trout.

Clark Fork of the Columbia
Upper Drainage

Albicaulis Lake. A high (elevation about 7950 feet) cirque lake, 16.6 acres, 57 feet deep and dammed at the outlet with as much as 14 feet of seasonal water level fluctuation and lots of submerged timber around the shores. It is reached by a poor jeep road 3 miles up Racetrack Creek past Danielsville, and then 2 miles in timbered country up the North Fork to the lake. This is real good water for 8 to 12 inch rainbow trout and rainbow-cutthroat hybrids.

Alder Creek. About 8 miles long with 500 feet of drop per mile in steep, heavily timbered mountains, Alder Creek debouches to Rock Creek on the opposite side from the road. It is reached by a private bridge across Rock Creek and followed upstream to the headwaters by a good USFS trail. This stream is really 'liquid ice,' has innumerable good holes that are overflowing with zillions of 4 to 7 inch cutthroat and is easily but unfortunately, because of the size of the fish and inaccessibility of the stream, seldom fished — except a little right around the mouth for rainbow and Dolly Varden spawners up from Rock Creek.

Alpine Lake. Twenty-two and a half acres, dammed at the outlet with 29 feet of water level fluctuation and lots of water-killed timber around the margins. Alpine is a clear, subalpine (elevation 7850) lake whose immediate shore is about a third talus and boulders. It is reached by quite a few fishermen via a poor jeep road 1 mile past Albicaulis Lake. There are a few nice 12 inch to 4½ pound-or-better rainbow here, but they are hard to catch.

Altoona Lakes. Big (or North) and Little (or South) Altoona, 2½ and 1¼ acres by 15 feet deep at the deepest; about an eighth of a mile apart in rocky, timbered country reached by a pickup road 2½ miles above the Brooklyn Mine on the Boulder Creek road. Big Altoona is mostly steep-offshore except for a shallow shelf on the northeast side, and is good fishing at best (not so hot at times) for 8 to 13 inch cutthroat up to 1 pound. Little Altoona is reached by a trail directly south from Big Altoona, and also contains cutthroat trout.

American Gulch. A small gulch through which flows an equally small stream that is paralleled by a new gravel road for about 3 miles from its junction with Brown's Gulch. It is poor fishing in occasional beaver dams for small resident cutthroat trout.

Antelope Creek. Reached at headwaters by county and private roads 10 miles west from Philipsburg, then 5 miles downstream by trail to its mouth on Rock Creek 45 miles above U.S. 10. The lower 1½ miles flow through open, but private land and are fair fishing for 6 to 8 inch cutthroat trout.

Barker Lakes. See Big and Little.

Basin Creek. Flows for 12 miles from the Basin Creek Reservoir to Blacktail Creek within the city limits of Butte, and is followed all the way by a county road. It is poor fishing for 5 to 7 inch brook trout.

Basin Creek Reservoirs.
Two, Upper and Lower, reached by an oiled road up Basin Creek 12 miles south from Butte. They are the city water supply and are seldom fished — mostly in "the dark of the moon." There are willow and lodgepole around to hike in, but be careful, the game warden is on his mettle here. The fish? — mostly 12 inch to 5 pound cutthroat and a few fat rainbow trout. The danger? — Well, there *have* been a few murders — but never more than two at a time.

Bear Creek.
Is crossed at the mouth by U.S. 10 at Bearmouth and followed upstream for 8 miles by a good gravel road. Back in the "good old days" it was excellent fishing for native cutthroat, but has since been extensively placer mined and is now virtually barren.

Bearmouth Access Pond.
A 1½ acre barrow pit on the Clark Fork floodplain just off U.S. 10 northeast of Bearmouth that was designed as a recreational pond after completion of the highway. You can get to it by hiking a hundred feet or so from the Rest Stop. It is planted every other year with 4 to 6 inch rainbow which grow like mad (the water is full of shrimp) until they're caught, which is pretty quick because the fishing pressure is high. Right now they'll run around a foot long, and there are a few in the 18 to 20 inch class, holdovers from an earlier plant.

Beaver Creek.
A small stream flowing through swampy, beaver dammed country to the North Fork of Rock Creek about ¾ of a mile west of the West Fork Guard Station. Beaver Creek is followed by a logging road and is fair fishing in the ponds for 8 to 9 inch cutthroat and brook trout.

Beefstraight Creek.
A very small tributary of German Gulch that is crossed at the mouth by a county road 6 miles south of Gregson Hot Springs and is excellent fishing for small, 5 to 7 inch brook and cutthroat trout that thrive in its numerous falls and pools.

Bell Creek.
A wee-small stream right in the city of Butte. It pretty well dries up in the summer but is planted now and then with rainbow for kids' fishing, derbys, and the like.

Big Altoona Lake.
See Altoona Lake.

UPPER BASIN CREEK RESERVOIR
Courtesy Montana State Fish and Game

Courtesy Danny On

KE SO!
Courtesy Ernst Peterson

Big (or Upper or East) Barker Lake.
A ½ mile hike and a 400 foot climb "around and over the ridge" southeast from Little Barker. It's a kidney-shaped lake, 11 acres and deep along the west shore with steep dropoffs beneath steeper, rocky cliffs, but shallower along the timbered west side. It is reportedly fair fishing and quite popular for cutthroat trout that will run between ½ to ¾ of a pound.

Big Hogback Creek.
Take the Rock Creek road from U.S. 90 for 31 miles to the Rock Creek campground, the another ½ mile to the mouth of Hogback Creek. This is a very small stream whose lower reaches flow through private meadowland and are easily fished for a few 6 inch-or-so cutthroat, brook and Dolly Varden trout.

Big Pozega (or Deep) Lake.
A high (elevation 7600 feet), 52 acre cirque lake in sparsely timbered, mountain goat country; reached by a good trail 3 miles from the end of the Modesty Creek road or you can drive to it up the Modesty Creek jeep road (if you are brave, or foolish). It is over a hundred feet deep in places and dammed, with water level fluctuations of as much as 18 feet and lots of dead timber around the shores. The fishing pressure is heavy for fair catches of 5 to 22 inch rainbow and cutthroat trout, and rainbow-cutthroat hybrids. The hyprids especially run towards the big end of the scale.

Big Spring Creek.
A small stream flowing southward to Rock Creek across from the road and about 1 mile upstream from the Rock Creek Campground. There is about ½ mile of trail up this brushy creek, but it is seldom fished for the few small cutthroat it contains.

Big Thornton Lake.
A 30 acre cirque lake, perhaps 25 feet deep; in timbered country on the headwaters of Racetrack Creek, but most easily reached by a good USFS trail a mile above the end of the Modesty Creek jeep road. Big Thornton is moderately popular and good fishing too for 8 to 10 inch rainbow trout.

Bison Creek Reservoir.
Take the Boulder highway (U.S. 91) 6 miles north from Butte and then a side road for 1 mile east to the reservoir in Elk Park meadowland. It is really two ponds that total about 10 acres, private, stocked with rainbow and brook trout, and reportedly excellent fishing.

Blacktail Creek. A small stream that flows for about 10 miles from Pipestone Pass to Butte where it joins Bison Creek to form the headwaters of the Clark Fork. It is poor fishing (our of town) for a few 6 to 7 inch rainbow and brown trout, but has been planted annually in town with 10 to 14 inch trout for the kids.

Blum Creek. Take the Gold Creek road 5½ miles west from U.S. 90 to the mouth of Blum Creek near Lingenpelter. It is a very small stream but fair fishing for 8 to 12 inch trout in a few beaver ponds along the lower half mile.

Bob Lea's Pond. A shallow, half-acre reservoir practically on the back porch of the Bob Lea place 4 miles up the Snowshoe road from U.S. 10, three miles east of Avon. There is excellent fishing here, with permission only, for brown trout that will average 10 to 14 inches and range up to 2 feet.

Bohn Lake. A 25 acre lake with 15 foot maximum depth (but there are steep dropoffs all around), in a cirque at the head of Dempsey Creek, in timbered country at about 7100 feet elevation above sea level. It is reached by a jeep road up the creek 8½ miles above the Perkins' ranch. It's slow fishing for 11 to 16 inch brook and cutthroat trout.

Boulder Creek. A small tributary of Flint Creek that is crossed at the mouth by U.S. 10A a mile south of Maxville, and followed upstream by a good gravel road for 7 miles and then a pickup road for 3 more to headwaters. It is good fishing for 8 to 10 inch cutthroat trout.

Boulder Lakes. Three in all, of which only the upper and lower (largest) ones contain fish. They are reached by 3 miles of very poor USFS trail from the Boulder Creek road about 2 miles above Brooklyn, are in alpine country, about 9 acres each and real deep. The fishing is good above for 8 to 10 inch cutthroat and spotty below for cutthroat trout that will average maybe 10 inches and range from minnows to 17 inches or better.

Bowles Creek. A small stream that rises in timber and flows eastward through meadowland for about 4 miles to the West Fork of the North Fork of Rock Creek. It is reached by a good gravel road 3.1 miles from the Sand Basin road 1.3 miles from High 38 about 25 miles west from its junction U.S. 10A and is good fishing in the lower reaches for 8 to 10 inch cutthroat trout.

Bowman's Lakes. Take the Bielenberg Canyon road for about 12 miles west, up from the old Quinlen school (or where it used to be), and you can drive right by all three of these lakes. They are in high, sparsely timbered, rocky country, will total about 20 acres, and are generally good, although sometimes temperamental, fishing for 8 to 13 inch cutthroat and rainbow trout.

Brewster Creek. Take the Rock Creek road 9 miles south from U.S. 10 to the mouth of this stream, then a gravel road for 6½ miles south and 3000 feet up along it to headwaters. It is a small stream in timbered country, and is fairly good fishing for 3 to 9 inch cutthroat trout.

Browns Gulch. Is reached at the mouth by U.S. 10 near Silver Bow, a few miles west of Butte, and followed upstream for about 10 miles by good county roads. This stream flows mostly through private pastureland in its lower reaches and is the site of numerous beaver dams above. It is generally good fishing for pan-size rainbow and brook trout.

RP CREEK
Courtesy USFS

COPPER CREEK NO. 1
Courtesy Montana State Fish & Game

Bryan Creek.
Flows to Telegraph Creek about 10 miles south of Elliston, and is followed for 1½ miles upstream by a county road. It is a small creek, about 3 miles long, mostly in timber but with a few meadows and some beaver ponds below. The fishing is excellent in the ponds for small cutthroat trout, but there is not enough water to stand much fishing pressure.

Burns Slough.
A group of small sloughs totaling about 25 acres, that used to extend for 1½ miles along both sides of the railroad tracks above Bearmouth. However, they have been partly filled in and reoriented by road construction so that their present status is debatable. They used to be lightly fished but excellent warm water fishing for Largemouth Bass that ranged all the way from 9 inches to 5 pounds, and there were also some nice brownies here. It is also real good duck hunting country.

Butte Cabin Creek.
A small stream flowing through a densely forested canyon for 5 miles to Rock Creek at the Butte Cabin Campground about 20 miles above (south of) U.S. 10 and followed by a road for a couple of miles. There is much willow, alder and brush along the banks and its tough fishing for the few fingerling rainbow and Dollies it contains.

Cable Creek.
A small tributary of Warm Springs Creek; crossed at the mouth by the Georgetown highway 2 miles east of Silver Lake and followed by a county road for 5 miles through timbered country to headwaters. It is fair early-season fishing for pan-sized rainbow, brook and Dolly Varden trout.

Carp Creek.
A beautiful little stream in the Anaconda Pintlar Wilderness Area; flowing west for 7 miles through timbered mountains to the Middle Fork of Rock Creek. Carp Creek is reached at the mouth by the Middle Fork road about a mile above Moose Lake, and is followed by a trail to headwaters — and you can also hit the middle reaches via the good East Fork road to the Meadow Creek road. A few big Dollys spawn up this stream in the fall and a few fingerling live in the upper reaches. Otherwise it is, oddly enough, barren for all practical purposes.

CLARK FORK of the Columbia (Upper)

Carp Lake. A shallow, 25 acre lake in a high (elevation 7750 feet) cirque reached by a fair trail a couple of miles up from the Meadow Creek road where it meets the creek. It is overpopulated with 3 to 6 inch cutthroat fingerling.

Carpenter Creek. Is reached near the mouth by a county road 2 miles east from Avon. This is a very small stream that was dredged for gold in the 1930's and is poor-to-fair fishing in a small 2 acre pond at its mouth for 8 to 10 inch brook and brown trout. The pond is on private land with fishing by permission only.

Carruthers Lake. A fairly deep, 22 acre lake reached by a jeep road 3 miles above (west of) Bohn Lake. It is reportedly good fishing for 6 to 9 inch cutthroat trout.

Cinnabar Creek. Take the Threemile Creek (in the Bitterroot Drainage) road to the Bitterroot-Rock Creek divide, then go south by jeep road along the divide for 1 mile to the headwaters of Cinnabar Creek and on downstream for another mile to the end of the road. It is a very small, inaccessible stream that flows through mountain meadows and timbered gulches to Welcome Creek, and is fair fishing for 6 to 8 inch cutthroat trout.

Clear Creek. A tiny stream flowing through timbered country to Beefstraight Creek. There is no trail and it is seldom fished because of its inaccessibility but contains a fair number of 7 to 10 inch cutthroat trout.

Conley's Lake. Take the Pioneer highway 2 miles west from Deer Lodge, then the Montana Prison Farm road for 4½ miles to the lake; which is about 3 acres, fairly shallow, bordered by brush and willow. It has been rehabilitated and stocked with rainbow trout.

Copper Creek. A small stream flowing northward for 5 miles from the Copper Creek Lakes to Boulder Creek about 2 miles above Princeton, and followed by a USFS trail to headwaters. The lower reaches contain a few small cutthroat.

Copper Creek. An Anaconda-Pintlar Wilderness Area stream flowing northeastward to the Middle Fork of Rock Creek about a mile below Moose Lake and followed by jeep road a mile or so and then by trail for 10 miles to headwaters. This is a mostly open stream except for willow and brush along the middle reaches. It is heavily fished and reportedly good for 8 to 11 inch cutthroat with lesser numbers of brook and Dolly Varden trout and whitefish.

Copper Creek Lakes. Three high (elevation around 7250 feet) lakes. No. 1 (to the south) is 14.3 acres; no. 2 is 3.7 acres; and no. 3 is 12 acres. All are real deep but with gradual dropoffs all around, and have good gravel bottoms. No. 1 (also called Crystal Lake) is reached by a good hikers' trail ½ mile north from a jeep road 4 miles east from Granite (which is west from Philipsburg). Nos. 2 and 3 string out ⅛ mile apart down the drainage. All are excellent fishing for 6 to 10 inch brook trout, but no. 3 also has lots of rainbow-cutthroat hybrids and a few of their progenitors.

Cottonwood Creek. A pretty little stream flowing to the Clark Fork River at Deer Lodge and followed by county road and trail 12 miles to headwaters. It is mostly de-watered for irrigation in the summer but the lower reaches are poor early-season fishing for rainbow, cutthroat and brook trout.

Cougar Creek. About 25 miles up Rock Creek to its mouth on the east side, good fishing for pan-size (exceptionally to 10 inches) cutthroat and brookies for a couple of miles up this little creek in heavily timbered mountain country. It gets fished a fair amount in early spring, and not much thereafter.

JUNCTION HIGHLINE & MIDDLE FORK TRAILS
ANACONDA-PINTLAR WILDERNESS
Courtesy USFS

MOUNTAIN GOAT & ELK COUNTRY
Courtesy USFS

Cramer Creek. A small tributary of the Clark Fork River, crossed at the mouth by U.S. 10 between Beavertail Hill and Bonita, and followed for 10 miles by county and jeep roads to the headwaters. The lower reaches are heavily fished for small cutthroat, rainbow, and brook trout and a few whitefish.

Crevice Creek. Flows eastward for 5 miles from the Goldberg Reservoir to Gold Creek near Lingenpelter and is followed by cow trails for a mile or two upstream. This is a small stream that is mostly overlooked by all save local fishermen for whom it produces good catches of 6 to 8 inch trout.

Crystal Lake. See in Copper Creek Lakes.

Davis Reservoir. See Dog Lake.

Dead Man's (or Dead) Lake. An 8.3 acre by 27 foot maximum depth, alpine lake in a small cirque about ¼ mile and 100 feet below (elevation 7850 feet) Albicaulis Lake; reached ¼ mile across the spillway and down the trail from Albicaulis Lake. It is fair fishing for cutthroat trout that average around a foot or better and sometimes grow to a foot and a half in length.

Deep Creek. A little bitty stream, about 5 miles in length, that flows to Bear Creek at "Old Bear Town" 7 miles upstream from Bearmouth on U.S. 10. In BIWP (Before It Was Placered) days Bear Creek was an excellent trout stream and Deep Creek still supports a few small cutthroat trout but they're very seldom fished for.

Deep Lake. See Big Pozega Lake.

Dempsey Creek. About 18 miles long from its headwaters in several high lakes, to its mouth on the Clark Fork 6 miles south of Deer Lodge, Dempsey Creek is followed by county and jeep roads, and a USFS trail throughout its entire length. The lower reaches are mostly de-watered for irrigation and are only poor fishing, but the upper reaches just below the lakes are generally excellent for 7 to 8 inch trout.

Dog Creek. A small stream flowing northward to the Little Blackfoot River near Avon. The lower 4 miles flow through a narrow, rocky gorge and are followed by cattle trails. The upper 10 miles flow through open pasture and timberland and are followed and crossed by county and private roads and (incidentally) by the old

CLARK FORK of the Columbia (Upper)

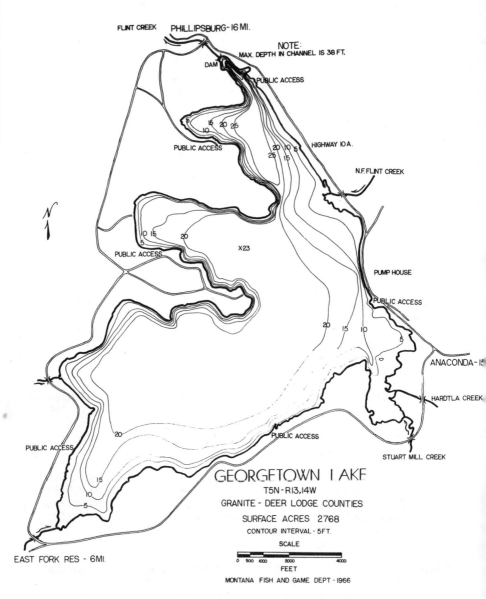

abandoned Fort Benton highway. This is a fair early-season stream for 2 to 9 inch cutthroat trout, and in late season produces some fair size browns and whitefish from its lower reaches — plus lots of suckers.

Dog Lake. Eleven acres, deep, with real black water and a 10 foot dam but little or no drawdown, in open rangeland at the head of a deep gorge on Dog Creek; reached by county and private roads 4 miles south from Avon. Dog Lake is lightly fished by local residents for 7 to 14 inch cutthroat, brook and a few brown trout. Other denizens of this lake now include 2 families of beaver and one mink.

Dolus Lakes. Four alpine cirque lakes ranging in area from 3.4 to 14.3 acres, about ¾ of a mile apart, between 7800 and 8000 elevation. The lower and middle lakes on the main drainage are reached by a good trail 3 miles from Rock Creek Lake and are slow-to-fair fishing for 7 to 10 inch rainbow-cutthroat hybrids and 9 to 16 inch Loch Leven. The upper lake is fair fishing for 6 to 11 inch cutthroat trout. The remaining lake (a little to the north of Middle Dolus) is barren.

Dora Thorn Lake. A high (elevation 7350 feet) alpine lake, 6.4 acres, fairly deep with gradual dropoffs and much aquatic vegetation around the margins. Dora is reached by a good trail a mile from the end of a jeep road from Granite (which is west of Philipsburg). It is a spotty lake where you get skunked more often than not as there is too much feed, but you can occasionally hit 'em for fair, mostly cutthroat trout that will range from 6 to 17 inches.

Dutchman Creek. A 5 mile "pick up" stream between the Lost Creek and Warm Springs Creek fans; never more than a mile from a good county road in marshy farmland. Dutchman is good fishing for 8 to 14 inch rainbow, brook, brown trout and grayling. It is usually subjected to moderately heavy fishing pressure by local residents, and occasionally to extreme pressure by way of a D-8 Caterpillar in the act of blocking off a section or two while effecting ditch repairs.

East Barker Lake. See Big Barker Lake.

East Fork of Brewster Creek. A very small stream reached by a gravel road 5 miles up from Brewster Creek and followed by a jeep road for ½ mile above the mouth. This is a brushy stream in timbered country. It is seldom fished but contains a few pan-size cutthroat and Dolly Varden trout in the lower end.

DOG LAKE
Courtesy R. Konizeski

EAST FORK RESERVOIR
Courtesy R. Konizeski

East Fork Reservoir. Near the middle of the East Fork Rock Creek Drainage; reached by a good county road 12 miles from the mouth on Montana Route 38, or 6 miles from Georgetown Lake. The East Fork Reservoir covers 500 acres, is 75 feet deep with shallow dropoffs near the dam and inlet, and straight off elsewhere. It has about 50 feet of water level fluctuation, in spite of which it is good winter fishing for 12 inch to 8 pound Bull trout and is good in the summer before de-watering, for 10 to 14 inch rainbow and brook trout. It's moderately popular.

East Fork Rock Creek. Twenty miles long, crossed at the mouth by Montana Route 38, followed by a county road for 12 miles to the East Fork Reservoir and thence by a good USFS trail for 6 miles to headwaters. The East Fork flows through partly timbered and partly open country, is fairly popular and produces good catches of 8 to 12 inch whitefish, cutthroat, brook and a very few rainbow trout.

Echo Lake. Take the Philipsburg highway ¾ of a mile from Georgetown Lake, then turn right on a county road for 1½ miles to the Echo Lake Campground 6650 feet above sea level. This is a summer resort lake, about 75 acres and around 20 feet deep with a mud bottom. It gets real warm in the summer and there is a lot of swimming, water skiing, quite a few summer homes and a lodge (or rather the remains of a lodge). It is fair fishing and heavily fished for rainbow and brook trout that range from 9 to 10 inches up to 1½ pounds.

Edelman Creek. A very small, northward flowing stream reached by the Brewster Creek road about 4 miles above the mouth. A brushy creek in timbered country, it contains a few pan-size, and seldom fished for, cutthroat trout for the first ½ mile upstream. There's no trail.

Edith Lake. Take a good USFS horse trail 6 miles to this lake from a point on the Middle Fork of Rock Creek road about 3½ miles above Moose Lake. Edith is a deep, 40 acre, cirque lake in alpine country. It is good for 9 to 15 inch rainbow trout but you have to work for 'em.

Eight Mile Creek. Is reached by the Harvey Creek road and trail 8 miles upstream and followed to headwaters by a good horse trail. It is a scenic stream in timbered country, and easily fished for good early season catches of 6 to 8 inch cutthroat, rainbow and brook trout, but it is seldom fished because you have to walk. 'Tis tough tick'n.

Elbow Lake. A high alpine lake about 12 acres, no more than 10 feet deep anywhere; reached by trail 4 miles from the end of the (one-dammed-poor) Rock Creek jeep road near the upper end of the Rock Creek Reservoir. This lake is poor-to-fair fishing for 6 to 13 inch rainbow trout.

Elk Creek. A small swampy stream flowing to the Ross Fork of Rock Creek about 7 miles above the East Fork road, and paralleled by a jeep road to headwaters. It is only occasionally fished but produces excellent catches of small cutthroat trout.

Elk's Tin Cup Creek Pond. On the Montana State Prison ranch about 6 miles west of Deer Lodge at the end of the prison road immediately west of the brick "residential" building. It is a shallow, 5 acre lake, dammed, with about 5 feet of draw-down, and good fishing for public and residents for 8 to 10 inch brook trout.

Elliot Lakes. See Upper and Lower.

Falls Fork Creek. Is crossed at the mouth by the Middle Fork of Rock Creek trail and followed upstream for 4 miles by a good USFS trail to Johnson Lake. Falls Fork is lightly fished through open timber country for good catches of 8 to 10 inch cutthroat, and a few Dollys in the fall.

Finlen Creek. See Gilbert Creek.

Fisher (or Jones) Lake. A high (elevation 7500 feet) cirque lake in partly timbered alpine country, dammed with 20 feet of draw-down, about 30 acres maximum area, 36 feet deep in the center with shallow margins. Fisher is barely reachable by jeep 1 mile from the Racetrack Creek road 5 miles above Danielsville, or 3 miles from Fred Burr Lake, where a brand new road is a-building. It is poor-to-sometimes-fair fishing for 7 to 12 inch cutthroat trout. Not many folks bother it.

Five Mile Barrow Pit (or Warm Spring Gravel Pit). About 5 acres of long, narrow ponds in open meadow just south of the Anaconda-Warm Springs highway (Montana 48) and about a mile west of U.S. 10. The Anacondites have fished it quite a bit in late years for excellent catches of rainbow trout planted by the State Fish & Game Department on a put-and-take basis. There are a couple of johns here — in case it rains.

Flint Creek. A large tributary of the Clark Fork River, Flint Creek flows northward from Georgetown Lake through mostly farm and meadow land for about 50 miles to its mouth at Drummond and is followed all the way by U.S. 10A. It is heavily and for the most part, easily fished for good catches of cutthroat, rainbow, brook, brown trout and some whitefish. Cutthroat and rainbow trout are the predominant species in the upper (above Philipsburg) reaches, rainbow and brook trout in the lower reaches. The cutthroat, rainbow and brook trout average around 9 to 10 inches, the whitefish and browns 14 to 16 inches.

Foster Creek. A small tributary of Warm Springs Creek in steep timbered country, crossed at the mouth by the Georgetown highway about 8 miles west of Anaconda and followed by road and trail for 10 miles to headwaters. Foster Creek is heavily fished without too much success for small cutthroat, brook and Dolly Varden trout. There is a campground about 1½ miles above the mouth.

GOLD CREEK RESERVOIR
Courtesy Phyllis Marsh

FISHER RESERVOIR
Clurtesy Phyllis Marsh

Four Mile Basin Lakes. Two deep, 5-acre lakes a few hundred yards apart; and three shallow ponds in alpine cirques reached by a good trail a couple of miles from the end of the Twin Lakes Creek jeep road. Lower Four Mile is usually good fishing for 8 to 10 inch brookies. Upper Four Mile supports a good population of 7 to 11 inch golden trout.

Fred Burr Creek. A small open stream flowing west from the Flint Creek Range to Warm Springs Creek about 2½ miles south of Philipsburg. The lower reaches, which are mostly on private land, are easily accessible by gravel road and produce fair catches of 7 to 9 inch cutthroat. It is fished mostly by Philipsburgians and local ranchers. You will need permission here.

Fred Burr Lake. A high (elevation 7600 feet), 161 acre by as much as 88 feet deep alpine lake that is nip and tuck getting into by a poor jeep road 7 miles from Philipsburg. Fred Burr is dammed and used for the city municipal water supply. Although closed for all fishing, it reportedly contains some real nice 13 to 14 inch rainbow trout and rainbow-cutthroat hybrids.

Fuse Lake. Take the Skalkaho highway 2½ miles east of the pass, then a good USFS trail about 1½ miles north to this lake which lies in a high (elevation 7950 feet) cirque, is 13.2 acres and 37 feet deep at the most with gradual dropoffs all around. This is a real pretty, easily accessible (after you stumble across the clear cuts) and moderately popular 'green algae' lake that is good-to-excellent fishing for mostly 5 to 12 inch — but with some up to 15 inch — arctic grayling.

Georgetown Lake. A large (4½ square mile), high (6350 feet elevation) lake (really a drowned meadow converted into a reservoir); easily accessible by state highway 10A twenty miles west from Anaconda. There are two excellent campgrounds, a boat livery and many summer residences. It is a popular recreational lake for boating, waterskiing and fishing, frequented mostly by vacationers from Butte, Anaconda and Deer Lodge, and heavily fished both summer and winter for 10 inch-or-so rainbow trout, quite a few kokanee or coho salmon, an occasional brook trout and once in a blue moon a grayling. This is probably the most popular body of water for fifty miles around.

German Gulch. A small stream flowing to the Clark Fork in Silver Bow canyon about 2½ miles south of Gregson; and followed upstream for 3 miles by a trail and then by a jeep road to headwaters. It was formerly excellent fishing but was extensively placered in the late 1800's and is now only poor-to-fair for small cutthroat especially in the occasional beaver ponds.

Gilbert Creek. Also known locally as "Finlen" Creek after the owner of the property just above the mouth. Take the Rock Creek road 4 miles west from U.S. 10, then a private road for 3 miles up the creek, and finally a trail for another 2 miles to headwaters. It is easily and heavily fished near the mouth for good catches of small brook, rainbow and cutthroat trout. There are 4 little artificial ponds: 2.6, 2.8, 2.9, and 3 miles above the mouth, the upper 3 about ½ acre each, the lower one about 6 acres. All have been planted and are used as private fishing ponds. The upper two miles of this stream are generally inaccessible without permission, but are overpopulated and underfished for long, snakey cutthroat, Dolly Varden and brook trout.

Glover Lake. See Tamarack Lake.

Goat Lake. An 18 foot deep by 12 acre cirque lake in alpine (elevation 8075 feet) country; reached by ½ mile of trail from the Dempsey Creek jeep road ½ of a mile above Carruthers Lake. It has an open shore, lies below open talus slopes to the south but is in timber to the north, is easily fished and produces good catches of 7 to 13 inch cutthroat and rainbow-cutthroat hybrids.

Goat Mountain (or Thompson) Lakes.
Six little alpine lakes in cirques on the upper flanks of Goat Mountain. Five are about 5 acres each and are reached by trail from the 6th and largest (about 15 acres) which is called Thompson Lake and is accessible by a jeep road and trail up Rock Creek. All 6 are overpopulated and excellent fishing for 8 to 9 inch brookies.

Gold Bar Lakes.
From Anaconda take the Georgetown highway for 10 miles west, and then turn north up the Warm Springs Creeks road for 7 miles, and thence about 1 mile cross-country to 4 small (1 to 3 acres) shallow lakes in heavily timbered mountain country. They are hard to find and poor fishing when you do, for small brook and rainbow trout.

Gold Creek.
Site of the first major gold strike in Montana, Gold Creek flows to the Clark Fork across from U.S. 10, eleven miles south of Drummond, and is followed by a good gravel and jeep road past the old ghost town of Pioneer (now completely vandalized) for 20 miles to headwaters at the Gold Creek Lakes. It is an easily fished stream that flows through meadows below and heavy timber above, and produces good catches of 7 to 10 inch rainbow trout and just enough big browns to keep the fisherman on his toes.

Gold Creek Lakes.
Two high (elevation about 7250 feet), 25 and 5 acre, cirque lakes on the western slopes of Rose mountain; reached by the Gold Creek logging road 20 miles from U.S. 90. Both used to be good fishing for foot-long rainbow trout, but hordes of suckers are taking over.

Green Canyon Lake.
A deep, cirque lake of about 8 acres by more than 50 feet deep, just east and below Whetstone ridge; approached to within ½ mile by a logging road (now closed) off the Copper Creek road above Moose Lake. This is a "hard-to-catch-'em-cause-there-are-not-many-to-be-had" lake for rainbow trout in the 17 to 20 inch class.

Green (or Pozega No. 3) Lake.
A high (elevation 7850 feet), 34 foot deep, 13 acre cirque lake southwest of Deer Lodge in the Flint Creek range, it is also known locally as Pozega No. 3. Green Lake is reached by a jeep road and trail 12 miles up Modesty Creek, or 18 miles up Racetrack Creek via Fisher and Meadow Lakes. It is partly dammed with water-level fluctuations of as much as 6 feet, and is fair-to-good fishing for rainbow, cutthroat, and rainbow-cutthroat hybrids ranging from 6 to 18 inches.

Grizzly Creek.
A very small, brushy stream flowing to Ranch Creek at the Grizzly Peak Campground. This is a hard, exasperating stream to fish and is mostly passed by, but does contain fair numbers of 6 to 8 inch cutthroat-rainbow hybrids. Dolly Varden and brook trout.

Haggin Lake.
A real deep little lake, about 8 acres, in a cirque just north of Mount Haggin in the Anaconda range 8250 feet elevation above sea level. Take the Georgetown highway 3 miles west from Anaconda, then a jeep road for another 3 miles up Big Gulch and thence proceed by foot trail for 2 miles to this lake, which is good fishing for 10 to 15 inch rainbow trout.

Hail Columbia Creek.
A kids' creek, tributary to Brown's Gulch and followed through open country for about 5 miles by county roads. It contains a few 6 inch and smaller brook and rainbow trout.

CLARK FORK of the Columbia (Upper)

ELBOW LAKE
Courtesy Montana State Fish & Game

JOHNSON LAKE
Courtesy USFS

Harvey Creek. A real pretty little stream that flows through low, open hill country to the Clark Fork about a mile north of Bearmouth and across the river from U.S. 90. It is followed for 2½ miles by a jeep road, and for 14 miles by a poor USFS trail to headwaters. This one is mostly skipped by all but local anglers, but is easily fished and excellent for 6 to 8 inch cutthroat and brook trout, plus an occasional rainbow and footlong Dolly.

Hearst Lake. This high (elevation 8200 feet), 18 acre lake occupies a barren rocky cirque just north of Mount Haggin, about 5 airline miles southwest of Anaconda, and is reached by a 2½ mile hike over a rough trail from the old city reservoir at the head of Ice House Gulch. It is real deep and excellent fishing for cutthroat and rainbow trout that range from 10 to 16 inches. Lower Hearst is a tiny pond about ⅛ of a mile below Hearst Lake, and reportedly contains a very few 12 to 14 inch cutthroat trout, but not enough to make it worthwhile fishing.

Hidden Lakes. Five (discounting Trask Lake, see listing) little alpine lakes with elevations between 8000 and 9500 feet and about 5 acres each; reached by 5 miles of good USFS trail from the end of the Rock Creek jeep road, or 2 miles by a good trail from Alpine Lake. They are all overpopulated and excellent fishing for 8 to 9 inch brook and some cutthroat trout. There is a special 10 pound and one fish limit here.

Hogback Creek. Take the Rock Creek road from U.S. 10 for 27 miles to the Rock Creek campground, then another ½ mile to the mouth of Hogback Creek. This is a very small stream whose lower reaches flow through private meadowland and are easily fished for a few 6 inch or so cutthroat, brook and Dolly Varden trout.

Hoover Creek. A small tributary of the Clark Fork River; crossed at the mouth by U.S. 10 at Jens; and paralleled by a gravel road for 12 miles to headwaters. Miller's Lake at the mouth of Elk Swamp Creek about 6 miles upstream, has been heavily fished for 6 to 9 inch rainbow and brook trout, but Hoover Creek itself (which is also fair fishing) is generally neglected. The present owner has put a 12 foot dam on the lake and posted the land.

Hope Creek. Flows from Faith and Charity Gulches for 4 miles through alpine meadows to Spotted Dog Creek. It is quite a small stream that is fair fishing, although almost never bothered, for 8 to 9 inch cutthroat, brown and brook trout.

Hunters Lake. This high (elevation 7850 feet) alpine lake lies in a broad timbered cirque, measures 8.7 acres, is dammed with 5 feet of drawdown, and is reached by jeep road up Modesty Creek. It has been fair fishing for 7 to 12 inch rainbow trout but reportedly froze out the winter of '73.

Ivanhoe Lake. An alpine cirque lake, 7 acres by 45 feet deep, on the headwaters of the Middle Fork of Rock Creek; reached by a jeep road 4½ miles above Moose Lake, then a steep switchbacked USFS horse trail half a mile to the shore. It's only fair fishing for 8 to 12 inch rainbow trout.

Jacobus Ponds. Two little half-acre private ponds 2 miles west of Elliston just south of the Little Blackfoot highway in open meadows. They are fair fishing for 6 to 8 inch cutthroat and brook trout.

Johnson Lake. This 70 acre by 68 foot deep lake is the first one in the Johnson Lake trail into the Pintlar Wilderness Area, a couple of miles south of Edith Lake. It has a nice campground, esceptionally scenic surroundings, and fair, skinny cutthroat fishing (along with lots of suckers).

Jones Lake. See Fisher Lake.

Jones Pond. A private 2 acre earth-dammed spring-fed reservoir right by the house 2 miles north of Warm Springs and a mile west of U.S. 10. It is stocked with browns and is very good fishing indeed for 12 to 24 inchers, with the owner's permission only!

Kaiser Lake. Twelve acres, 22 feet deep in conifer-covered mountains; available by 1 mile of pick-up road from the Middle Fork of Rock Creek road 7 miles above the mouth. Kaiser is poor fishing at best for small cutthroat, Dolly Varden, rainbow and hordes of suckers.

Lake Abundance. A high alpine lake in heavily timbered country 4 miles south of Condon Peak. Lake Abundance is only 3.6 acres, long, narrow and shallow (but does not freeze out), with swampy margins and much aquatic vegetation — really sort of a dammed-up mountain meadow. It is excellent fishing for 6 to 10 inch cutthroat trout, but is only hit occasionally inasmuch as it is 14 miles by horse from the end of the Ross Fork of Rock Creek road. It's in real good elk hunting country too, but rough!

Lake Ellen at YMCA Camp. A shallow beaver pond in a ground moraine outwash plain reached by the Little Blackfoot road 3 miles south of Elliston. It's good 6 to 14 inch brook trout fishing for the kids at summer camp.

Lake of the Isles. A high (elevation 8150 feet) lake just north of the Continental Divide in the Anaconda range 11 miles west of Anaconda; reached by what is left of an old, unmarked trail a mile east of Twin Lakes. Lake of the Isles lies in a timbered basin, is about 3 acres, fairly deep in the upper end, and only poor-to-fair fishing now for cutthroat trout that range up to 2 pounds.

Lea's Pond. See Bob Lea's Pond.

Little Altoona Lake. Altoona Lakes.

CLARK FORK of the Columbia (Upper)

LOWER WILLOW CREEK RESERVOIR
Courtesy USDA, SCS

UPPER & LOWER ELLIOT
Courtesy Montana State Fish & Game

Little (or Lower or West) Barker Lake.
Take the Georgetown highway 8 miles from Anaconda to the mouth of Barker Creek, then a rough jeep road for 3½ miles to Little Barker in a small timbered cirque at 7850 feet (above sea level). It is about 10 acres, mostly less than 5 feet deep but with 2 small deep holes about a fourth of the way out towards the middle from the inlet, which are probably what keeps the fish from freezing out. It's quite popular and good fishing (maybe a little better than Big Barker) for 10 inch cutthroat trout, plus an occasional one to 14 or 16 inches.

Little Blackfoot Rover.
A good sized mountain stream but small for a river, it flows for 50 miles from Electric peak, elevation 8100 feet, through mostly a broad flat-bottomed agricultural valley to eventual junction with the Clark Fork at Garrison, elevation 4650 feet, and is paralleled all the way by good roads. This is a heavily fished stream whose flow is sometimes entirely diverted for irrigation, but that is somehow consistently good fishing for 9 to 12 inch browns and whitefish below Elliston, and fair for cutthroat and brook trout above.

Little Fish Lake.
A mostly shallow, 8.9 acre mountain lake with much vegetation around the shores; reached by trail a mile from the South Fork of the Ross Fork of Rock Creek trail, and good to excellent fishing, although seldom bothered, for small (6 to 10 inch) cutthroat trout.

Little Hogback Creek.
A tiny, brushy stream flowing north for 3 miles to Rock Creek between the Cougar Creek and Hutsinpilar campgrounds. It's too small to be of much note, but does contain a few small Dolly Varden near the mouth.

Little Pozega Lake.
A high (elevation 7650 feet) alpine lake about an eighth of a mile above Big Pozega by jeep in real good mountain goat country. Little Pozega has a maximum area of about 13 acres, is as much as 35 feet deep in places, dammed with 8 feet of water-level fluctuation and lots of drowned timber around the margins. It is fished a fair amount for this neck of the woods, and is good too, for 5 to 15 inch rainbow trout.

Little Racetrack Lake.
Reached by a ¾ of a mile hike from Racetrack Lake, in a high (7450 feet elevation) alpine cirque. Little Racetrack is 5.8 acres, 27 feet maximum depth, dammed with 5 feet of drawdown and lots of boulders and drowned timber around the edges. It is lightly fished and no better than poor-to-fair for 10 to 11 inch rainbow and a few rainbow-cutthroat hybrids. Would probably be better except for the excessive drawdown.

Little Thornton Lake. About 500 yards north and east of Big Thornton, 3.7 acres by 42 feet deep, and real slow fishing for 14 to 18 inch rainbow.

Lost Creek. A small, easily wadeable stream flowing for about 15 miles through open meadow-and-crop land (just over the hump north from Anaconda), and for 5 miles in a heavily timbered mountain canyon. Lost Creek is easily accessible for its entire length by good county roads and is heavily fished for good catches of small brook, cutthroat, Dolly Varden and brown trout, with an occasional 3 to 4 pounder from the lower reaches.

Lower Barker. See Little Barker.

Lower Carp Lake. A shallow, 15 acre, alpine meadow (elevation 7750 feet) lake that is reached by a good USFS trail 2½ miles from the Carp Creek road. It is good fishing for overpopulated 6 inch cutthroat trout.

Lower Elliot Lake. A real high (elevation 8500 feet) cirque lake in alpine country; reached by a good USFS trail ½ mile to the right beyond the end of the Dempsey Creek jeep road (you can make it if you're a wild man). Lower Elliot has an area of perhaps 63 acres and is more than 80 feet deep with steep dropoffs, dammed but with only 3 feet of draw-down — just enough to drown the lower fringe of trees around the edges. It is seldom visited but a good producer of 6 to 9 inch cutthroat trout.

Lower Willow Creek. A small jumpable stream flowing for about 12 miles through farmland in the Flint Creek valley west of Hall, and easily accessible all the way by county roads. It is heavily fished by local ranchers for good catches of 7 to 9 inch cutthroat, brook and brown trout.

Lower Willow Creek Reservoir. A 170 acre, 85 foot maximum depth reservoir behind a 940 foot long, earth-fill dam in open grassland at the junction of the North and South Forks of Willow Creek; reached by good county

YEAH!
Courtesy Montana State Fish & Game Department

CLARK FORK of the Columbia (Upper)

MARTIN LAKE
Courtesy USFS

MEADOW LAKE NO. 4
Courtesy Montana State Fish & Game

roads about 9 miles north from Hall. The water rights to this reservoir are all taken for irrigation, — the thing can be drained *completely* dry and therefore is *not* stocked by the State Fish and Game Department. However, it does provide an excellent habitat for the native cutthroat trout from the creek — and is sometimes excellent fishing for 14 to 16 inches. It's also used a moderate amount for waterskiing.

Mallard Creek. A very small stream flowing for 3 miles north to Rock Creek about 4 miles below the West Fork Guard Station. The lower mile is on private land with fishing by permission only, is easily fished and contains goodly numbers of 6 to 7 inch cutthroat trout.

Marshall Creek. An open, easily accessible stream flowing for about 6 miles through mostly farmland to Flint Creek 3 miles north of Philipsburg. It is only fair fishing for 6 to 7 inch cutthroat trout.

Martin Lake. A real high (8700 feet elevation) cirque lake in barren "way back" country at the headwaters of Tin Cup Joe Creek; reached by horse trail 2½ miles from Carruthers Lake, or 5 miles from the end of a jeep road up Elk ridge. It is as much as 50 feet deep in spots, 45 surface acres when full, and so far back and high up that it is seldom visited, much less fished, but does support a small population of 7 to 13 inch cutthroat trout.

Meadow Creek. A small, easily wadeable tributary of the East Fork of Rock Creek; reached three miles above the mouth by the East Fork road and followed to headwaters by 11 miles of good road. Here is an open, muchly beaver-dammed stream that is fair fishing for 7 to 10 inch cutthroat trout.

Meadow Creek. This little stream flows from some springs just below the Continental Divide north of Mullan pass, for 2 miles through alpine meadows to Spotted Dog Creek. It is fair-to-good fishing for small brook, cutthroat and brown trout in some beaver ponds about ½ mile above the mouth.

Meadow Lakes. Six alpine lakes (elevation between 7650 and 8650 feet) at the headwaters of Racetrack Creek in sparsely timbered, rock-and-mountain-meadow country; reached most easily by a 1½ mile pack trip east and then south around the hill from Fisher Lake to Meadow no. 3. No. 4 is a few hundred yards on to the southeast; no. 5 about 500 hards to the southwest of no. 4 in fairly flat country; and nos. 2 and 1 about 800 and 1500 yards to the northeast in small cirque like depressions. No. 1 is 12.9 acres by 50 feet deep with shallow dropoffs and considerable plant life around the shore. It is the easternmost of the six and receives its water from Lake no. 2, is seldom fished but produces a few 13 to 14 inch rainbow on occasion. Lake no. 2 is 8.7 acres, by 38 feet deep with gradual dropoffs and lots of rocks, trees and some brush around the shore. It was planted in the '50s and still contains a small population of 10 to 14 inch rainbow. Lake no. 3 is 6.9 acres, by 25 feet deep with gradual dropoffs and lots of rocks, marsh and some trees around the shores. It is fished a little for 10 to 12 inch rainbow but very few are taken. Lake no. 4 is about 9.7 acres, by 20 feet deep at most with gradual dropoffs and a rocky to muddy bottom. It partially winterkills and now contains only a few (if any) 16 to 18 inch rainbow. Lake. no. 5 is only 5.1 acres, by 12 feet deep at the deepest, and supports a very few 10 to 14 inch rainbow. There are some other lakes above (to the southwest) but they're barren.

Meadow Lakes.
Five little alpine lakes around 7300 to 7500 feet above sea level at the headwaters of Rock Creek; reached by 5 miles of jeep trail west from the upper end of Rock Creek Lake; or 2½ miles by jeep trail south from the Gold Creek Lakes. Only West Meadow contains fish. It lies below (east of) a 350 foot high, mostly open rocky ridge but is in timber to the north and east; is 15½ acres, fairly deep, and pretty good fishing for 6 to 14 inch rainbow-cutthroat hybrids.

Medicine Lake.
A 7000 foot elevation, 72 acre lake in heavily timbered mountains; reached by a jeep road 4½ miles from the junction of Elk Creek to the Ross Fork of Rock Creek; but you can also get in by hiking a couple of miles from the end of the Sand Basin road. It is about 35 feet deep with gradual dropoffs, a muddy bottom, lots of lilies and a beaver dam at the outlet. The fishing used to be and still is no less than excellent — for 6 to 14 inch cutthroat trout.

Middle Fork Rock Creek.
A clear mountain stream flowing northward from its headwaters in the Anaconda range for about 25 miles to its mouth, and followed for the first 20 miles upstream by a road, and the remaining 5 miles by a good USFS trail. The Middle Fork is an easily wadeable stream that is heavily fished for excellent catches of 6 to 10 inch rainbow, whitefish, Dolly Varden and a few cutthroat trout.

Mike Renig Gulch.
A small stream paralleled from its headwaters on Jericho mountain for 8 miles to its mouth on the Little Blackfoot at Elliston by good gravel and jeep roads. Mike Renig is generally poor fishing except for a string of beaver ponds in meadowland about 3 miles above its mouth; from them are caught good messes of 6 to 8 inch brook and cutthroat trout.

Mill Creek.
A clear mountain stream that rises in the Anaconda range and flows eastward behind the barren slopes of Mount Haggin for about 15 miles to the Deer Lodge valley and thence for 5 miles through meadowland to the Clark Fork River. The lower and middle reaches are followed by road and are easily and heavily fished for fair catches (if you could call it that) of 5 to 6 inch cutthroat, brook and rainbow trout.

Mill Creek Lake. See Miller Lake.

Miller (or Mill Creek) Lake.
Really just a small (2½ acres) shallow, swampy pond, in subalpine, scattered timber and meadow country near the head of Mill Creek, just below and north of the Continental Divide. Miller Lake is reached by 8 miles of very poor foot trail from the end of the Mill Creek road, is very seldom visited but excellent fishing for 6 to 10 inch native cutthroat trout.

CLARK FORK of the Columbia (Upper)

MILLER LAKE
Courtesy R. Konizeski

UPPER PHYLLIS LAKE
Courtesy Ernst Peterson

Miller's Lake. See *in* Hoover Creek.

Minnesota Gulch. A very small (2½ miles) tributary of Beefstraight Creek; followed about half way upstream by a trail. It's fair fishing for pan-size cutthroat and brook trout.

Monarch Creek. A tiny (4 miles) tributary of Ontario Creek. Followed by a road for 2½ miles above the mouth, and fair fishing for small population of small trout — 6 to 7 inch cutthroat.

Moose Creek. This small stream flows for 5 miles through boggy, floating turf in the Moose Meadows country to the Ross Fork of Rock Creek. It is crossed at the mouth by the Ross Fork road and followed upstream by a jeep road (THAT YOU SHOULD STAY ON IF YOU DON'T WANT TO GET STUCK). If you stray far from the creek you could easily break through and provide food for the fish, which are mostly 6 to 8 inch cutthroat and hungry.

Moose Lake. A beautiful little 21 acre by 26 foot deep lake; reached by a good road a mile from the Copper Creek campground. There is only limited public access and the fishing is poor anyway for a few good-sized Dolly Varden that migrate up from Rock Creek via the Copper Creek drainage, plus a very few small brook trout and hordes of suckers.

Mosquito Lake. This lake lise ¼ mile north of the Copper Creek Lakes at the head of Swamp Gulch and a tough hike in by either of the Fred Burr, Copper Creek, or Swamp Gulch trails. It is in heavily timbered mountains, about 5 acres with a muddy bottom but clear water in which live numerous salamanders and a fair population of 6 to 8 inch rainbow trout plus a few BIG (to 8 pounds or better) cutthroat; all that are left of an abortive planting many years ago.

Mountain Ben Lake. Reached by a good trail ½ mile from Caruthers Lake about a mile south of Racetrack Peak (a nice view from here at 9524 feet and a pleasant hike after the morning's fishing). Mountain Ben is 32.7 acres, over 50 feet deep, in timbered country but with rocky talus slopes to the west, and slow fishing for 7 to 15 inch rainbow-cutthroat hybrids.

Mount Powell Lake. See Upper Elliot Lake.

Mud Lake. A muddy little 2 acre pot-hole at the top of Tin Cup Joe's moraine, reached by a good gravel road 2 miles west of the Deer Lodge prison farm. It's poor fishing for 7 to 10 inch brook trout, and a few sunfish.

Mud Lake. Proceed west on the Skalkaho highway for ¾ of a mile past the Crystal Creek campground to about a mile east of the pass, and then hike a few hundred yards south of the road across the North Fork of Rock Creek to Mud Lake which is 4½ acres, swampy, and up to 34 feet deep in the middle so it doesn't freeze out (although it's mostly shallower). It's slow fishing for the blackest pan-size cutthroat you've ever seen, plus an occasional Dolly.

Mud Lake. Is reached by a poor logging (jeep) road 2 miles north from Big Racetrack Lake. It's about 12 acres, mostly shallow but up to 17 feet deep in places, and slow fishing for 16 to 17 inch rainbow trout.

North Fork Flint Creek. Flows from the southern slopes of the Flint Creek Range to Georgetown Lake and is followed by a good gravel road for 7 miles from its mouth to headwaters. The North Fork is a small, heavily fished stream for limited catches of 7 to 8 inch cutthroat and brook trout.

North Fork Gold Creek. A small tributary of Gold Creek; crossed at the mouth by the Gold Creek road a half mile south of Jones Mountain, or over-the-top a couple of miles from the end of the Douglas Creek road which takes off from U.S. 10A 2½ miles south of Hall. It isn't much for fishing but does support a few 5 to 7 inch trout.

North Fork Lower Willow Creek. Is easily accessible by 11 miles of county road from Hall, or by 14 miles of private and county roads from Bearmouth. There are about 7 miles of open, easily fished water that is good early season kids' fishing for 6 to 8 inch cutthroat and brook trout.

North Fork Rock Creek. The North Fork is an easily wadeable, very popular stream that flows through heavily timbered mountains and is followed for 15 miles from its headwaters near the Skalkaho pass to its mouth 2 miles west of the West Fork campground. It is a consistent producer of small cutthroat trout.

Ontario Creek. Is followed by a good gravel road through heavy timber for 8 miles northwestward between Nigger and Bison mountains to the Little Blackfoot River about 8 miles south of Elliston. Ontario gets real low in late summer and its trout "take to the holes" which are fair fishing then for 6 to 8 inch cutthroat and brookies.

Ophir Creek. A wee-small stream flowing for about 3 miles to Carpenter Creek 4½ miles northeast of Avon, and crossed at the mouth by the Carpenter Creek road. Ophir "picks up" irrigation water during the summer months and supports a few 6 to 10 inch cutthroat, browns, and brook trout, but on the whole it is poor fishing and very seldom bothered.

Perkin's Reservoir. A dammed-up, 4 acre pothole on the toe of the Dempsey Creek moraine. Perkin's has a maximum depth of about 20 feet, goes almost dry in late summer and somehow supports a fair crop of 8 to 12 inch, muddy tasting brook trout. It hasn't been fished much since some enterprising individuals tried to seine it out several years ago.

CLARK FORK of the Columbia (Upper)

ROCK CREEK ABOVE GRIZZLY CREEK
Courtesy USFS

RACE TRACK LAKE
Courtesy Montana State Fish & Game

Peterson Creek. A small stream flowing for 15 miles westward to the Clark Fork at Deer Lodge. The lower reaches are mostly de-watered for irrigation but there is fair fishing for small trout in an occasional pothole here and there upstream. A kid's creek that is followed by county roads to headwaters.

Pfister Pond. A private, ½ acre pond a mile east of Elliston on the south side of the highway. Pfister was planted years ago and is presently fair fishing, but seldom visited for 8 to 10 inch brook trout.

Phyllis Lakes. Take a good USFS trail for 4 miles above the end of the Middle Fork Rock Creek road, or ½ mile by foot path below and north of the Highline trail in the Anaconda-Pintlar Wilderness Area, to this pair of high alpine lakes which are about 10 acres each, very deep with steep dropoffs and rocky bluffs to the south and timber on around. The upper lake is spotty for 8 to 10 inch cutthroat to about 1 pound. The lower lake is very poor fishing for LARGE (reportedly to 11 pounds) cutthroat trout.

Powell Lake. Measures 8.5 acres, 40 feet deep in at least two places, in timbered country on the eastern flanks of the Flint Creek range. Powell is reached by a foot trail ½ mile from the end of a jeep road from the Deer Lodge prison farm west of town. It is seldom fished, and only poor fishing anyway, for 11 to 15 inch brown trout and hordes of suckers.

Pozega Lakes. See Big and Little.

Pozega No. 3 Lake. See Green Lake.

Racetrack Creek. An easily wadeable fishing stream that flows from Racetrack Lake for 12 miles down a deeply glaciated, heavily timbered canyon, then for 2½ miles through the great hummocky Racetrack terminal moraine, and finally for 6 miles across the wide open meadows of the Racetrack-Dempsey Creek alluvial flat to the Clark Fork River 2 miles north of Galen. It is easily accessible through its entire length by good county roads, jeep roads, and USFS trails. The middle and upper reaches are heavily fished for good catches of 6 to 9 inch rainbow, cutthroat and brook trout, a few browns and Dolly Varden.

Racetrack Lake. A high (elevation around 7700 feet above sea level) alpine lake in deeply glaciated country just north of Twin Peaks. Racetrack is reached by a rough jeep road (if and when its open to traffic) from the end of the North Fork of Flint Creek road. Little Racetrack lies about ¾ of a mile beyond. Racetrack is about 80 feet deep, 35 acres, dammed with perhaps 14 feet of drawdown, and slow fishing for 11 to 14 inch rainbow trout.

Rainbow Lake. This high (elevation 7200 feet), 20 acre lake is up to 35 feet deep in spots, in heavy timber at the headwaters of Gold Creek about an eighth of a mile below Gold Creek Lakes. It is reached by a jeep road 8 miles above the old ghost town of Pioneer, is moderately popular, and used to be excellent fishing for 10 to 30 inch rainbow trout — but there are danged few left.

Ranch Creek. Take the Rock Creek road for 10 miles from U.S. 10 to the mouth of Ranch Creek, then up that creek for a mile to the Grizzly campground, another mile on a private road (through a locked gate) to the Spink ranch, another mile of poor road (with permission), and finally 12 miles of trail to headwaters. Here is a beautiful mountain stream flowing through steep timbered country above and open meadows below, that provides excellent fishing for mostly small cutthroat and brook trout, plus a few rainbow, Dolly Varden and whitefish near the mouth. It is heavily fished near the campground but seldom bothered above.

Rock Creek. Drains as many as 19 alpine lakes and runs eastward for 5 or 6 miles through heavily timbered, glaciated canyons to Rock Creek Lake, thence for another 5 miles down the great Rock Creek boulder moraine, and finally for 3 miles through a narrow, rocky gorge and out across the Clark Fork flood plain to the river near Garrison. It is readily accessible by road to the lake, and by trail above, and is lightly fished in the lower and middle reaches for good catches of 7 to 12 inch whitefish, plus a few brook and rainbow trout and lots of suckers.

Rock Creek. Probably the most famous and heavily fished stream in Montana west of the Continental Divide. Rock Creek flows for 50 miles from the junction of its North and Middle Forks to the Clark Fork River 20 miles east of Missoula and is readily accessible all along by oiled and graveled roads, not exactly a super highway but there's none of it a caddy couldn't readily negotiate. A small, easily wadeable stream in open rangeland above, it quickly becomes a large, swift and often unfordable stream for most of its length through heavily timbered hills between the Long John and Sapphire mountains. It is good summertime fishing for "one and all" in its upper reaches for 8 to 10 inch rainbow and cutthroat trout; excellent fishing below by the experienced fisherman for mostly 10 to 16 inch rainbow trout, and lesser numbers of brook, cutthroat, brown and Dolly Varden; and winter fishing for 10 inch to 5 pound whitefish that could stand a lot more pressure than they get.

Rock Creek Lake. One hundred and eighty acres, deep, with a concrete dam and 30 feet of drawdown, Rock Creek Lake is easily accessible by 12 miles of good gravel road northwest from Deer Lodge, and has been a popular boating, swimming and water-skiing area. It was poisoned in 1959, but with only a partial kill and has since been planted repeatedly with rainbow trout. As of last reports it is only fair fishing for rainbow and whitefish, and the rough fish are rapidly coming back. It is now under new ownership and its recreational future is somewhat dim.

Ross's Fork Rock Creek. Flows for about 20 miles from its headwaters at Lake Abundance to the North Fork or Rock Creek a couple of miles east of the West Fork Work Station and is followed for about 4 miles by jeep road and then trail on up to the lake. Ross's Fork flows mostly through timber and swampland, is easily wadeable and is considered to be fair fishing for 9 to 10 inch cutthroat trout, and a few brookies, Dolly Varden and whitefish in season. The lower end is mostly on private (posted) land.

Ryan Lake. A little old reservoir lake, 5½ acres by 10 feet deep (almost); reached by trail 1¼ miles from the end of the Bielenberg Canyon Road. It's slow fishing for big brooks in the 15 to 18 inch class.

Sallyann Creek. A very (very) small stream crossed by the Telegraph Creek road and consisting mostly of a few beaver ponds which drain to Bryan Creek, and in which thrive some real nice fat brook trout.

Sand Basin Creek. A small, 2½ miles, open meadow stream flowing to the West Fork of Rock Creek and reached by the Sand Basin road 6 miles up from the main stem. It is easily fished for fair catches of pan-sized cutthroat trout, but is reportedly soon to be placer mined for "rare earths."

Sauer Lake. Take the (good) East Fork Rock Creek trail for 6 miles above the reservoir, then a poor, unmaintained foot trail for another 2½ miles to this lake in high (elevation 8050 feet) timbered country. It is about 8 acres, used to be good-to-excellent for 6 to 9 inch cutthroat trout, but is now "reportedly" barren and seldom if ever fished.

Sawmill Creek. A darned good little stream for 6 to 10 inch brookies and Dollys too, for a couple of miles up its timbered canyon from the mouth of Brewster Creek (from the sawmill access on Rock Creek) up the main stem for 1½ miles to an old cable crossing.

Schwartz Creek. Flows east from the Bitterroot-Clark Fork drainage divide for 7 miles to the Clark Fork River about 3 miles below the Rock Creek Lodge; and is followed by good logging roads all the way. It is fished near the mouth by kids for a few small cutthroat and rainbow trout.

Senate Creek. A small open stream that drains Kaiser Lake for 1 mile to the Middle Fork of Rock Creek. The lower ½ mile is fair fishing for 6 to 9 inch cutthroat trout.

Senecal Ponds. Two artificial ponds of about 1 and 2 acres, at the mouth of Trout Creek canyon. Take U.S. 10A three miles east from Avon, then the Snowshoe Creek and county roads for 7½ miles to these private ponds below the old Senecal residence. Fishing is with owner's permission only and is reportedly good for 8 to 14 inch cutthroat and rainbow trout.

Sidney Lake. Reached by trail ½ mile northeast from Dora Thorn Lake, 14.7 acres, over 50 feet deep and almost barren except for a very few large rainbow trout, holdovers from an old plant.

Silver Lake. An A.C.M. reservoir about a mile long by ½ mile wide; in timbered country 1 mile east of Georgetown Lake and reached by the highway 13 miles west from Anaconda. Silver Lake is lightly fished (in favor of Georgetown) and usually poor at best for 8 to 10 inch rainbow trout and sockeye salmon, plus a few Dolly Varden.

Smart Creek. A small open stream flowing north to Flint Creek 6 miles above Hall and followed by road and trail for 10 miles to headwaters. It is not much for fishing but "looks good," is easily accessible and is subjected to moderate fishing pressure for 6 to 7 inch cutthroat trout.

Snowshoe Ponds. Two scenic "used-to-be" fish-rearing ponds of about 3 and 4 acres each, 100 feet apart in scattered timber and open meadowland hills at the mouth of Snowshoe Canyon; reached by U.S. 10A three miles east from Avon, then the Snowshoe Creek road for 6 miles and a private road ¼ mile west to the ponds. The owner's home is located between the ponds, which used to be fabulous fishing "with permission only" for 9 to 12 inch rainbow trout, they were practically fished out during the owner's absence a few years ago and its anybody's guess as to their present status.

South Boulder Creek. Flows north through steep, timbered country (the northern slopes of the Flint Creek range) to Boulder Creek 2 miles above Maxville, is reached at the mouth by the Boulder Creek road and followed upstream for 5 miles by a road, and thence, 2½ miles by a trail to headwaters, at Stewart Lake. This stream is easily accessible and good fishing for 9 to 12 inch cutthroat, rainbow, brook and a few Dolly Varden trout.

South Fork of Gilbert Creek. Here is a small stream that flows through brushy, timbered country to Gilbert Creek at the old Finlen ranch. The lower couple of miles are followed by a trail and could provide fair catches of cutthroat, brook and Dolly Varden trout, but are generally inaccessible to the public and seldom fished.

South Fork of Mill Creek. A tiny stream flowing for a couple of miles through a willow-brush-beaver-dammed flat a few hundred feet west of the Anaconda-Big Hole highway. It contains quite a few 5 to 7 inch cutthroat trout.

South Fork Ross Fork Rock Creek. This small, open stream is followed by a good USFS trail for 5 miles from its mouth past headwaters and for 1 mile beyond the Frog Pond basin. It's good fishing, but not all that popular for 7 to 9 inch cutthroat trout.

WEST FORK ROCK CREEK
Courtesy USFS

TAMARACK LAKE
Courtesy USFS

CLARK FORK of the Columbia (Upper)

TWIN LAKES CREEK
Courtesy USFS

UPPER TWIN
Courtesy USFS

Spotted Dog Creek. A small, open, easily fished stream followed by the Mullan Pass highway for 8 miles from just east of Elliston to Blossburg, and well fished by local residents for 6 to 10 inch cutthroat, brown and brook trout.

Spring Creek. Take the Rock Creek road for 6 miles from U.S. 10 to the lower pastureland reaches of Spring Creek on the Handley ranch (fishing with permission only) then on upstream by a real poor USFS trail through a timbered gorge for 5 miles to headwaters. The lower reaches are heavily fished for mostly 8 to 10 inch (and a few to 14 inch) rainbow, cutthroat and Dolly Varden trout; the upper reaches are lightly fished for small cutthroat and Dolly Varden.

Stewart Lake. A 20 acre mountain lake in heavy timber; reached by car 4 miles northeast from Philipsburg. Stewart was rehabilitated in the '50s and has since been planted several times with rainbow trout which now average about 9 to 11 inches. It is quite popular and fair-to-good fishing.

Stony Creek. The outlet of Stony Lake, flowing for 10 miles through steep, heavily timbered country northeastward to Rock Creek at the Squaw Rock campground, and followed upstream by a road for 5 miles and then a USFS trail for 6 miles to headwaters. Stony is an open, easily fished stream that is quite popular for good catches of 7 to 9 inch cutthroat trout.

Stony Lake. A 24 foot maximum depth, 11½ acre alpine cirque lake; reached by a real steep USFS trail 6 miles up Stony Creek from the end of the road; or you can hike in 6 miles from the Crystal Creek campground a couple of miles east of Skalkaho Pass. The fishing is slow here for 6 to 12 inch cutthroat trout — you have to work for 'em.

Storm Lake. Fifty-five acres, about 8700 feet above sea level, in a cirque on the northern slopes of Mount Tiny in the Anaconda range; reached by road and trail 7 miles south from the Anaconda-Georgetown highway at a point ½ mile east of Silver Lake. Storm Lake is over 100 feet deep with steep dropoffs all around. It was originally barren but has been planted a number of times and is now fairly good fishing for cutthroat, rainbow and Dolly Varden trout that will range from 9 to 18 inches. It's a good weekend lake.

Storm Lake Creek. The small, fast and clear outlet of Storm Lake; flows northward down its steep timbered canyon and is followed by the Storm Lake road for about 7 miles to Warm Springs Creek ½ mile east of Silver Lake. It's not fished much but does contain (here and there along its course) a few 6 to 10 inch cutthroat and brook trout.

Stuart Mill Creek. A 200 yard long inlet to the northwest end of the Georgetown Lake that is heavily used by spawning rainbow trout in the spring, and sockeye salmon in the fall. It is closed until late July, and open for sockeye snagging from October through December.

Sydney Lake. This is a very deep, 8 acre alpine lake that is reached by 1½ miles of good trail (past Dora Thorn and the Copper Creek Lakes) from the end of a jeep road from Granite. Sydney is good fishing the year around for rainbow and cutthroat trout that will average 10 to 12 inches, along with a few that will range up to better than 2 pounds.

Tamarack (or Glover) Lake. Is reached by 4 miles of good trail and one mile of lousy rough trail from the Meadow Creek road in the Anaconda-Pintlar Wilderness Area. Tamarack is a high lake in a timbered cirque about a mile west of Warren peak. It covers about 15 acres, is only of moderate depth with shallow dropoffs all around and swampy with quite a few aquatic plants around the shores. It is seldom fished but reportedly contained small cutthroat at one time although they may have frozen out by now, and probably have. Not recommended unless you need the exercise.

Telegraph Creek. A small tributary of the Little Blackfoot River, Telegraph is followed by good gravel roads for 10 miles from its headwaters to its mouth near Elliston. The lower 3 miles flow through meadow-and-willow land and are fair fishing for 6 to 8 inch brook, cutthroat and Dolly Varden trout.

Thompson Lakes. See Goat Mountain Lakes.

Thornton Lake. A 30 acre "cirque" lake in timbered country on the headwaters of Racetrack Creek, but most easily reached by a good USFS trail a mile above the end of the Modesty Creek jeep road. Thornton is moderately popular and good fishing for 8 to 10 inch rainbow trout.

Three Mile Creek. A small stream that is mostly a series of meadowland springs above, and flows for 1 mile through a flat-bottomed canyon below to its junction with the Little Blackfoot River 2 miles west of Avon. The lower reaches are followed by a private road and are good fishing for 8 to 12 inch browns.

Tin Cup Joe Creek. A small stream flowing eastward from the Flint Creek Range for 8 miles across state (Deer Lodge) prison farm meadowland and the Deer Lodge Country Club to the Clark Fork at Deer Lodge. The lower reaches are easily fished (mostly by kids) for poor catches of 5 to 6 inch rainbow and cutthroat trout, and trash fish.

Trask Lake. The largest (10 acres) of the Hidden Lake group, reached by a trail 5 miles above the end of a jeep road past Rock Creek Lake, or about 2 miles by a good trail from Alpine Lake. It is overpopulated with 8 to 9 inch brook trout, and some cutthroat. There is a special 10 pound and 1 fish limit here.

Trout Creek. About 5 miles long in open ranchland easily accessible by private and county roads from its junction with Flint Creek 5 miles south of Philipsburg—to headwaters. The fishing here is real good for pan-size brook and cutthroat trout, but the stream is too small to stand much more than local pressure.

Trout Creek. A small stream flowing for 10 miles through private ranchland northward to the Little Blackfoot River 2½ miles east of Avon. The lower reaches are used for irrigation but are good in early season for 8 to 10 inch brookies, cutthroat and brown trout.

Twin Lakes Creek. A real fast, steep stream flowing down a steep timbered canyon for 8 miles from Twin Lakes to Warm Springs Creek ½ mile east of the Spring Hill campground (3½ miles east of Silver Lake), and followed all the way by a logging road and USFS trail. The first 4 miles are fair for small cutthroat and brook trout, but are seldom fished.

Twin Lakes. Reached by a good trail a mile from the end of the Twin Lakes Creek jeep road, in subalpine country. Lower Twin is small, shallow and contains few if any fish. Upper Twin lies about ⅛ mile above Lower Twin, is 18 acres, fairly shallow and is bordered by scattered timber, alpine meadow and talus slopes. It is usually good fishing for 8 to 12 inch cutthroat and rainbow trout.

Tyler Creek. Take U.S. 10, twelve miles south from the Rock Creek Lodge to Byrne, then cross the Clark Fork River to the mouth of Tyler Creek on private property. It is followed by a private jeep road through a locked gate and for 5 miles up the creek, which is brushy and hard to fish for the limited number of small cutthroat it contains.

Upper Barker Lake. See Big Barker Lake.

Upper Elliot (or Mount Powell) Lake. A high (elevation 8500 feet) cirque lake 3½ miles east of Racetrack Peak; reached by 1½ miles of good USFS trail from the end of the Dempsey Creek jeep road, or by 1 mile of trail above Martin Lake by way of the Elk Ridge road and trail. Upper Elliot is 73 acres, 55 feet deep with steep dropoffs and about 5 feet of water level fluctuations due to an irrigation dam at the outlet. It is seldom fished but good for camp-fare cutthroat trout.

Upper Willow Creek. A pretty little stream flowing from heavily timbered hill country in the Long John mountains for 20 miles to Rock Creek 4 miles above the Squaw Rock campground and followed by a road all the way except for maybe the last two or three miles. The lower reaches flow through meadowland, are wadeable, and easily and heavily fished (with owner's permission) for excellent catches of 8 to 12 inch cutthroat, rainbow and some brook trout. Most of the creek is on private land.

Wahlquist Creek. A real small, brushy stream in a heavily timbered draw, flowing to Rock Creek on the opposite side from the road, 3 miles above the Harry's Flat Guard Station. It CAN be reached at the mouth, but only by a cable crossing and so is but seldom fished, for the plentiful small cutthroat and brook trout it contains.

Wallace Creek. A small (4 miles) stream flowing through heavily timbered hills to Rock Creek 10 miles west from U.S. 10, and followed by a logging road to headwaters. The lower reaches contain a few 8 to 10 inch brook and cutthroat trout.

Wallace Reservoir. About 9 acres by 10 feet maximum depth behind a 300 foot long earthfill dam in a partly open draw a couple of miles east (by good gravel road) of U.S. 10 at Clinton. It is sometimes drawn clear down to nothing for summertime irrigation but has been known to somehow harbor a fair population of nice fat 6 to 15 inch cutthroat that are not often fished for. They must hole up in the inlets during the "dry" periods. A close to town "sleeper."

Warm Springs Creek. Heads in heavily timbered mountains at the southern end of the Flint Creek Range and flows southward for 9 miles (past the Upper and Lower Warm Springs campgrounds) to the Warm Springs valley and thence for 20 miles eastward through open meadow and farmland past Anaconda to the Clark Fork at the town of Warm Springs. Here is an easily accessible (by state highway and county roads), well stocked and easily fished stream that is just poor-to-fair fishing for 6 to 10 inch rainbow, brook and a few brown, Dolly Varden and cutthroat trout. It is "kids' fishing only" in and near Anaconda.

Warm Springs Creek. A small tepid stream flowing through limestone terrain for 10 miles to the Clark Fork River; crossed near the mouth by U.S. 10, three and a half miles south of Garrison and followed upstream all the way to headwaters by a good gravel road. The upper reaches are overpopulated with 3 to 5 inch fingerling and trash fish, the lower reaches flow through private (no fishing allowed) land and are excellent fishing for 9 and 14 inch browns, and a few rainbow and Dolly Varden trout.

Warm Springs Gravel Pit. See Five Mile Barrow Pit.

Warm Springs Hospital Pond. A shallow, 2 acre pond on the hospital grounds 100 yards west of U.S. 10. It is usually planted on an annual basis with keeper-size trout for the kids and hospital patients. This is a good place to take pictures of geese from the bird farm next door.

Welcome Creek. A small stream followed by a good USFS trail for 11 miles up a heavily timbered canyon from its mouth on Rock Creek opposite the Dallas Creek campground, to headwaters (the middle 3 miles or so is hazardous "foot" trail only). It is seldom fished (probably because its on the opposite side of Rock Creek from the road) but fairly good for pan-size cutthroat, plus a few Dolly Varde, rainbow trout and whitefish.

West Barker Lake. See Little Barker Lake.

West Fork Cramer Creek. Take the Cramer Creek road for 1½ miles from U.S. 10 to the mouth of the West Fork, then a dirt road for 2 miles and finally cow trails for 3 miles on to the headwaters. It's easily and heavily fished for good early-season catches of 8 to 10 inch cutthroat trout.

West Fork Lower Willow Creek. Take a county road for 10 miles west from Hall to the mouth of this creek and then for 3 miles on upstream. A timber and meadowland creek, it's only fair fishing for small cutthroat, brook and Dolly Varden trout.

West Fork of Rock Creek. A small, open stream in mountain peak and timber country, the West Fork flows for 8 miles to the North Fork of Rock Creek about 4 miles east of the Skalkaho Pass, and is followed by 4 miles of good road and then trail to headwaters. This is a popular, easily fished stream that produces good catches of 4 to 11 inch cutthroat trout and whitefish, a few 5 to 10 inch Dolly Varden and fewer still brook trout.

West Fork Willow Creek. About 2½ miles of creek, mostly in swampy meadowland of the Willow Glen ranch, 6 miles east of Anaconda. The West Fork provides fair meadow fishing for 6 to 8 inch brook trout with permission only.

Willow Creek. A small, jumpable stream flowing to the Clark Fork River a mile north and on the opposite side from Garrison; and crossed 4 miles upstream by the old Pioneer highway 10 miles north from Deer Lodge. It is fair-to-good fishing through the meadowlands (about 2½ miles) of the Willow Glen ranch, for 6 to 7 inch brook and cutthroat trout.

Willow Creek. Eleven miles of grassy bank, open meadow stream flowing to the Clark Fork River a mile northeast of Opportunity. This one is real easy to fish and a popular kids' creek that consistently produces excellent catches of eat'n size brook and cutthroat trout.

Wyman Creek. Take the Rock Creek road from U.S. 10 for 2½ miles above the guard station, then a federal cable car across Rock Creek to the mouth of Wyman Creek, and thence a good trail 3 miles upstream. It is a small, ice cold mountain stream that flows through a heavily timbered canyon and produces fair catches of 4 to 7 inch scarlet-red cutthroat and brook trout which, because of their small size and the inaccessibility of the water, are very seldom fished for. Down around the mouth you will pick up mostly Dolly Varden and rainbow come up from Rock Creek.

Wyman Gulch. A tiny stream flowing for 5 miles to South Boulder Creek; paralleled by a road from mouth to headwaters. The lower couple of miles are mostly in meadow and are easily fished for fair catches of pan-sized cutthroat trout.

UPPER ROCK CREEK NEAR PHILLIPSBURG
Courtesy Ernst Peterson

Courtesy Cary Hull

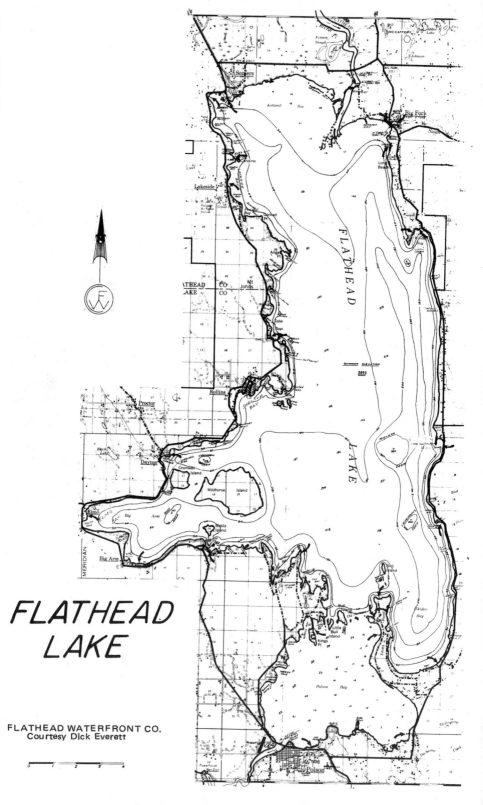

Behold, The Flathead!

When a Montanan says he is going to fish the "Flathead," he usually means the lake. However, there is also excellent fishing in the stretches of river above the lake to its three forks (North, South and Middle), and below the lake to its junction with the Clark Fork near Paradise.

Flathead Lake provides some of the most magnificent natural beauty found in America. The lake is large: 28 miles long by an average 6 in width. It is encircled by paved highways, resorts, summer homes and several towns and is heavily used, especially during the summer months, for water skiing, sailing, swimming and picnicking, as well as year-round fishing.

Here is a typical example of the fishing you'll find. Just after the first of the year I had a call from an old friend vacationing at Big Mountain and enjoying some wonderful skiing. He wanted to do some winter fishing for lake trout and Dolly Varden. At that time of year most large mackinaw and Dolly's congregate in relatively shallow water at the south end of the lake where there is abundant feed for the large fish. Using heavy rods, 30 pound steel line and T-50 Flatfish bouncing on the bottom, we managed to pick up four fish with a total weight over 60 pounds.

As the weather warms, the large mackinaw move to their spawning grounds, two of which are between Wild Horse and Melita islands, and further north around Angel Point. To catch a trophy mackinaw (or lake) trout, the best bet is to fish in August or September when the fish are most active in these areas. Here they lie about 80 to 100 feet down and steel line is used to bounce the Flatfish or other lure along the bottom. No weight is used; only a snap swivel between the lure and the steel line.

In April the Dollys start their migration toward the north end of the lake and as high water comes, they head up the river. During this northward migration, the Dollys tend to move close to shore. They are best caught in April or May, using a floating line, a ten foot-8 pound leader and any favorite hardware. Hooking a big Dolly on a light outfit and the ensuing battle to land him is great fun. More often than not, the fish wins, but there's always another in the next cove.

At this same time, the flat (or native cutthroat) trout also move in close to shore to feed. It is fast, exciting fun to locate a school, stop the boat and spin cast with a small lure and monofilament line. If a hatch is on and the fish are feeding on the surface, a dry fly will "limit a fellow out" before the motor is cold. The fish will run around 1 to 2 pounds each.

Flathead Lake also supports a large population of kokanee salmon. Although not large, these fish in the 12 to 14 inch class are easy to catch and wonderful to eat. Kokanee feed at different levels and once you have the correct depth, the rest is easy. Until acquainted with this summer sport, cow bells are probably the best way to attract the fish, although most local, experienced fishermen use leaded line, with about 15 feet of 4-pound-test leader and a small lure such as a Needle-Fish or Triple-Teaser.

In the main river above the lake, fishing generally begins around mid-July. At this time casting with lures will produce some tackle-busting experiences; be prepared for fish up to 30 pounds! It would be unforgivable not to mention the wonderful fly fishing also available, where the primary target will be beautiful native cutthroat (flats).

While Flathead Lake is a natural body of water, it is blocked at the south end by Kerr Dam, with resulting ten foot maximum annual fluctuations of water level – up in the summer and down in the winter. Shortly ahead of the dam, the river again takes form as it leaves the lake. Below the dam are some fantastic prospects for trout, largemouth bass, and for those in the know, Northern Pike in the 20 pound class. You can't go wrong fishing the Flathead!

LYLE JOHNSON

Courtesy Lyle Johnson

Flathead River Drainage

Abbot Creek. A small stream flowing through timber and pastureland to the Flathead River at Martin City. It's paralleled by a good road and provides about 2½ miles of good kid's fishing for 6 to 8 inch brookies.

Agency Creek. A little stream flowing from McLeod peak for 8 miles northwest towards the Jocko River. The lower, farmland reaches near Old Agency are accessible by county roads and are occasionally fished by kids for 5 to 8 inch cutthroat.

Alder Creek. A very small stream, crossed at the mouth by the Good Creek road 15 miles west of Lower Stillwater Lake. Its followed all along by a road (you'll need permission to tresspass for the first mile) and is fair fishing for 3 to 9 inch cutthroat trout.

Allentown Pond. A three-quarter acre pothole just behind the Allentown Motel on U.S. 93 near the Nine Pipe Reservoir. This one used to be stocked annually with rainbow for the motel guests, strictly on a put-and-take basis — but is no longer. Its present status is probably not much, maybe a few largemouth bass and sunfish but that is just a guess.

Ashley Creek. The outlet of Ashley Lake, this stream flows for 35 miles through Lone Lake, Lake Monroe, Smith Lake and eventually past (just south of) Kalispell to the Flathead River 5 miles southeast of town. It is all easily accessible by U.S. 2, county and farm roads. The lower reaches below Kalispell are polluted and the upper reaches are mostly on private land so it's not very heavily fished, although it contains a fair population of 6 to 10 inch cutthroat, brook, and some rainbow trout.

Ashley (or Big Ashley) Lake. Four and a half miles long by almost 2 miles wide at the west end, over 200 feet deep with mostly steep dropoffs but 40 or 50 acres of lilies and rushes; reached and followed all around by good county roads 12 miles west from Kalispell. There are 50 or 60 summer homes and a state campground (on the north shore) of this popular, dammed lake which has a water-level fluctuation of about 5 feet. It is excellent fishing for pencil-shaped sockeye, fair for 2 to 5 and occasionally up to as much as 17 pound cutthroat, and some of the nicest yellow perch in the state.

ATER SKIING ON FLATHEAD
urtesy Ernst Peterson

KEEP'N SIZE
Courtesy Dallas Ecklund

Ashley (or Hidden) Lake.
Take U.S. 93 to the north side of St. Ignatius (to where it narrows at the intersection), then drive 4 miles east on a county road, turn a couple of miles north on a truck road, and finally take a logging road up Ashley Creek as far as your vehicle will carry you. The lake is 1½ miles on up the drainage at about 5200 feet above sea level. It is 30 acres, 68 feet deep at the deepest — and used to be barren. So 'twas stocked with small cutthroat in 1968 and — it looks like its barren again. A good try. P.S.: Don't mess with the pond below, it's barren too.

Basham Lake.
Take the Creston road 4½ miles north from Big Fork, then a secondary (Echo Lake) road ½ mile east, and finally a poor logging road for another ½ mile to Basham on the right hand (east) side of the road. This is a 5 acre pothole in logged-over land. It used to be stocked off and on with rainbow but no longer, and inasmuch as there is no reproduction, you're wasting your time unless you like to fish for small yellow perch.

Bead Pond.
A 1 acre pothole in open country; reached to within ¼ mile by a gravel road about 1 mile east of Allentown on U.S. 93. Bead is planted annually on a more or less put-and-take basis for rainbow, a few of which will run up to 16 inches but they will mostly average between 10 and 12 inches. It is lightly fished.

Beaver (or Big Beaver) Lake.
A deep, 106 acre lake with lilies and aquatic plants along the north side; in rolling timbered country accessible by road 5 miles west of Whitefish. It was poisoned out in 1961, planted with rainbow in 1964, and is real fine fishing now for 1 to 5 pounders.

Big Ashley Lake.
See Ashley Lake.

Big Beaver Lake.
See Beaver Lake.

Big Knife Creek.
A short, muchly beaver-dammed tributary of the Jocko River, very steep in its canyon above and mostly de-watered for irrigation below; crossed at the mouth by the river road 5½ miles west from Arlee and more or less accessible upstream by a trail. It is seldom fished but the beaver ponds contain small brook and cutthroat trout.

Big Lost Creek.
Twelve miles long from its headwaters in the hills a few miles north of Ashley Lake to its lower end in the Flathead Valley where it sinks about 5 miles northwest of Kalispell; reached and followed all the way by good county roads. Big Lost Creek is mostly fished by local people, for good catches of 8 to 10 inch cutthroat and a few brook trout.

Billings Pond.
A moderately popular 3 acre, 20-foot-deep pothole reached by road a short way east from Allentown on U.S. 93. It is planted with rainbow on a put-and-take basis and is good early spring and late fall fishing for 10 to 12 inchers.

Lake Blaine.
A popular summer home-water-skiing-picnicking lake with one resort and a private campground, 372 acres in timber and farmland right at the foot of the Swan Range; reached by good county roads 10 miles east from Kalispell. Blaine is fair fishing for 1 to 4 pound largemouth bass, some 12 to 14 inch rainbow, an occasional large lake trout and reportedly a few whitefish. Has been that is. It was recently stocked with kokanee and the plans call for replants every four years so the fishing should take a turn in that direction. Oh yes! — there are twenty zillion wash tubs full of squaw fish around too.

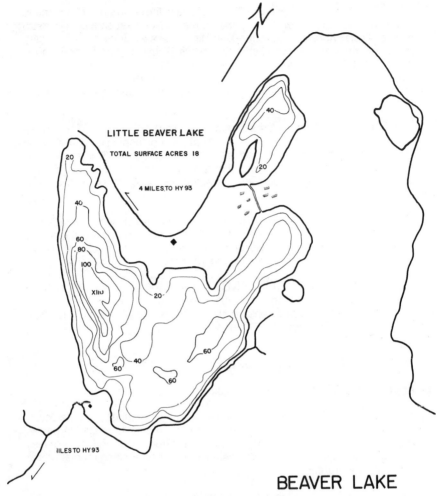

Blanchard Lake. A 147 acre body of water; accessible by road a couple of miles south of Whitefish. Blanchard is up to 32 feet deep in places with clear water and a mossy bottom. It gets pretty warm in the summer and is good fishing from boats for mostly 10 to 12 inch (and some recorded to 4 pounds deep down in the weeds) largemouth bass, 8 to 14 inch yellow perch and lots of "keeper size" sunfish. There is public access at the north end.

Blasted Lake. See Crater Lake.

Blue Lake. An 8 acre pothole, 40 feet deep, in the hills ¼ mile north of Blackie's Bay on Echo Lake; reached by road 3 miles east of the Creston Highway 6 miles north of Bigfork. This one never was bothered much but used to be stocked periodically on a put-and-take basis, and was good fishing for 10 to 12 inch largemouth bass and zillions of sunfish. It may still be fair — no better.

Blue (or Green) Lake. Misnamed Green Lake on some maps, about 12 acres, the entire south side of which lies along the edge of a big rock slide and is real deep straight off from shore; in rolling, timbered country accessible by a poor road 1 mile south from Stryker. It is spotty but sometimes good fishing for 1 pound brookies and millions of shiners.

Bock Lake. One of the pothole lakes — just east of (barren) Kohler Lake, 6 miles north from Bigfork on the Creston road, then a mile east on a gravel road, and finally ¼ of a mile hike across farmer's fields to the shore. It's about 5 acres, right on the edge of the timber, was planted with rainbow in 1960 and is reported to be fair fishing now for 12 to 16 inchers.

Bootjack Lake. Sixty-five acres with fairly steep dropoffs but swampy on the south end; up to 6 feet deep, in timbered hills 7 miles west of Whitefish and reached by an old logging road 4 miles east of the Tally Lake Ranger Station. Most of the shore is under private ownership but there *is* some public access. It was rehabilitated in the 60's and is very good fishing now for 10 inch to 5 pound cutthroat.

Boyle Lake. Forty-one acres, moderately deep, in conifers on the north side of the railroad near the west end of the tunnell a mile west of the northwest end of Whitefish Lake. Boyle contains only sunfish now.

Bull Lake. Almost all access to this lake is across private land, but you can hike into it at the southeast end a few hundred yards from the southwest end of Stryker Lake, or, if you can get permission, take U.S. 93, fifty miles north from Kalispell to about 3 miles below (southeast) of Dickey Lake, then turn north on a good gravel road for 2 miles and you can drive right to the north end. Its in timbered mountains, 106 acres, over 70 feet deep and with mostly steep dropoffs all around. There are summer cottages here and there, a boat livery (and all that implies) but only poor fishing for 10 to 12 inch cutthroat trout, and too many sunfish. May be a rehab coming up.

Bugress Lake. A shallow, 30 acre lake; reached by road 1½ miles south of U.S. 93 three miles west of Perma. It sometimes winter-kills, and summer-kills too, but is planted off and on and is good fishing (or was the summers of '73-'74) for 12 to 18 inch rainbow trout.

Burnt Lake. Four and a half acres by 40 foot maximum depth, planted with cutthroat in 1969 and reportedly pretty fair fishing now. To get there take off cross-country for two miles east from the Sunday Creek road a mile northwest of Wall Lake. Good luck!

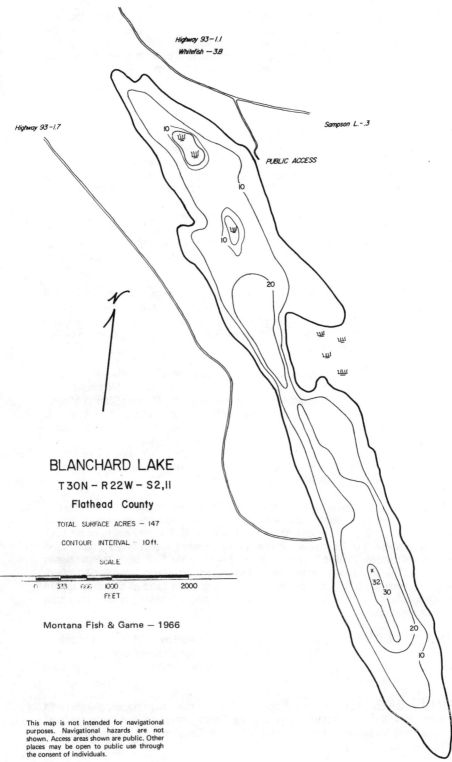

Cabin Lake.
A pretty little 16.8 acre pothole, 50 feet deep, in timber a couple of miles north of Echo Lake; reached by road about 10 miles north from Bigfork. It never was much of a fishing lake and what with the recent summer home development in that area, it likely never will be.

Canyon Lake.
About three-eighths of a mile long by 200 yards or so wide and pinched to a narrow, shallow channel in the middle so that it's almost two lakes; reached by a ½ mile (a steep drop) hike from the road about 2½ miles north of Hubbart Reservoir, on the Flathead Indian Reservation. It's in sparsely timbered country, is fairly deep in spots, was rehabilitated in 1973 and should be good fishing now for rainbow and cutthroat trout. Let's hope it doesn't get taken over again with trash fish.

Carter's Pond.
Six acres, 15 feet deep, in open meadows 2 miles south of the Nine Pipe Reservoir with a county road right by it. It's private but not posted and wonderful kid's fishing for sunfish and small bullheads.

Cedar Creek.
A small, brushy stream in an old burn, Cedar Creek flows south for 10 miles towards the Flathead River and eventually sinks about a mile north of Columbia Falls. Its impounded north of town for the municipal water supply and is closed in that area. However, it is open above the impoundment and fair fishing for 6 to 7 inch cutthroat trout. There's a road along much of its length.

Cedar Lake.
Really a shallow, 15 acre peat bog in brushy burned-over country; accessible by a logging road to within a couple of hundred feet from the shore, 3½ miles north of Columbia Falls. It is seldom fished (a boat is a must) but is fair for 8 to 10 inch cutthroat.

Chinook Lake.
A seldom fished, deep, 8 acre lake in timbered country; reached by a logging road 2 miles north and 2 more southeast from the Tally Lake Ranger Station. It's full of fresh-water shrimp, was planted with cutthroat in 1960 but they mostly died out. The D O is just too low for trout. Montana State Fish and Game fishery biologists are talking of maybe trying it with grayling.

Courtesy USFS

Courtesy USFS

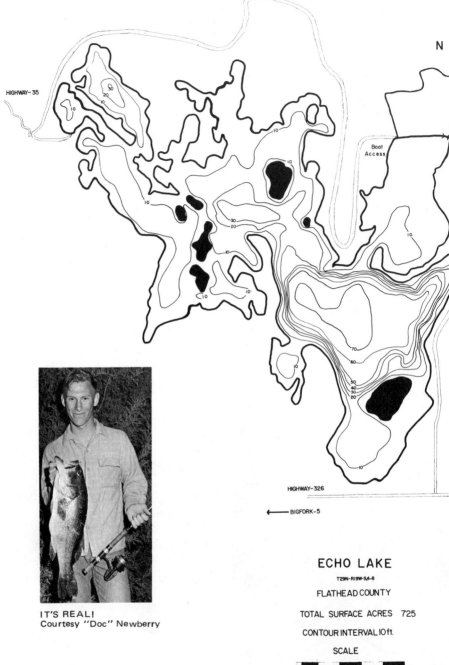

IT'S REAL!
Courtesy "Doc" Newberry

ECHO LAKE
T29N-R19W-SA-8

FLATHEAD COUNTY

TOTAL SURFACE ACRES 725

CONTOUR INTERVAL 10 ft.

SCALE

FEET

MONTANA FISH & GAME—1967

This map is not intended for navigational purposes. Navigational hazards are not shown. Access areas shown are public. Other places may be open to public use through the consent of individuals.

Circle Lake. A 2 acre pothole in the "Many Lakes" housing development (used to be the "pothole district") in logged-over country; reached by road a half dozen miles north of Bigfork. It used to be planted off and on with rainbow and was good fishing, but has now gone private, is no longer planted and likely has little or no fishing potential.

Clark Lake. Take the Creston highway 2 miles north from Bigfork, then turn east on the Swan highway for a couple of miles, and finally ½ mile off the road through open fields and some timber to the shore. This one is about 3½ acres and full of weeds in the summer. It has been planted with rainbow in years gone by and is fished occasionally (with the owner's permission) through the ice, but is marginal at best.

Cliff Lake. See in First Lake.

Corduroy Creek. A tiny, brushy tributary of Good Creek; crossed at the mouth by the Good Creek road 10 miles west of the Stillwater guard station, and also accessible by the Martin Creek road at headwaters. It's not fished much but is fair for 6 to 10 inch brookies in some beaver ponds about ½ mile above the mouth.

Crater (or Blasted) Lake. Three acres, 20 feet deep, in an old rocky burn; reached to within 200 yards by road 3 miles south from Whitefish. Crater was poisoned out in 1961 and is now planted every third year with cutthroat and grayling, so is fair-to-good fishing at times.

Crazy Fish Lake. Five acres, deep, in dense timber on the headwaters of the Jocko River 1½ miles northeast of McLeod peak and reached by good USFS trails either 8 miles west from the Jocko River road, or 5 miles east from the end of the Agency Creek road. Crazy Fish is on tribal land and closed to all but Indians; but is good fishing for 8 to 10 inch cutthroat trout.

Cree Lake. A 30 foot deep, 7 acre pothole in timbered hills (the "Many Lakes" housing development); reached by road 1½ miles east from Creston. It has been planted off and on in the past, but no more. It once supported a fair population of brook trout but what will be there in a year or two is anybody's guess.

NORTHERNS
Courtesy USDA, SCS

ECHO LAKE
Courtesy Phyllis Marsh

Creston Lake (or Jessup Mill Pond). Just east of, and supplies the water for, the Creston Hatchery; reached by paved highway either 10 miles east from Kalispell, or an equal distance north from Bigfork. It lies in an east-west direction, is about ⅝ of a mile long by a couple of hundred yards wide, in mostly cut-over timber but open on the west end. It is spring fed, very fertile, has been planted now and again with this and that, was poisoned out, and is no longer stocked but presently fair fishing for nice fat brook trout.

Crow Creek. From the junction of its North and South Forks at the base of the Mission Range, Crow Creek is readily accessible by county roads as it flows for 15 miles through mostly open farmland to the Lower Crow Reservoir and eventually the Flathead River. A pretty stream, its lower reached have been polluted by sewage from Ronan (via Spring Creek) and are accordingly only lightly fished for 8 to 10 inch rainbow and brook. The upper reaches are fished mostly by kids from Ronan and are good in early spring for 8 to 14 inch brookies.

Crystal Creek. A small "kids" creek, tributary to Cedar Creek in timbered, brushy country; accessible by road 5 miles north from Columbia Falls. There are about 3 miles of fair fishing here for 6 to 7 inch cutthroat trout.

Crystal Lake. A private, commercial fishing pond in open pastureland on the north side of U.S. 93 about ½ mile north of Fortine. It's excellent fishing (after you pay your money) for 16 to 18 inch cutthroat and rainbow trout. There is also a 9 hole golf course here and a private landing field.

Crystal Lake. A private, commercial fishing pond in open pasture on the north side of U.S. 93 about ½ mile north of Fortine. It's excellent fishing (after you pay your money) for 16 to 18 inch cutthroat and rainbow trout. There is also a 9 hole golf course here and a private landing field.

Disappointment Lake. See in First Lake.

Dog Creek. The outlet of Dog Lake, flows for 3½ miles through timbered country parallel to and just west of U.S. 93 to the Stillwater River about a mile above Lower Stillwater Lake. The lower mile contains a few small brookies; it's marginal at best.

Dog Lake. See Rainbow Lake.

Dog Lake. A 99 acre lake that is only about 25 feet deep at most, in rolling timber just west of U.S. 93 about 25 miles north of Whitefish. It has a shallow, narrow neck in the middle, is fairly deep at the north end and marshy at the south end. Mostly an early spring and winter lake, 98% of the local fishermen claim there's nothing in it, but the other 2% report that it's good if you know it for brook trout that average 12 inches and range up to 5 pounds.

Dollar Lake. A deep, 8 acre pothole in rolling timberland a half mile west of Whitefish Lake and a mile north of Beaver Lake; accessible by a logging road 5 miles from U.S. 93. Here is a moderately popular pond that is fair-to-good fishing for 8 to 16 inch rainbow and grayling. Because it *is* such a nice *little* pond, luxury liners aren't allowed here, only car top boats, canoes, etc.

Double Lake. A fairly deep, 18 acre pothole in rolling timbered hills reached by road a little over a mile east from the Bigfork road 1¾ miles southeast from Creston. Double Lake used to be good fishing, with the owner's permission but is now infested with bullheads.

Dry Fork (or Lone Pine) Reservoirs. Upper and Lower, 200 and 322 acres, connected by about 4 miles of pastureland irrigation ditch; reached by county and private roads (DON'T fortet to close the gate, Richard) 1½ miles west of Lone Pine. Both reservoirs are nearly drained in summer, have shallow dropoffs and real muddy bottoms, and the lower one is choked with weeds. It is also unique in that it is one of the few lakes in this neck of the woods that is a northern pike fishery: good-to-excellent for 2 to 27 pounders as well as yellow perch and a very few cutthroat trout. The upper reservoir is fair cutthroat fishing.

Dry Lake. Take the Creston road 6 miles north from Bigfork, then turn east on a farm road for 1¼ miles, then north on the Cabin Lake road for ½ mile, and finally a private road for ⅛ of a mile to Dry Lake in rolling timberland. This is a VERY shallow, 1½ acre pothole that is mostly less than THREE feet deep but has some nice springs that keep it from winter killing. It has been good fishing in early winter only — for yellow perch that will average 14 inches and reportedly range up to 21 inches, but when I tried it last winter, I got skunked.

Ducharme Creek. A kids' creek, the lower couple of miles in open pastureland reached by Montana 35, five miles west of Polson and fair fishing for 6 to 8 inch brook.

Duck Lake. Really just a wide, brushy spot in the Stillwater River (about 60 acres) 1½ miles above Upper Stillwater Lake and right beside U.S. 93, twenty-nine miles north of Whitefish. It's poor fishing (and then mostly in winter and spring) for 8 to 10 inch brook and cutthroat, plus a few Dolly Varden that will range up to a couple of feet. P.S.: It's probably better know locally as a darned good water fowl hunting area.

Duncan Lake. A barren lake in open scrub timber below McDonald Peak 7 airline miles east of St. Ignatius. It's about a mile due east up the drainage from Ashley (or Hidden) Lake but a good 2000 feet above — with no trail. You just take off up the ridge until you get there, if you don't miss it, because it's only 9.9 acres but good and deep, 47 feet to be exact. The Fish and Game boys surveyed it with a helicopter, liked what they saw and dumped a load of cutthroat in it the following summer (1967), but they didn't survive. You could get lonesome here.

Dunsire Creek. Go from Whitefish to Tally Lake, then take first the Logan Creek road and next the Sheppard Creek road for 5 miles above the Star Meadow guard station to Dunsire Creek which is mostly too small to fish but is fair in some beaver ponds near the mouth — for 6 to 10 inch brook and native cutthroat trout.

East Bass (Skyline or Rainbow) Lake. A private, 4 acre pothole ¼ mile east of Sawdust Lake in the "Many Lakes" housing development; has been good fishing for largemouth bass and rainbow trout but is no longer stocked and probably isn't much.

East Fork Whitefish (or Swift) Creek. The inlet of Upper Whitefish Lake; about 6 miles long in a narrow, flat-bottomed timbered canyon followed along the east side for 4½ miles by road. It is fair fishing for 8 to 10 inch cutthroat, small Dollys and 12 to 14 inch whitefish.

Echo Lake. A 725 acre, mud-bottomed lake of most uneven outline; over 100 feet deep in spots, in what used to be excellent deer hunting country; reached by U.S. 95, five miles north of Bigfork. Echo Lake has one small inlet (Echo Creek) but no surface drainage, and is mostly spring fed. It is moderately popular (there are lots of summer homes and more a-building all the time), with the water skiers and

attendant activities more in evidence all the time. It used to be good winter fishing for whitefish ranging from 18 inches up to 10 pounds or so, spring fishing for 12 inch to 5 pound largemouth bass, and summertime fishing for yellow perch and sunfish. However, some nut slipped northern pike in three or four years ago and they are taking over. If it turns out as expected they will run up to 25 pounds or so and eat the rest of the fish out of house and home. Let's hope they don't get into any of the other western Montana waters.

Emmet (or Sun Fish) Slough, or Horseshoe Bend.
Makes almost a complete circle ¾ of a mile in diameter, in open farmland with cottonwood along the banks; reached by road ½ mile east and then about 3 miles north from Somers. Emmet is good fishing for largemouth bass that are reported up to better than 5 pounds, and also contains a large population of yellow perch and sunfish.

Estes Lake.
Take the east lake highway 5 miles south from Bigfork and then a washed-out rocky mountain road 1½ miles east through heavy timber to Estes; which is about 3 acres, not too deep and is fair-to-good brook trout fishing for mostly 10 inchers, plus a few recorded to 1½ pounds.

Evers Creek.
A small swampy stream, about 6 miles long with lots of beaver ponds that are known locally as "Short's Meadows." Evers flows to the north end of Tally Lake just east of the ranger station and is mostly accessible by a rough, muddy logging road, and a USFS all purpose road to the "Meadows" that connects with the Logan Creek road. It is hard to fish but contains a fair number of 6 to 8 inch brookies.

Fenan Slough.
See Russell Slough.

Finger Lake.
See Wall Lake.

Finley Creek.
A small stream emptying to the Jocko River near the U.S. 93 crossing 1 mile north of Arlee, is accessible by U.S. 93 and county roads, for 13 miles to headwaters. Mostly a "pastureland" creek, it is only lightly fished, but excellent for 6 to 8 inch brook, some cutthroat and a very few rainbow trout.

Finley (or Schley) Lakes.
Two little shallow mountain lakes just below and southwest of Murphy Peak. Take U.S. 93 about 5 miles north from Evaro, then a jeep road east for ¼ mile and thence up a very poor unmaintained trail up the East Fork for 5 miles to Lower Finley, and another ¾ of a mile to Upper Finley. The lower lake contains a very few nice fat 14 to 18 inch rainbow that you can sometimes see but seldom catch, and quite a few 10 inch cutthroat. The upper lake is overpopulated with 6 to 10 inch, half-starved, bug-eyed cutthroat that bite like mad. Neither lake is fished much. They're on Flathead Indian Reservation land.

Fire Lakes.
Three, about 7, 7 and 13 acres, in an old burn accessible to within 1½ miles by the Sunday Creek road, then cross-country to the east over sharp, brushy ridges and draws. They are lightly fished in winter, almost never in summer, but are good for brooks that average a pound and grow bigger.

First Lake.
Is the lowest (at 6400 feet above sea level) of a group of 5 glacial lakes just east of McDonald Glacier and below (west of) the crest of the Mission Range. There is no trail, and any way in, other than flying, is nothing but rugged. Perhaps the 2 easiest routes, if there are any, are: cross-country a couple of miles up over the top of the range from Island Lake in the Swan drainage, or, from the McDonald Reservoir-Summit Lake trail. From the top of the switchbacks about 3 miles up from the reservoir, you take off to the south up the creek, over talus and

through brush for 2½ miles to First Lake. It's 57.8 acres, a good 65 feet deep and will taste like champagne by the time you get there. Another ¼ of a mile and 300 feet up the drainage through more brush and scrub timber brings you to 14.6 acres by 50 feet deep Disappointment Lake. Now if you're still in the mood and the body is willing, Cliff Lake is another 500 feet up and ¼ of a mile beyond (surrounded by cliffs, naturally) and is 45 acres by 150 feet deep. All three were barren but were planted with cutthroat fingerling the spring of '68 and are pretty good fishing now. The lakes above are barren, and just as well because not many folks make it that far.

Fish Lake. See Stryker Lake.

Fitzsimmons Creek. Followed for its full length, 10 miles, by a good logging road down a steep timbered gorge on the west side of the Whitefish Range, to Bull Lake. Fitzsimmons is fair-to-good spring fishing for 8 to 16 inch cutthroat, and some brook and rainbow, but is too small to hold up in the summer.

Flathead (or Spring) Creek. A nice, clear, open-meadow stream that flows from a few miles north of Arlee for 3½ miles along the east side of U.S. 93 to the Jocko River 3½ miles south of Ravalli. It's planted regularly, heavily fished, and fair for 6 to 14 inch rainbow and brook trout, plus whitefish and an occasional loch leven spawner.

Foy Lakes. Three: Lower (20 acres); (Middle (35 acres) and Foy Lake (273 acres and over 130 feet deep); about ⅛ of a mile apart in open country reached by paved road 2 miles southwest from Kalispell. The Lower and Middle Lakes are barren. The Upper lake is heavily fished for excellent catches of 10 inch to 2 pound rainbow trout. What with all of the summer homes building around and about it will soon be just like fishing in your own bathtub.

Frog Lake. See in Summit Lake.

Garlick Lake. A lightly fished but used to be good put-and-take, 3.7 acre, shallow pothole in the "Many Lakes" housing development north of Bigfork. It has been planted off and on in the past but no longer. Best you forget it.

CLEAR 'N' COLD
Courtesy Harley Yeager

Courtesy USDA, SCS

Gilbertson Lake. Another fairly deep, 5 acre pothole in the "Many Lakes" housing development that has geen a moderately popular rainbow fishery in the past but is no longer being stocked and might just as well be struck off your books.

Good Creek. A fair size stream emptying into the Stillwater River 3 miles below the lower lake and fishable either from shore or boat. There are about 20 miles of fishing water clear to the mouth of Plume Creek, and it is followed most of the way by a good logging road west from the Stillwater Ranger Station at the north end of Lower Stillwater Lake. There are many beaver ponds along the middle reaches, a half mile of box canyon near the mouth. The entire stream is reported to be good for 3 to 11 inch brook and cutthroat trout.

Grayling Lake. Another pothole lake such as Gilbertson and Garlick that used to be a fishery but is no longer being managed by the State Fish and Game Department. It's about 5 acres, fairly shallow and lies about 2¼ miles east of the Bigfork-Creston highway.

Grayling Lake No. 2. A half mile from Grayling Lake, and of the same general fishing potential.

Green Lake. See Blue Lake.

Griffin Creek. From a point on the Sheppard Creek road 5 miles above the junction with the Logan Creek road, take a good USFS all purpose road (the Ingalls Mountain road) for 20 miles or the entire length of the creek. The lower couple of miles past some beaver ponds and then 8 miles up a steep, rocky gorge are good fishing with the catches pretty evenly divided between 8 to 9 inch cutthroat, brook trout and whitefish.

Half Moon Lake. A fairly deep, 54 acre lake in relatively flat, timbered country; reached by county road 4½ miles from Coram, or 2½ miles from West Glacier. Half Moon was rehabilitated and planted with cutthroat in 1960, thereby eliminating a super-abundant pumpkinseed and sucker population. It is heavily fished for fair catches (at times) of 10 to 12 inch trout, especially in the winter months.

Hand Creek. A small tributary of Griffin Creek, in timbered mountains but not too steep a gradient to prevent its being an excellent producer of 6 to 7 inch native cutthroat. There are about 3 miles of fishing, reached at the mouth by the Griffin Creek trail and followed by a trail and logging road to headwaters.

Hanson (or Pike) Lake. Drive ¾ of a mile west from Sunrise Lake on the Tally Lake road, and then a mile south on a single track dirt road to this shallow, 5 acre pond in rocky, mostly logged-over country. It was stocked with brookies in 1966 and is pretty good fishing for 10 to 12 inchers now. N.B. Somebody caught a pike here years ago, hence the alternate (and older) name.

Harbin Lake. A 2 acre, muddy bottomed pothole in timberland reached by a poor road 1¼ miles east from the Creston Hatchery. This one is fair fishing on a put-and-take basis for 10 to 14 inch rainbow, plus some yellow perch and a few largemouth bass.

Harry Creek. See Herrig Creek.

Haskill Creek. About 10 miles long, in timber above and ranchland below; a part of the whitefish municipal water supply. Haskill flows to the hitefish River, a few miles below town and is crossed here and there by county roads for 6 miles of fair 6 to 7 inch cutthroat and brook trout. "Kid's" fishing.

FLATHEAD RIVER

This map is not intended for navigational purposes. Navigational hazards are not shown. Access areas shown are public. Other places may be open to public use through the consent of individuals.

LAKE FIVE
T31N-R19W-S9,10
Flathead County

TOTAL SURFACE ACRES 235

Contour Interval — 10 ft.

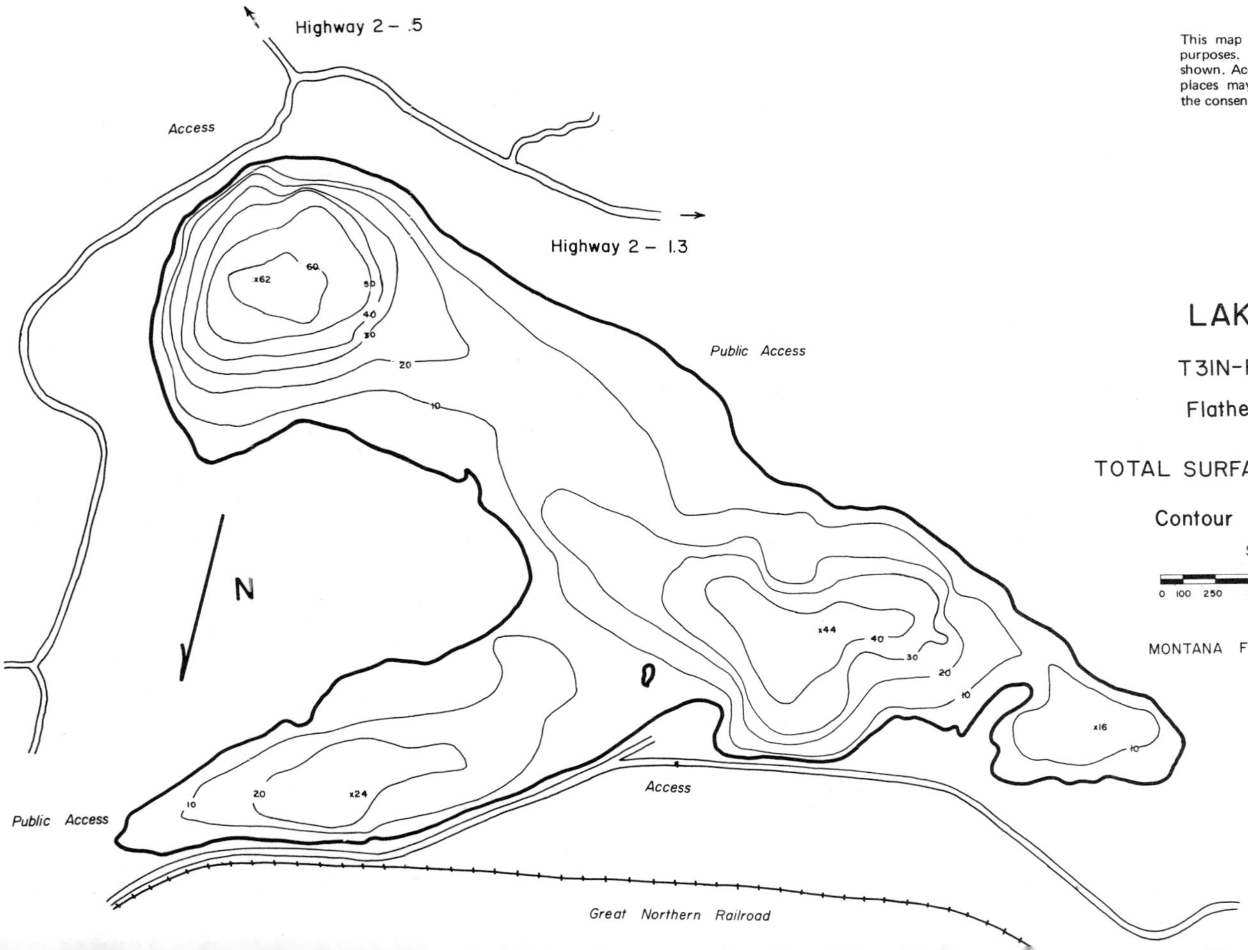

Scale
0 100 250 500 1000
Feet

MONTANA FISH & GAME - 1964

Herrig (or Harry) Creek. Drains from Herrig Lake on the southern slopes of Pleasant Mountain for 5 miles to the north end of Little Bitterroot Lake and is followed along the lower reaches by a good gravel road. Herrig is mostly posted but reportedly good for 10 to 20 inch cutthroat and rainbow during the spawning season.

Hewolf Creek. Very small, flows for 5 miles through mostly private (posted) land to Valley Creek. The lower 2 miles are followed by a road of sorts. Hewolf is extensively beaver-dammed, and fair fishing for 6 to 7 inch brook, plus a few cutthroat and rainbow trout.

Hidden Lake. See Ashley Lake.

Hole-in-the-Wall Lake. From U.S. 93 at Mock (Radnor) take a logging road west across the Stillwater River and then northwest for 1 mile to about the middle of the west side of Duck Lake; now park your car and hike a few hundred yards to Wall Lake, then southeastward ¼ of a mile over timbered rocky ridges to Hole-in-the-Wall which is more or less round, about 3 acres, deep, and relatively unknown. It's *loaded* with 10 inch brookies and is heavily fished during the winter months.

Horseshoe Bend. See Emmet Slough

Hubbart Reservoir. Four hundred and thirty acres but drawn down very low at times; in open country reached by a good road 12 miles north of Niarada. Hubbart was rehabilitated in 1959 and again in 1973, was planted in the fall of that same year with rainbow trout and should be excellent fishing now. Fact is, the reservoir has been moderately popular in the past, and the probable reason is that the trout get nice and fat and large — in the 15 to 18 inch class.

Jessup Mill Pond. See Creston Lake.

Jette (or Sheridan) Lake. Used to be a commercial fish hatchery, 24 acres; reached by a good road 8 miles northwest from Polson with a GREAT BIG SIGN at the entrance on U.S. 93. The owner used to raise a few rainbow and brook trout for his own use (until he was killed in a plane crash) and to the best of my knowledge it's gone commercial once again.

DUCKLINGS
Courtesy USDA, SCS

NOT MANY LIKE THIS
Courtesy USDA, SCS

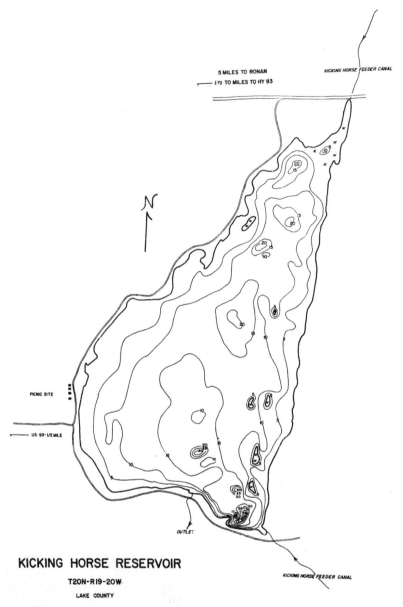

KICKING HORSE RESERVOIR

T20N-R19-20W.

LAKE COUNTY

TOTAL SURFACE ACRES - 800

CONTOUR INTERVAL - 5 FT.

MONTANA FISH & GAME - 1965

This map is not intended for navigational purposes. Navigational hazards are not shown. Access areas shown are public. Other places may be open to public use through the consent of individuals.

KICKING HORSE RESERVOIR
Courtesy Ernst Peterson

FLATHEAD MACKINAW
Courtesy Lyle Johnson

Jocko River.
A nice-sized stream, about 18 miles long from the junction of the Middle and South Forks to the Flathead River at Dixon, and followed by good roads all the way. It flows for a few miles through timbered mountains, but mostly through open farmland down the Jocko Valley past Arlee to Ravalli, and finally through wooded bottomland to Dixon. It is planted annually with rainbow, and as a whole is fairly good fishing for 12 to 14 inch rainbow, with a scattering of cutthroat, brown, brook, Dolly Varden and whitefish.

Johnson Lakes.
Three pretty little potholes — North, South, and East, in cut-over timber and farmland. From Bigfork go 2 miles north on the Creston road to the "Little Brown Church," then turn east for 1 mile and thence hike about ¼ mile north to South Johnson and follow on around the others which are all one during high water, and cover a combined total of about 60 acres. They have been planted at times with cutthroat (and reportedly some rainbow) but are fairly shallow and may have frozen out.

Kicking Horse Reservoir.
About 800 acres, in open farmland ½ mile east of U.S. 93 and a mile from Nine Pipe Reservoir. It is moderately popular but not quite as good as Nine Pipe although some of it is better if you really know the water. It contains largemouth bass to 16 inches, yellow perch, sunfish, and a few 2 to 4 pound rainbow which are seldom caught and then mostly by trolling. Note! this reservoir was almost dry in 1967 but is planted each fall with BIG (4 to 8 pound) average spawners that really kick up a fuss come the following summer.

Kila Lake. See Smith Lake.

Lagoni Lake.
In rolling timberland reached by a lousy-rough, almost nonexistent logging road 1½ miles south from Mock (Radnor) on U.S. 93 down the west side of Upper Stillwater Lake. Lagoni is 20 acres, deep on the northeast side with steep rocky cliffs on the south side and timber to the north. It should be fished from a boat (if you're going to fish). Those that claim to know the facts report that it is seldom bothered and only poor-to-fair for 14 to 18 inch brook and cutthroat, plus loads of trash fish.

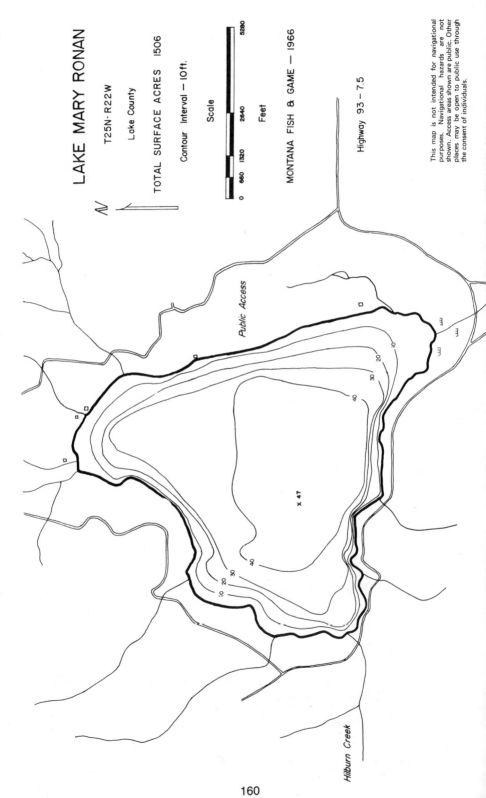

OPENING DAY — LAKE MARY RONAN
Courtesy Meir's Studio

Lake Five (or Spruce) Lake.
Two hundred and thirty-five acres, up to 62 feet deep in spots with a gradual dropoff and gravel bottom, reached by a good county road 3 miles from West Glacier, or 4 miles from Coram. It has been good fishing but was taken over by trash fish, rehabilitated in 1969, restocked with cutthroat trout in 1970 and provided some fair-to-good fishing for awhile, but the sunfish and suckers have taken over again. A losing battle.

Lake Five (or Spruce Lake).

Lake Mary Ronan.
A large lake (1506 acres) in rolling, timbered mountains; reached by a good county road 8 miles northwest from Dayton on U.S. 93 across the bay from Wildhorse Island in Flathead Lake. Mary Ronan is a popular recreation lake with a good state campground, three resorts, some private cottages, and the usual boating and picnicking activity. It is excellent fishing for 14 to 16 inch Kokanee, 8 to 20 inch rainbow, and 3 to 4 pound largemouth bass. And lots of crayfish too. Furthermore, this one should continue to be good because the Montana State Fish and Game Department keeps a close watch on it and keeps it well stocked.

Lake Monroe (or Lower Ashley Lake).
Monroe is connected with Lone (or Middle Ashley) by a swampy but passable-by-boat area so that the two lakes are almost one. They are reached by a good road 4 miles south from Ashley (Big Ashley) Lake in timbered country. They contain lots of small yellow perch, a few really nice cutthroat and grayling, and a super abundance of trash fish.

Lake of the Woods.
A pothole lake, 62 acres and fairly deep (for a pothole); reached by road three miles east from the Creston highway 6 miles north of Bigfork. It has been planted off and on for years by the State Fish and Game Department and has been fair fishing for 12 to 14 inch cutthroat trout along with a few brooks.

Lazy Creek.
A small, swampy stream with lots of beaver ponds; in brushy timber above Whitefish Lake; accessible by a logging road from the northwest end of the lake for about 6 miles of good, rough, hard fishing for 12 to 16 inch brookies.

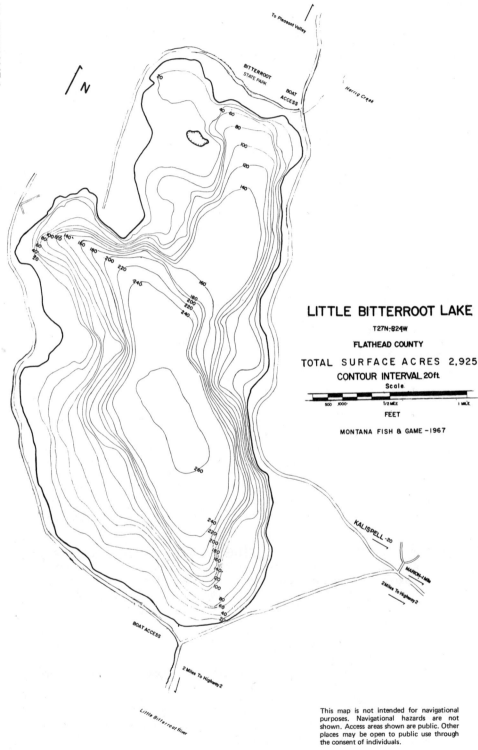

FALLS on LITTLE BITTERROOT RIVER
Courtesy USFS

Courtesy Dallas Ecklund

Lebeau Creek. A small, brushy stream in a rocky, beaver-dammed canyon. LeBeau sinks about ½ mile above its mouth and rises into Upper Stillwater Lake. It is reached in the middle reached by the Martin Creek logging road 8 miles west from the Stillwater Ranger Station. There are reported to be about 4 miles of good fishing here for 10 to 12 inch brookies.

Leech Lake. 9.6 acres, 25 foot maximum depth, in timber all around; reached by a mile hike through foothill country north of Upper Martin Lake. Was stocked with cutthroat in 1969 and the results are so far indefinite. Who knows, perhaps they're hiding in the weeds.

Little Beaver Lake. About 23 acres and almost an extension of the north end of Big Beaver Lake, but separated by a narrow swampy area. Little Beaver is reached by road through rolling timberland 5 miles west from Whitefish. It was poisoned with Big Beaver in 1961, subsequently planted with rainbow in 1964, and has been good fishing up to now for 12 to 16 inchers.

Little Bitterroot Lake. Mostly in timber; about 3 miles long by 1½ miles wide; fairly deep, dammed with 4 feet of water-level fluctuation and is reached by good roads 25 miles southwest from Kalispell (1½ miles west of Marion). This is a moderately popular recreational lake; there is a state park at the north end, a couple of resorts and 60 or 70 summer homes with all of the attendant activities. The fishing is mostly good for 10 to 13 inch rainbow (maximum recorded to 4 pounds).

Little Bitterroot River. A small stream that was once fair sized but is now almost de-watered in places for irrigation. It heads in Little Bitterroot Lake and is followed southward by good county roads for 11 miles to Hubbart Reservoir, and then another 45 or so to the Flathead River 15 miles north of Dixon. The lower reaches, below the reservoir, are fairly falt and mostly poor-to-fair early spring fishing; the upper reaches are steep and generally fair to good for 8 to 10 inch rainbow and cutthroat, plus occasional concentrations of brookies and whitefish. There are also quite a few northern pike from Lone Pine down.

Logan Creek. About 25 miles long and followed by a logging road for 15 miles of good fishing to its mouth at the north end of Tally Lake 1 mile west of the Ranger Station. There are lots of beaver ponds in Star Meadows, along the upper reaches, and they are fished almost entirely by rubber boat (a lot by dudes from the Dude Ranch there), for mostly 8 to 14 inch rainbows and brooks, a few cutthroat, and still fewer whitefish.

Lone (or Middle Ashley) Lake. Four miles south of Ashley Lake in timberland reached by the Ashley Creek road; 200 acres, and all under 50 feet deep with shallow dropoffs, lots of vegetation and a muddy bottom. Lone Lake is connected with Lake Monroe (or Lower Ashley) by a swampy but passable-by-boat area. Both lakes were poisoned out and rehabilitated in 1963 and '64 and were for awhile good fishing, but the trash fish and yellow perch have mostly taken over again, although there are still a few nice grayling and a very few cutthroat.

Lone Pine Reservoir. See Dry Fork Reservoir.

Long Lake. See McCaffrey Lake.

Long Lake. See in Summit Lake.

Lore Lake. Take the "west valley" (Tally Lake) road 12 miles north from Kalispell and then a private lane ½ mile west where you'll find Lore right behind a ranch house. It's ½ mile long but only 100 feet or so across, below timber on the west, open ranchland to the east, and so shallow as to freeze out over exceptionally hard winters (like 1972). There is quite a lot of aquatic vegetation and right now it's fair fishing for small brook trout — if they didn't get it in the neck last winter.

Lost Creek. Flows north for about 3 miles to some 90 foot falls, and then for another mile almost to the south end of Tally Lake where it makes a 135 degree turn and flows to the southeast for 7 miles to the Flathead Valley where it finally sinks about 5 miles northwest of Kalispell. The lower reaches are not worth fishing but it is very good for a couple of miles (available by a poor logging road from Beaver Lake) above the falls for 6 to 10 inch brook trout.

Lost Lake. One-third of a mile northeast and across the road from Turtle Lake; Lost Lake is a private, 3 acre and deep pothole that was poisoned out in 1966 and has not been replanted.

Lost Sheep Lake. Drive to the end of the Middle Fork Jocko road at the mouth of Deep Creek. Now take off cross-country up the ridge along the west side of the creek for about a mile until you hit the old Jocko trail. Follow it up around the headwaters of Deep Creek and then for another 1¼ miles until you come to the first fair-size stream. Lost Sheep Lake is just out of sight about 200 yards up the creek. It's 9 acres, in alpine timber, only poor fishing for fair-size cutthroat, rarely visited and even more rarely fished. However, the State Fish and Game Department stocked it with cutthroat a couple of years ago and are awaiting the results with "baited" breath.

Lower Ashley Lake. See Lake Monroe.

Lower Crow Reservoir. An irrigation reservoir (subject to excessive draw-down and water level fluctuation); about 700 acres in a set of wooded breaks with open grainfields all around; reached by a county road 6 miles west from Ronan. Unfortunately this reservoir has in years past been polluted with sewage from Ronan and probably still is. (See Crow Creek.) It is fair at times for 16 to 20 inch (2 to 3

LOOKING EAST TO LUCIFER LAKE
Courtesy USFS

FLATHEAD DOLLIES
Courtesy John Stewart

pounders) rainbow, 10 to 12 inch whitefish, and a large population of 6 to 8 inch yellow perch, sunfish and bullheads.

Lower Jocko Lake.
One mile long by ⅛ of a mile wide; dammed with 75 feet od drawdown, and followed along the north side by the Jocko River road 22 miles east from Arlee. It's all muddied up, full of debris, and has practically no fishing value. There are reportedly a very few 10 inch cutthroat and rainbow come in from the river — but they are nothing to get excited about.

Lower Stillwater Lake.
In a timbered flat alongside U.S. 93, seventeen miles north of Whitefish; Lower Stillwater is about 248 acres, with mostly shallow dropoffs. It is polluted by a lumber mill and generally very poor fishing except for small yellow perch; except in early spring directly below the mouth of the Stillwater River where it's goof for 16 to 18 inch rainbow, Dollys and squawfish.

Lower Sunday Lake.
Is 8 acres in size, 30 feet deep at the most, in wooded foothills with some steep sided cliffs on both sides of the lake; reached cross-country a ¼ mile hike east of Upper Sunday. It's got zillions of pumpkin seeds and that is all.

Lower Wall Lake.
See Wall Lake.

Lucifer Lake.
As you drive north over the hill (on U.S. 93) by the National Bison Range from Ravalli, look ahead and to your right and you will see the Mission Creek (or Lucifer) Falls at the top of a great rock-walled glacial cirque on the southwest end of the Mission Range. Drive on to St. Ignatius, turn right for 3 miles on a county road to the Mission Reservoir, on around it and ¼ of a mile above to the end of the road. The best way in used to be by taking off to the northeast up the trail for a couple of miles until you can look back down to your right at the falls and then breaking off cross-country to Lucifer Lake in lodgepole about ⅛ of a mile above. Now there's a good trail along the creek and up the cliff to the right (north) side of the falls. It's 41.4 acres, 110 feet deep and did have trout years ago. Moreover, the State Fish and Game Department restocked it with cutthroat fingerling in 1968 and it's pretty good fishing for fair size cutthroat right now.

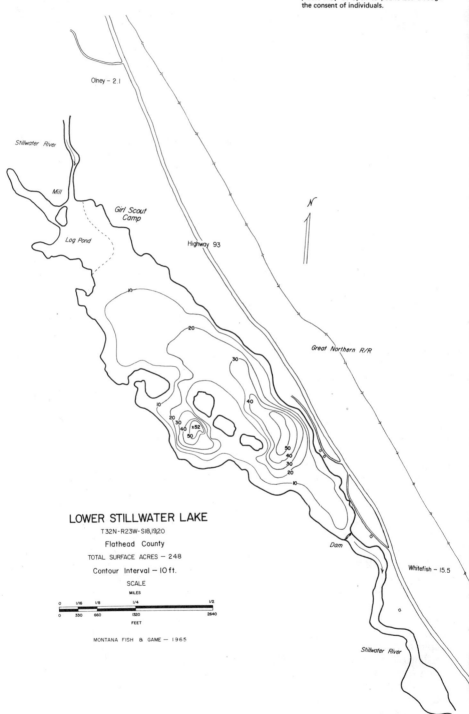

McDONALD LAKE
Courtesy Ernst Peterson

MISSION RANGE FROM RED BUTTE
Courtesy USFS

Lupine Lake. Is 13 acres, steep off-shore and deep in the middle; in a timbered mountain pocket reached by trail 1½ miles from the Griffine Creek road. Lupine is moderately popular, there are a few rafts and it's good fishing for 12 to 18 inch cutthroat trout.

Magpie Creek. Crosses U.S. 10A (or would if it didn't sink) 6 miles east of Perma and is followed by a road for 4 miles to headwaters. The lower reaches go dry but the upper reaches (above Brown's ranch) are reportedly fair early-season fishing for pan-size rainbow and brookies.

Martin Lakes. Two lakes in timbered hills reached by a cross-country hike 1 mile north from the Martin Creek logging road about 3 miles west of Lower Stillwater Lake. Upper Martin is in a steep rocky canyon, about 13 acres and very deep but marshy around the shore. It is fair fishing for 8 to 10 inch brookies, a very few rainbow, and millions of trash fish. Lower Martin lies at the head of a meadow and is fed with seepage from Upper Martin. It is about 6 acres, shallow, has a muddy bottom and is fair ice fishing for about the same as in the upper lake.

Martha Lake. Is only 2 acres, deep, in alpine country (there's not even any brush around it) on Birch Creek; reached by a mountain climber's nightmare of a trip ½ mile below Birch Lake. This is almost never fished but contains a fair population of 12 to 14 inch rainbow and rainbow-cutthroat hybrids.

McCaffery (or Long) Lake. Long and narrow, 27 acres, in timberland ½ mile south of Peterson Lake and reached by the Echo Lake road 5 miles north and east of Bigfork. McCaffery is (or rather has been) moderately popular and good fishing for 12 to 20 inch largemouth bass, and lots of pumpkinseeds. It's now in the "Many Lakes" housing development, is no longer being managed by the State Fish and Game Department, and likely won't retain its fishing potential.

FLATHEAD RIVER

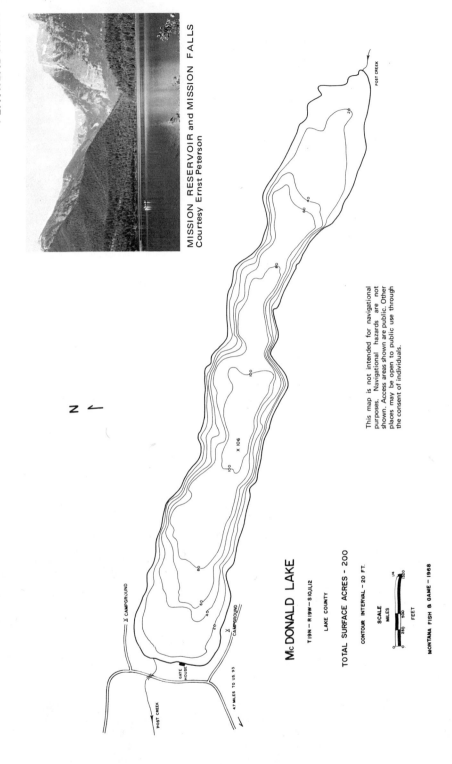

MISSION RESERVOIR and MISSION FALLS
Courtesy Ernst Peterson

McDONALD LAKE

T19N – R19W – S10,11,12
LAKE COUNTY

TOTAL SURFACE ACRES - 200

CONTOUR INTERVAL - 20 FT.

MONTANA FISH & GAME – 1968

This map is not intended for navigational purposes. Navigational hazards are not shown. Access areas shown are public. Other places may be open to public use through the consent of individuals.

McDonald Lake. Two hundred and fifty acres behind a 30 foot earthfill dam in a rock-walled canyon on the precipitous western slopes of the Mission Range; reached by good county roads 5½ miles west of U.S. 93 one mile south of the Nine Pipe Reservoir. McDonald is moderately popular and has been excellent fishing at times for 8 to 10 inch rainbow. There is an Indian Service campground at the lower end.

McGilvray Lake. A most irregularly shaped, 34 acre pothole in rolling timberland; reached by the Creston road 6½ miles north from Bigfork, and then take secondary roads another 1½ miles in to the shore. This pothole, like many of the others, has been planted off and on with rainbow trout and is presently good fishing for 12 to 20 inchers, plus a few largemouth bass. It was rehabilitated in 1974 and planted with rainbow again; should be good now.

McWineger Slough. About 50 acres, right beside U.S. 2 four miles east of Kalispell. The water here is too cold for most fish although there are a few largemouth bass and yellow perch and lots of black bullheads. It's really used mostly by duck hunters. There is a well-used picnic area just off (north of) the road.

Meadow Lake. A small lake on tribal land in the burned over, 1960 Belmont fire area; reached by a jeep road up the South Fork of the Jocko River. It's open only to Indians, who report that the fishing is good for 1 to 3 pound cutthroat.

Meadow Lake. See Thornburg Lake.

Meinsinger Creek. Heads just north of Haystack Mountain and flows rapidly down the steep western slopes of the Mission Range until it sinks just above the Pablo Feeder canal. It's reached by logging roads 4 miles east from U.S. 93 (6 miles south from Polson), is small and cold, and fair kid's fishing for pan-size brookies.

Middle Ashley Lake. See Lone Lake.

Middle Crow Creek. A small, snowfed stream above; the lower reaches flow through mostly timberland along the base of the Mission Range for about 4 miles of excellent 8 to 10 inch brook trout fishing. Middle Crow is reached at the mouth by county roads a couple of miles southwest from Ronan, and is followed by an unimproved jeep road.

Middle Fork Jocko River. Twelve miles long from its junction with the main stream 11 miles west of Arlee, and followed for half its length by a good gravel road. It is a moderately popular stream, though quite small, and good late season fishing for mostly 7 to 9 inch cutthroat, a few Dolly Varden, and still fewer brook trout. There are 2 public campgrounds.

Mission Creek. A small, brushy, pastureland stream, the outlet of Mission Reservoir and a tributary of Post Creek. Mission Creek is followed and crossed by county roads along its entire 10 mile course across the Flathead Valley near St. Ignatius. It is moderately popular with local residents and fair fishing for 8 to 10 inch brook and rainbow trout.

Mission Reservoir. Is 289 acres, in a steep, rock-walled canyon behind a 30 foot earth-fill dam on the lower slopes of the Mission Range; reached by a good county road 3½ miles west from St. Ignatius and there is a poor road clear around it. Mission Reservoir is very good fishing at times for pan-size rainbow, and also contains a few large ones and a very few Dolly Varden spawners running up the creek.

FLATHEAD RIVER

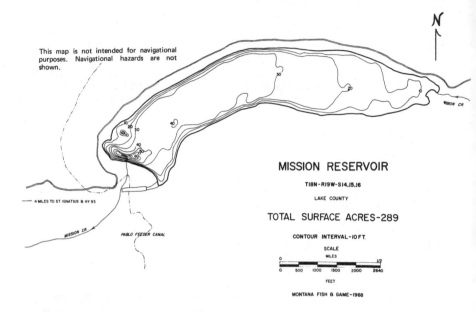

Moon (or Round) Lake. See in Summit Lake.

Moon Slough. See Russell Slough.

Morigeau Lakes. Upper and Lower, or East and West, 15 and 8 acres by 36 and 65 feet deep, 450 yards apart at 6200 and 5800 feet elevation above sea level in timber and rocky cliff country on the steep west face of the Mission Range. Reached by a cross-country mile through timber and downfalls from the end of the Morigeau Branch of the South Crow Creek trail 2½ miles above the end of the road. Both were originally barren but were stocked by the State Fish and Game Department with fingerling cutthroat in '68. They're doing fine now.

Morning Slough. Thirty acres, on the east side of the Flathead Valley; reached by good county roads 4½ miles north of Lake Blaine, and reportedly contains largemough bass and brook trout, but is better known as a duck hunting rather than a fishing slough.

Mud Creek. A tiny (but about 25 miles long) brushy, pastureland "kid's" creek crossed by U.S. 93 1½ miles south of Pablo and flows to the Lower Creek Reservoir. Mud Creek is easily accessible by county roads and is fair fishing in spots for mostly 5 to 10 inch brook trout, plus a very few cutthroat and rainbow.

Mud Lake. Five acres, near Lake Five, about 30 feet deep, and reached by a good county road 4 miles from Coram. Mud Lake and the potholes associated with it have been rehabilitated. It's good fishing now (especially in the winter) for 10 to 12 inch brook trout — or rather it was for the past couple of years, but the sunfish are taking over.

Mud Lakes. There are 14 in all but only the lower 4 have fish. They lie at 5500, 5750, 6300 and 6500 feet above sea level in an east-west string 200, 400 and 50 yards apart at the head of a great sparsely timbered, rocky, glaciated canyon on the precipitous western front of the Mission Range. Lower Mud (or Lake No. 4) is reached by pack trail, two miles from the end of a jeep road 7 erratic miles east from Pablo. You just hike on up the drainage to the others. No. 4 is 15 acres, a good 145 feet deep and swarming with 5 to 10 inch brook and some 7 to 15 inch rainbow trout.

NINE PIPES
courtesy Ernst Peterson

No. 3 is 10 acres, 70 feet deep and has a good population of nice fat little 8 to 9 inch rainbow plus some 7 to 12 inch cutthroat trout. No. 2 is 20 acres, all of 115 feet deep and provides good catches of nice fat 6 to 11 inch rainbow trout. No. 1 (or the uppermost of the fishable lakes) is 10 acres, only 34 feet at maximum depth, and used to be barren but was stocked with cutthroat in 1967, has been good since and should be good now.

Murray Lake. A cartop-boat only lake; 45 acres and deep, in rolling timberland a mile west of Beaver Lake; reached by a logging road a couple of miles north from U.S. 93, six miles north from Whitefish. Murray is heavily fished (mostly from boat) for good catches of 10 to 17 inch rainbow and because its planted off and on, it should remain so.

Nine Pipe Reservoir. An 1800 acre irrigation reservoir on a National Migratory Waterfowl Refuge away out in open, rolling farmland in the middle of the Flathead Valley right beside U.S. 93, nineteen miles south of Polson. As a waterfowl refuge it greatly enhances migratory bird hunting throughout the entire region. It is also a popular year-round fishery for 10 to 14 inch largemouth Bass, 10 to 12 inch yellow perch, good sized bullheads and lots of sunfish. There is a public campground right beside (west of) the highway.

North Fork Crow Creek. Ten miles long, the upper reaches on the steep western slopes of the Mission Range, the lower in brushy pastureland west of Ronan. It is easily available by county roads and excellent fishing for 4 to 10 inch brook and cutthroat trout.

North Fork Jocko River. Is really the main stream (quite small to call a river) above its junction with the South Fork; followed by a good road for 7 miles above its mouth, past the lakes, and over the divide to the Clearwater River drainage. It flows across Indian land but is open for good late-season fishing for mostly 7 to 9 inch cutthroat, plus a few Dolly Varden and still fewer brook and rainbow trout.

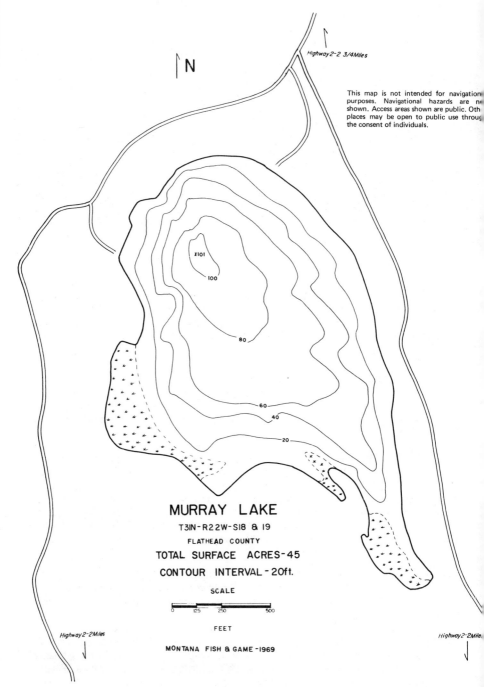

North Fork Valley Creek. A small, brushy, pasture-and-meadowland stream; crossed at the mouth by the Valley Creek road and followed by unimproved roads for 3½ miles of good early spring fishing. It contains mostly 6 to 9 inch rainbow, brook and a few cutthroat trout.

Olive Lake. One of the potholes, in hilly, logged-over country, now in the "Many Lakes" housing development. Reached by road about a ½ mile from Grayling No. 1 or 2. It is 6 acres, fairly shallow, and used to be lightly fished for yellow perch and a few largemouth bass. What it will be doing in a year or so is anyone's guess.

Pablo Lake. Drive 3 miles east from Pablo on gravel and logging roads, and then hike (climb) another 3 miles (and 3000 feet) up an old trail on the west face of the Mission Range along the ridge between Goat and Crow creeks to this one. It's at 6550 feet above sea level in a beautiful rocky cirque, is 30 acres and 45 feet deep and used to be barren. However, the State Fish and Game Department stocked it with cutthroat the summer of '68 and it has been and supposedly still is good fishing now.

Parker Lakes No. 1 and 2. Ten and 20 acres, in logged-over timberland now partly mantled with second growth; reached by a farm road ½ mile east of the Creston highway 5 miles north from Bigfork. No. 1 is open on the west side. Both are fabulous fishing for small largemouth bass (they are starving). You should ask the owner's permission on these.

Parker's Private Pothole. Is deep, clear, ⅛ of a mile long in an east-west direction by 100 yards or so across; in a wood lot just west of the ranch buildings ¼ of a mile west from the Creston road, 3 miles north from Bigfork. Parker's is closed to the public but supports a good population of 12 to 18 inch rainbow.

Peterson Lake. Eighty acres, with a very irregular shoreline, timbered on the east side and farmland on the west. Peterson can be reached by boat from Echo Lake during periods of high water, but is normally approached through a farmer's yard. Take the Creston road 3½ miles north from Bigfork, then a secondary road 2 miles west and finally a private lane a few hundred yards to the house. The lake is good fishing for largemouth bass, yellow perch and sunfish.

Picture Lake. Is 450 yards up the drainage from Lucifer Lake in a stupendous glacial cirque the likes of which you've never seen before. It's at 6450 feet above sea level in lodgepole timber below great 2000-to-3000 foot vertical rock walls all around, is 20.4 acres, 50 feet deep — and used to be barren. However it was planted with a goodly supply of cutthroat (left over from Lucifer) and has been fair-to-good fishing since for to to 12 inches, for the few fishermen who make it in.

Pike Lake. See Hanson Lake.

Plummer's Lake. A fairly shallow and warm (25 acre) pothole in cut-over timberland reached by a farm road (with the owner's permission) ½ mile east of the Bigfork highway 2 miles east from Creston. As of the early 60's Plummer's was good-to-excellent fishing for 1 to 3 pound largemouth bass, but the pothole district is now being "developed" for summer homesites and the good Lord only knows what's going to happen to the fishing.

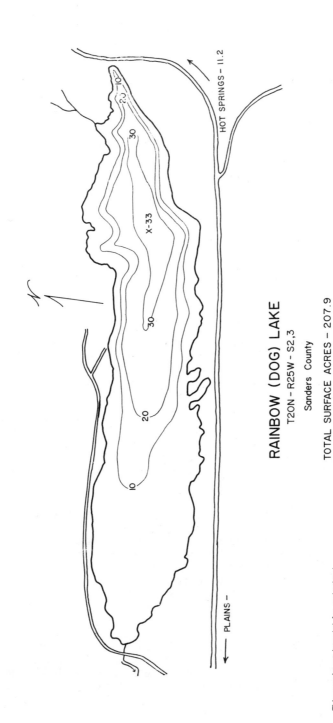

HEADWATERS OF POST CREEK
Courtesy USFS

RAINBOW LAKE
Courtesy Ernst Peterson

Post Creek. A small, farmland stream; the outlet of McDonald Lake and a tributary of the Flathead River 3 miles above Dixon. The upper reaches are readily accessible by county roads and are fair fishing for 6 to 12 inch rainbow, cutthroat and brook trout. The lower reaches are badly polluted by stock and farming practices and contain only rough fish.

Pratt Lake. One of the best potholes for warm water fish, Yellow Perch from 6 to 11 inches. It's about 10 acres, deep enough to keep from freezing out, and reached by an old skid road ¼ of a mile south from Double Lake (which is to say ½ mile south from Sawdust Lake).

Rainbow Lake. See East Bass Lake.

Rainbow Lake. In timbered (second growth) hills north of the road about midway between and ½ mile from Murray and Beaver Lakes. Rainbow is 40 acres, deep, moderately popular and good fishing (from cartop boats only) for 12 to 16 inch rainbow (what else?).

Rainbow (or Dog) Lake. From U.S. 10 near Plains take Montana 28 for 10 miles north to the lake shore, about 10 miles from Hot Springs. A 207 acre lake, full of beaver dams at the inlet and has been used for a mill pond. It was rehabilitated in 1966 with rainbow that furnish excellent sport now for 1 to 3 pounders. It's very popular and heavily fished (from boats in the summer and through the ice in the winter) but is reportedly being taken over again by sunfish.

Redmond Creek. A very small tributary of Briggs Creek, reached at the mouth by a road just south of Hubbart Reservoir; no trail. Redmond flows through a wide, timbered canyon and provides about 3 miles of fair 6 to 8 inch cutthroat fishing.

Revais Creek. A small stream whose lower reaches dry up and are crossed by U.S. 10A, two and a half miles west of Dixon. The upper reaches are in foothill rangeland; are followed by a road for about 3 miles of fishing; and are fair for pan-size brookies.

Rogers Lake. Is 237 acres, deep in spots but shallow around the margins; in rolling timberland reached by U.S. 2 about 20 miles southwest from Kalispell. Rogers was one of the original grayling lakes in this part of the country and is operated now as a rearing pond.

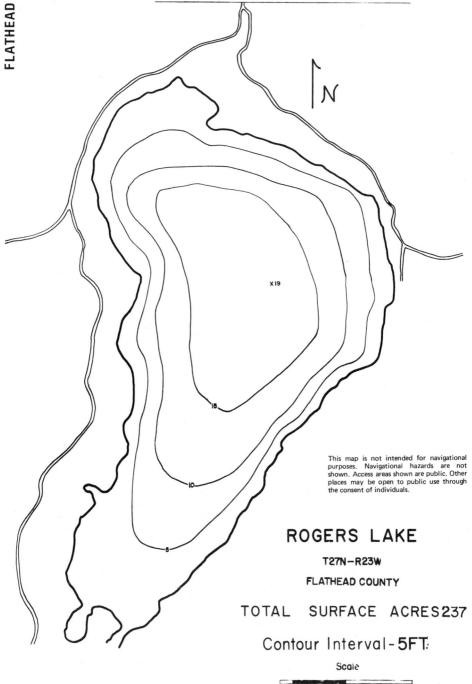

Round (or Moon) Lake. See in Summit Lake.

Russell (Fenan or Moon) Slough. Take the Somers road 4 miles west from Bigfork to the south end of this slough, which is about 150 yards or so across and forms almost a complete circle about ½ mile in diameter; in a cottonwood wood lot. It is moderately popular with local people and tourists from Bigfork and provides some of the best largemouth bass fishing in this area, from 1 to 5 pounds.

Sabine Creek. A tiny stream that rises just west of St. Ignatius from pickup irrigation water and flows through farmland and the northeastern corner of the National Bison Range to Post Creek. It is a favorite water hole for the buffalo but is all muddied up and of marginal value for fishing.

St. Mary Lake (or Tabor Reservoir). A 274 acre irrigation reservoir in heavy timber at the base of the Mission Range; reached by the Dry Creek road 10 miles east from St. Ignatius. St. Mary is a popular camping spot, but there is a very large drawdown and the fishing is no better than fair for small (8 to 10 inch) rainbow trout.

Schley Lakes. See Finley Lakes.

Senielen Lake. No trail to this one so...drive to the end of the logging road above the Mission Reservoir, gird your loins and take off up the creek for 1½ miles, and then take off again for a rugged climb up the right tributary an equal distance and that's it — Senielen Lake in a narrow rocky glacial canyon at 6700 feet above sea level, 20 acres by 40 feet deep. It used to be barren but was stocked with fingerling cutthroat the summer of '68, was fair-to-good fishing in the early 70's and should still be good now.

Sheppard Creek. A small, beaver-dammed, willowed-up, brushed-up stream that is reached at the mouth by the Logan Creek road 7½ miles above Tally Lake, and is followed upstream by a logging road for 7 miles of fishing. Sheppard is a tough one to fish unless from a rubber boat, but is a fair producer of 8 to 12 inch rainbow and brook trout.

Sheridan Lake. See Jette Lake.

Skaggs Lake. Forty-two acres, 17 feet deep, in timbered country at the head of Skaggs Creek; reached by a fair pickup road 2½ miles east and north of Proctor (4 miles east from Dayton on U.S. 93). This lake has been very good in the past, but partly winter-killed in 1963, '64, '66, and '67. However, it is planted with rainbow now and real good fishing again.

Skyles Lake. Thirty-seven acres, mostly shoal; reached by U.S. 93 three miles west from Whitefish. There's a State Fish and Game access area here, and it's close enough to town to be moderately popular, especially for winter fishing through the ice for 10 to 18 inch rainbow trout. It was rehabilitated in 1955 and has been stocked several times since — off and on as the need arises.

Skyline Lake. See East Bass Lake.

Smith (or Kila) Lake. About 300 acres, shallow, in a flat-bottomed mountain valley right beside the town of Kila on U.S. 2, nine miles southwest from Kalispell. Smith Lake is followed all the way around by a road, but is so marshy and boggy that about the only place you can get to the water is along the south side

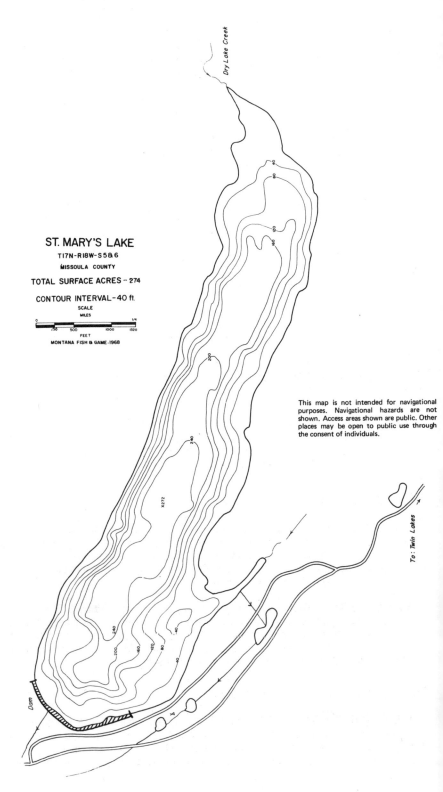

(there is Fish and Game Department access), and only the south half is fishable. It is excellent, though, for 6 to 12 inch yellow perch, a few largemouth bass, brooks, some small sunfish and a very few 8 to 12 inch rainbow trout.

Smokey Lake.
Ten acres, 20 foot maximum depth, a ½ mile cross-country hike west from Burnt Lake, in timbered foothills. Was planted with cutthroat in 1969 and may or may not have any left now. Give it a try.

South Crow Creek (or Terrace) Lakes.
Five beautiful alpine lakes in 3 timber-bottomed, rock-walled glacial cirques at the heads of the North, Middle and South Forks of South Crow Creek near the crest of the Mission Range. Take the South Crow trail a little over 3 steep miles (to about a mile beyond the Morigeau Lakes turnoff) to Terrace or Crow Creek Lake No. 2 at 6500 feet above sea level and ½ mile beyond to No. 1 (if you need the exercise and enjoy the scenery). It's barren. Or you can follow the North Fork ½ mile downstream from the Morigeau turnoff to the Middle Fork and then up that tributary for ½ mile to South Crow No. 4 at 5900 feet above sea level and a mile beyond to No. 3 at 6500 feet above sea level. No. 5 is at 6350 feet above sea level, and is reached ¼ mile down the Middle Fork from No. 4 and then ½ mile up the South Fork. Terrace (or South Crow No. 2) is 55 acres, 110 feet deep, has a few very nice 13 to 20 inch rainbow and 10 to 11 inch cutthroat, and lots of 8 to 10 inchers planted in '68. South Crow No. 3 is 11 acres by 35 feet deep and used to be barren but was also planted with cutthroat the summer of '68. South Crow No. 4 is 36 acres, 105 feet deep and chuck-a-block with skinny 8 to 14 inch cutthroat. South Crow No. 5 is 33 acres in size, 65 feet deep, and it too was planted with fingerling cutthroat in '68 that provide some real fine fishing now.

South Fork Crow Creek.
This little stream flows down a steep timbered canyon above, but is fairly flat below. It is reached by county roads 4 miles south from Ronan and is fair fishing for 8 to 10 inch brook, rainbow and cutthroat trout.

UP'S ON!
Courtesy USFS

READY TO GO
Courtesy USFS

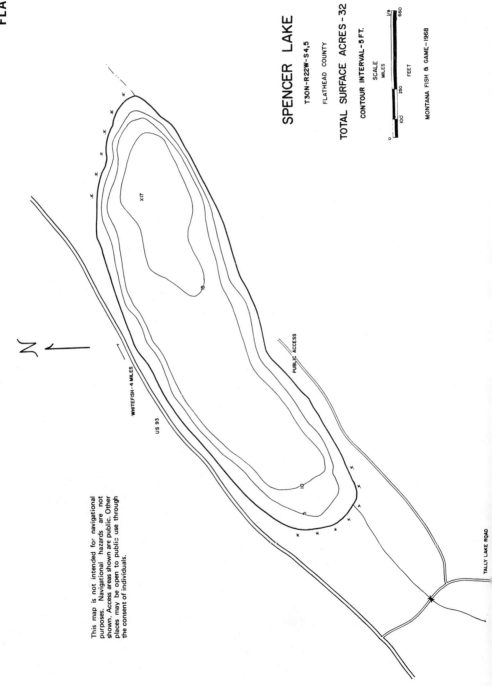

South Fork Jocko River. A small mountain stream followed by a logging road and trail for 15 miles through steep timbered mountains from its junction with the Middle Fork (11 miles west of Arlee) to its headquarters in lakes Sal-ol-sooth, Sudden, Wapiti and Crazy Fish. It is mostly on tribal land and closed to all but Indians, but is reported to be good fishing for 9 to 11 inch cutthroat, 10 to 16 inch Dolly Varden and a very few rainbow trout.

Spencer Lake. A put-and-take lake, rehabilitated in 1956 and now planted regularly with cutthroat. Heavily fished in winter and spring for mostly small trout, but there are a few good-sized ones too. Spencer is 28.5 acres and no more than 15 feet deep anywhere with mud bottom; found in timbered mountains available by road 4 miles west of Whitefish on U.S. 93, or 1 mile west of Skyles Lake.

Spill Lake. Just over the bank from Blackie's Bay on Echo Lake. It is 9.7 acres, has been good fishing for 10 to 12 inch rainbow but has gone private and is no longer stocked.

Spring Creek. See Flathead Creek.

Spring Creek. A tiny stream that rises about 3 miles northwest of Kalispell and flows to the southwest through private farmland to town (and the Stillwater River). It is fished mostly by kids the year around for fair-to-good catches of small brook and now and then a rainbow or cutthroat trout.

Spring Creek. A beautiful spring-fed stream above Ronan that is polluted on its passage through town and eventually debouches into Crow Creek. It is an open meadowland creek, easily accessible for its full length of about 6 miles by county roads, and is occasionally fished by kids for good catches of 7 to 8 inch brook trout.

Spring Lake. Two ponds: From Mock Radnor go 3 miles north on U.S. 93 to Point Of Rocks Cafe, then cross-country 1½ miles east up Spring Creek and then take the left fork to the Lower Lake, or up the east fork to the Upper Lake. Both are virtually unknown but the upper one is very good fishing for 1 to 1½ pound cutthroat (it was last stocked in 1972). The lower lake goes dry.

Spring Creek. Drains Morning Slough for 4 miles south through farmland to Lake Blaine and is accessible by county roads at the mouth and at intervals here and there upstream. The lower ¾ of a mile is fair fishing for largemouth bass that will run up to better than 5 pounds, and 6 to 8 inch brook trout.

Spruce Lake. See Lake Five.

Stillwater River. Flows to the Whitefish River 1 mile east of Kalispell and is followed by U.S. 93 through mostly rolling farm and ranchland for 45 miles to the northwest and then through steep timbered mountains for another 10 miles to headwaters. It used to be poor fishing due in large part to excessive siltation and a fish barrier at the mouth, but the dam was recently removed and the silt cleaned up so that the Stillwater is now fair-to-good fishing for 10 to 12 inch cutthroat and brook trout.

Stoner Creek. A very small stream, flowing for 5 miles eastward to Flathead Lake at Lakeside where it is crossed by U.S. 93. It is mostly too small to fish but there is about ½ mile of meadows in the upper reaches (available by road) where the creek has been impounded by beaver dams and here live a fair number of brook trout.

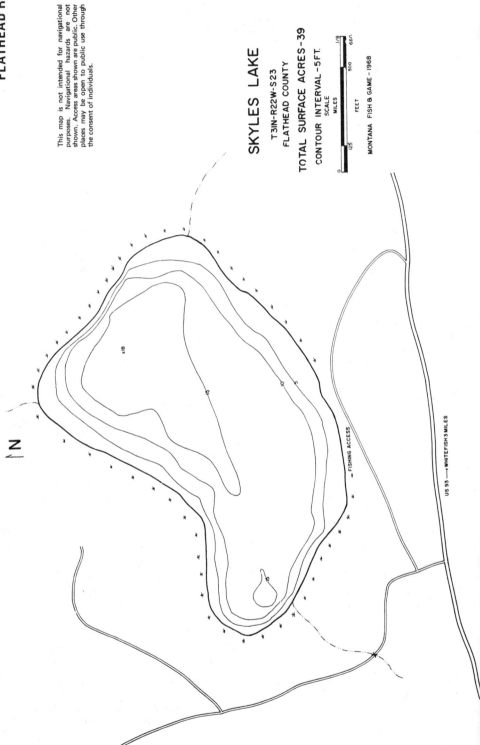

Courtesy Danny On

Strawberry Lake. From Kalispell take county roads east for 18 miles to Krause Creek, then the Krause Basin road for 2 miles and finally a steep U.S. Forest Service trail for another 2½ miles to the lake near the crest of the Mission Range. Strawberry is about 12.8 acres, 24 feet deep, in scrub (subalpine) timber. It is not fished much but is fair for 8 to 12 inch rainbow trout.

Stryker (or Fish) Lake. A deep lake, 32 acres, right next to Bull Lake in timbered hills; reached by road 3 miles northeast from U.S. 93, two and a half miles south of Dickey Lake (on the Kootenai River drainage). Stryker is moderately popular and excellent fishing for 9 to 10 inch brookies, and there are also quite a few nice ones — up to 3 pounds.

Summit Lake. From the end of the McDonald Lake road (at the lower, dam end of the reservoir) take a fair-to-middling trail up the north side of the lake for a moderately flat 1½ miles and then a steep switchbacked 2½ miles to ROUND (or MOON) LAKE which is in a brushy, steep-walled canyon, is 15.3 acres, mostly between 10 and 20 feet deep, and generally freezes out each winter but is restocked each spring with 6 to 10 inch (plus a few to 16) Rainbow by natural migration downstream from LONG LAKE about ⅛ of a mile to the north at 5800 feet above sea level. This one is ½ mile long in a north-south direction; 29.6 acres, shelves off rapidly from about 30 feet deep along the west side to as much as 46 feet below steep talus slopes along the east side, and is slow fishing (as a rule) for 14 to 16 inch rainbow. The State Fish and Game Department restocked it with cutthroat the summer of 1968 and it's good fishing now for 10 to 12 inchers. The trail follows along the brushy west side of LONG LAKE and beyond for another half mile to FROG LAKE (which is only about 6 acres and quite shallow but does sometimes contain fish that have migrated to it from above and below). Another half mile northwestward up the trail brings you to SUMMIT LAKE which is almost round, about ⅜ of a mile in diameter, 68 acres by 50 feet deep, beneath steep slopes to the northwest but a fairly flat swale to the southeast (where there is lots of room for your camp) and has been slow fishing for 12 to 14 inch rainbow. However, the State Fish and Game Department planted it with cutthroat in 1968 and they're in ascendency now. If you don't want to retrace your steps on the way out — you can proceed westward through Eagle Pass and on down the western slopes of the Mission Range along a steep switchback trail for 4½ miles to the Cheff Ranch 4 miles due east of the Allentown Motel on U.S. 93.

TALLY LAKE
Courtesy USFS

Courtesy Montana State Fish & Game

Sun Fish Slough. See Emmet Slough.

Sunrise Lake. A 10 acre beaver pond in broken-rocky-ledge-and-timber country (an old burn); reached to within ¼ mile by the Beaver Creek-Tally Lake road 4 miles southeast of Tally Lake. It was rehabilitated in 1962 and is now fair-to-good fishing for 12 to 14 inch brook trout and lots of small pumpkinseeds.

Sunday Creek. A small stream flowing for 2 miles northeastward through brushy, timbered mountains to Duck Lake. The lower reaches are paralleled by U.S. 93, the upper reaches are available by logging road and trail. There are a lot of beaver ponds up and down this creek and they are fair fishing for pan-sized brookies.

Sunday Lakes. See Upper and Lower.

Swartz Lake. A private, 5 acre, shallow swamphole with muddy margins surrounded by timber at the western base of the Mission Range, reached near the end of the South Crow Creek road 5 miles east and a little south from Ronan. It's good fishing if you're of a mind to, (and if you get permission) for 6 to 8 inch cutthroat and brookies.

Swift Creek. Twelve miles from Upper to Lower Whitefish Lake; all in timbered mountains and accessible to within half a mile by roads the full length. It is a moderately popular stream that is mostly good fishing for 8 to 10 inch cutthroat and Dolly Varden.

Swimming Lake. Eight acres, 15 feet deep, in logged-over land now included in the "Many Lakes" housing development. It is no longer managed by the State Fish and Game Department and the fishing (there used to be a large population of small largemouth bass and lots of sunfish) is in the laps of the Gods.

Sylvia Lake. A grayling lake; 22 acres, fairly deep but with gradual dropoffs, in rolling timberland reached by a USFS all purpose road a couple of miles south from Sheppard Creek. Sylvia is moderately popular with the local boys and good fishing for 5 to 16 inchers.

Tabor Reservoir. St. Mary Lake.

Tally Lake.
A nice-size (1326 acres), deep (over 492 feet at the maximum) lake with rock cliffs along the east side and here and there on the west; in rolling, glaciated timberland reached by a good road 12 miles west from Whitefish. There is an excellent USFS campground at the west end and the lake is fairly popular, although only poor fishing for 12 to 16 inch cutthroat, rainbow, Dolly Varden, brook and whitefish. Also it was planted with kokanee in 1967 and they now average 9 to 10 inches in length — what few there are of them.

Tamarack Creek.
A small, beaver-dammed stream in rolling grazing land, flowing east to the Little Bitterroot River 2 miles above Hubbart Reservoir, and accessible by a poor jeep road for about 1 mile upstream. All told there are about 2½ miles of poor-to-fair fishing here for small cutthroat and brook trout. It's a marginal stream.

Terrace Lakes. See South Crow Creek Lakes.

Thornburg (or Meadow) Lake.
From Olney (near the upper end of Lower Stillwater Lake), take U.S. 93 north for 1 mile and then turn east on a county road for 1½ miles to this private, 6.1 acre pond in open meadow ¼ mile north of the road with a ranch house right by it. It's reportedly fair fishing for 10 to 12 inch brooks (?) and maybe some pumpkinseeds.

Three Eagle Lakes.
Two; Lower and Upper, 6.3 acres by 78 feet deep and 10.8 acres by 82 feet deep at 5500 and 6175 feet in elevation, 500 yards apart in a cirque ¼ of a mile over the top of Three Eagle Mountain from (barren) Pilgrim Lakes (which can be reached by shanks mares 2 miles south by trail from Birch Lake and then another rough cross-country mile east to the shore). Both Three Eagles used to be barren, too, but they're good and deep with fertile water and plenty of feed so the Fish and Game boys planted them with cutthroat in '68 and they're better fishing than some others I know of...now.

Turtle (or Twin) Lake.
Take the east side Flathead Lake road 4 miles east from Polson, and then a county road 1½ miles south to this 44 acre, 35 foot deep reservoir lake in open country but with conifers along its east side. It was rehabilitated in 1962 and is sometimes excellent fishing for cutthroat and brook trout. Watch for special (Flathead Indian) regulations here.

Twin Lakes.
On the lower, southwestern flanks of the Mission Range at about 4500 feet elevation. They lie about 100 yards apart and are followed around the north side by the Tabor Feeder Canal road about 12 miles east-southeast from St. Ignatius. There's a campground between them. Upper Twin is 11 acres; Lower Twin 28 acres. Both are fairly deep with moderately steep dropoffs. The upper lake reportedly contains some fair-sized yellow perch but no trout. The lower lake is moderately popular and fair summer-time fishing for rainbow, cutthroat, Dolly Varden and brook trout. It's also reported to be excellent ice fishing when and if the access road is open.

Upper Jocko Lake.
A deep, 100 acre Indian reservation lake reached by the Jocko River road 25 miles east from Arlee, or 15 miles west from Seeley (on the Blackfoot River drainage). It is surrounded by dense timber, has a public campground on the upper end, and is only lightly fished for fair catches of 12 to 14 inch rainbow plus a few lunkers that are reported to go to better than 2 pounds. However, it is now being drawn away down for irrigation and the fishing is indeterminate.

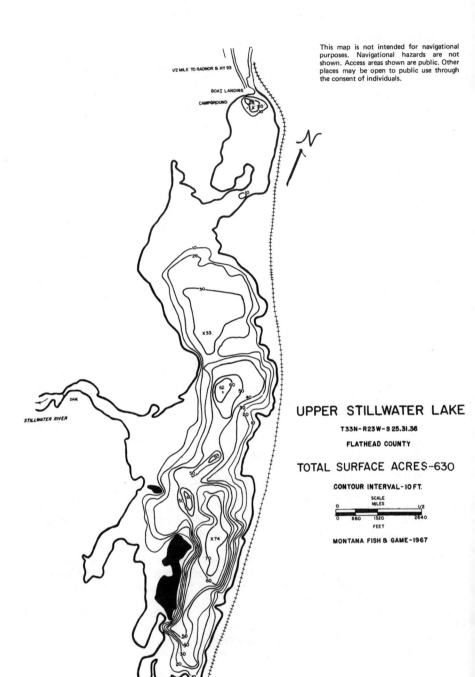

Upper Lagoni Lake.
Eight acres, less than 32 feet deep all over, in timber 1 mile cross-country west from Wall Lake, stocked with cutthroat in 1969 and reportedly good fishing now for 10 to 12 inchers, but could possibly have frozen out the winter of 1973-74.

Upper Stillwater Lake.
In flat timberland; reached by U.S. 93 twenty-five miles north from Whitefish, then a poor jeep road ½ mile southwest to the shore, and thence another 2 miles along its east side. This one is shallow, 630 acres, with marshy areas along the west side and south end. It is very lightly fished but fair for 12 to 16 inch Dolly Varden, rainbow, cutthroat and brook trout, lots of yellow perch and sunfish, and (of course) hordes of trash fish.

Upper Sunday Lake.
Lies in wooded foothills 4 miles by road southeast of Stryker, 6.4 acres, 20 feet deep at most and good fishing for 9 to 12 inch brook trout, 8 to 13 inch largemouth bass, and lots and lots of shiners and suckers.

Upper Whitefish Lake.
Is 88 acres, mostly less than 20 feet deep with quite a few logs and aquatic vegetation around the shore; in rolling timbered bottomland on the East Fork drainage, reached by road 20 miles above Whitefish Lake. Upper Whitefish is moderately popular but you need a boat to fish it, for fair-to-good catches of 9-10 inch cutthroat and Dolly Varden.

Valley Creek.
A small, brushy, pastureland stream that is followed by roads of sorts for its full length of 15 miles to its mouth on the Jocko River 2½ miles south of Ravalli. There is some posted land (and they MEAN it) but it is a moderately popular stream with the local people and fair-to-good fishing for small brook and a few cutthroat trout.

Wall (or Lower Wall or Finger) Lake.
In timbered rocky-ridge country; about 16.5 acres, ¾ of a mile long, narrow, and deep in places with steep, rocky shores; reached by trail 200 yards from the end of a logging road up the west side of Duck Lake. Wall is heavily fished in winter (but only lightly in summer) for good catches of 12 to 18 inch brook trout.

S REAL!
ırtesy "Doc" Newberry

UPPER WHITEFISH LAKE
Courtesy USFS

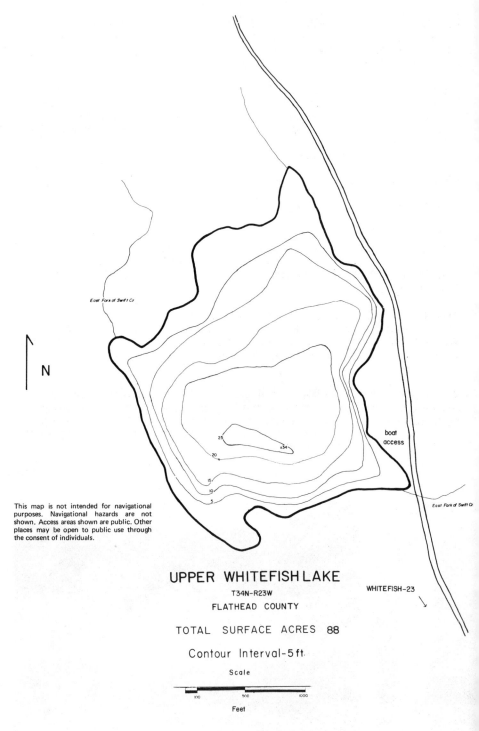

West Fork Swift Creek. A small stream with about 6 miles of fishing water, in a steep-sided, timbered canyon. The West Fork is reached near its mouth by a truck road 1 mile south of Upper Whitefish Lake and is followed upstream to headwaters. It is fair-to-good fishing for 10 to 12 inch cutthroat, and Dolly Varden that run as large as 10 pounds.

White Clay Creek. A small, spring-fed farmland creek flowing for 2½ miles to the Flathead River southwest of Polson. It's fished (mostly by kids and in the upper reaches only) for fair catches of 6 to 10 inch brookies.

Whitefish Lake. A popular resort lake with many summer homes and public recreational facilities. Reported by State Fish and Game personnel to provide the best Lake Trout (Mackinaw) fishing in the United States — for 9 to 30 pounders. Lake whitefish are also present in large numbers (naturally), but are seldom caught. It is also good summer trolling for 12 to 14 inch kokanee and cutthroat and produces a few largemouth bass, Dolly Varden and brook trout. Whitefish is well over 222 feet deep, 6 miles long by a mile wide, in the upper Flathead Valley with its south shore within the city limits of the town of Whitefish, and easily accessible around each end and along the east side by good county roads. There is a state park on the south end.

Whitefish River. Eighteen miles long from its head at Whitefish Lake to its juncture with the Stillwater River a mile east of Kalispell, and easily accessible by good county roads all the way. It is a good floating stream (for ducks), but is fished mostly from shore for good early season catches of cutthroat, and late season kokanee snagging. Unfortunately, it suffers from industrial pollution at the headwaters or it would probably be as good as the lake. There is also an overlarge population of trash fish here.

White Horse Lake. Is in an old burn, on Indian land, 1½ acres and deep; reached by a jeep road 1½ miles from the South Fork of the Jocko River 9 miles above the mouth. White Horse was planted with cutthroat in 1953 and has been excellent fishing ever since for skinny 8 to 10 inchers.

PTARMIGAN
Courtesy Danny On

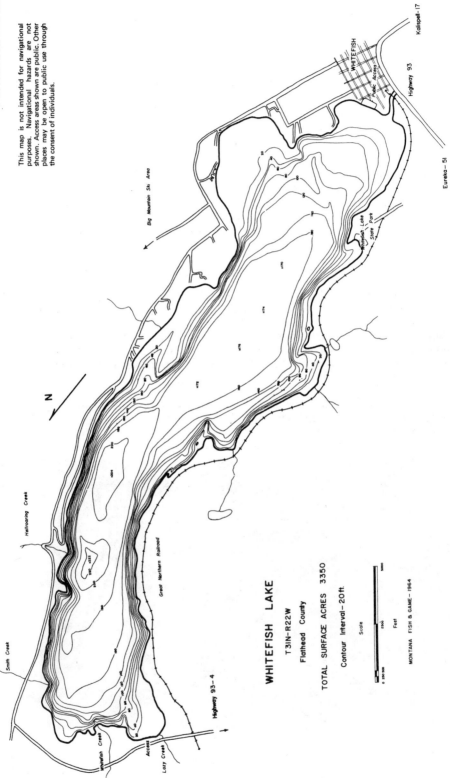

WHITEFISH LAKE FROM BIG MOUNTAIN
Courtesy Burlington Northern R.R.

FLATHEAD RIVER (Middle Fork)

BLACK BEAR CUBS
Courtesy USFS

Middle Fork of the Flathead Montana's Wildest River

The Middle Fork of the Flathead River is born along the rugged western slope of the Continental Divide in the Bob Marshall Wilderness Area. One of the last of America's truly wild rivers, it flows through beautiful timbered mountains and winds it way in narrow canyons for the most part. There are no roads and very few people inhabiting this region. Wild trout fishing, white-water floating, big game hunting, in combination with scenic, glaciated valleys and timbered mountains, are some of its outstanding recreational attributes.

Back-country wilderness visitors find all their requirements met by the upper forty mile segment of the river. Remoteness and solitude without highways, people or modern conveniences, combined with excellent fishing, provide outstanding recreational opportunities. The lower section which is some thirty-five to forty miles long, lies immediately adjacent to Highway 2 and is readily accessible by car at many points. Both sections provide ample white-water floating.

From Bear Creek downstream to West Glacier the lower section of the Middle Fork forms the southern boundary of Glacier National Park and is paralleled by Highway 2 and the Burlington Northern Railroad. Access is no problem although camping facilities are few, if any. This section of the river has some relatively swift white-water that can be floated by experts but is not recommended for the inexperienced boater. Fishing is best and floating the safest in July when the snow melt has diminished. The scenery is beautiful and the crystal clear water is like that seen only in Montana.

Access to the upper section of the river is provided by an airfield at Schaefer Meadows twenty-six miles from Highway 2. There are also trails into the area from the Spotted Bear Ranger Station or Java, but horseback or foot are the only means of transportation. The river at Schaefer Meadows is small. Float trips are somewhat hazardous and should be taken with caution and only by seasoned boaters. Fishermen are rewarded with some excellent cutthroat trout and Dolly Varden fishing.

The cutthroat trout generally range from 12 to 16 inches in size and average about a pound each, while the Dolly Varden must be 18 inches in length to keep. Trophy sized Dolly Varden in the 15 to 20 pound class are frequently caught.

The Middle Fork, like the North Fork, relies on Flathead Lake for most of the fish that are found there from June to September. Dolly Varden and cutthroat migrate from Flathead Lake up these tributaries to spawning

FLATHEAD RIVER (Middle Fork)

grounds creating an inter-dependency of the lake and its tributaries. Naturally, the summer time is the best fishing time in the Middle Fork. The spring run-off creates murky water until middle or late June and during this high water it is best to stick with live bait, spoons, wobblers, lures and spinning gear for cutthroat. However, when the water clears dry fly fishing comes into its own and remains excellent until fall. Dark patterns are favored in sizes eight to fourteen; however, in the more remote areas any fly works well. Dolly Varden are husky, scrappy and hard nosed and gear should be stout when these fish are sought. The use of medium and large spoons, as well as plugs, are recommended and bouncing the lure along the bottom can be very productive.

OTIS ROBBINS
Fisheries Biologist
Montana State Fish and Game Department

MIDDLE FORK FLATHEAD NEAR SPRUCE PARK
Courtesy USFS

Middle Fork Drainage

Basin Creek. Rises on the slopes of the Continental Divide at an elevation of more than 7000 feet and flows eastward for 6 miles to Bowl Creek (elevation 5800 feet) just below the Sun River Pass. It is followed by a trail all the way but is off the beaten path and seldom fished for the fair number of small cutthroat it contains.

Bear Creek. A fair-sized tributary of the Middle Fork of the Flathead River, Bear Creek is followed by U.S. 2 for 13 miles of fair fishing now, although it used to be good, down a brushy, timbered and some pastureland valley (Steven's Canyon). It was practically washed (scoured) out by the 1964 spring flood, and even now at this late date contains less than optimal numbers of 8 to 10 inch cutthroat and 14 to 20 inch Dollys (come up from the river).

Bergsicker Creek. The outlet of a small lake on Prospector mountain, Bergsicker is followed by a trail for 5 miles down its narrow, timbered canyon to the Middle Fork of the Flathead River about 8 miles above Nimrod. It is good fishing in the lower reaches for 8 to 10 inch cutthroat, and 14 to 20 inch Dolly Varden.

Bowl Creek. A high, snow-fed stream that heads on the Continental Divide at Teton Pass and is followed by a trail for 10 miles through timbered mountain past the Sun River Pass to the Middle Fork of the Flathead River 20 miles above the Schaefer Ranger Station; or — it is also reached by trail over Teton Pass 12 miles from the Trail Head above the West Fork (of the Teton River) Cabin. It is fished a moderate amount by parties in transit over the passes, and is fair-to-good from the mouth upstream — as far as Basin Creek, for 8 to 10 inch cutthroat and Dolly Varden.

Calbick Creek. A small stream, about 3 miles long in a narrow timbered canyon; followed by a trail from headwaters to mouth on the Middle Fork of the Flathead River, 4 miles above the Schaefer Ranger Station. It is mostly passed up for larger streams but the lower 2 miles are fair fishing for 6 to 8 inch cutthroat.

Challenge Creek. A very small headwaters tributary of Granite Creek; reached at the mouth by the Granite Creek trail, or 6 miles from Fielding on U.S. 2; and followed by a trail for 3 miles of fair 8 to 12 inch cutthroat, and 14 to 30 inch Dolly Varden fishing...if it wasn't closed for spawning.

Charlie Creek. The outlet of (barren) Cup Lake, Charlie Creek is followed by a USFS trail for 3½ miles through timbered, brushy country to the Middle Fork of the Flathead River, 6 miles above Nimrod. It is lightly fished for 8 to 10 inch cutthroat, and an occasional 14 to 20 inch Dolly Varden.

Clack Creek. The outlet of (barren) Dean, and Trilobite lakes; Clack Creek is followed by a trail for 6 miles down its narrow, steep-walled canyon to the Middle Fork of the Flathead River at Gooseberry Park. The lower 2 miles are fair-to-good fishing for resident 8 to 10 inch cutthroat, but it is an out-of-the-way stream and not bothered much.

FLATHEAD RIVER (Middle Fork)

Cox Creek. The outlet of Beaver Lake (no fish here), Cox Creek is followed for 9 miles through steep timbered mountains to the Middle Fork of the Flathead River 5 miles above the Schaeffer Ranger Station. It is not heavily or even moderately fished, but the lower 5 miles are good for 8 to 10 inch cutthroat.

Deerlick Creek. A small stream flowing down a steep timbered canyon above and through fairly flat pastureland below to the Middle Fork of the Flathead River opposite and about 2 miles below the Nyack Ranger Station. It is crossed at the mouth by U.S. 2 and is followed by a trail for 2 miles of fair, 6 to 7 inch brook and cutthroat fishing.

Dirtyface Creek. A short stream in a steep V-shaped, heavily timbered canyon below; but it levels out somewhat above. Dirtyface is reached at the mouth by the Elk Creek trail and is followed for 5 miles past the headwaters to Logan Creek and on to the Betty Creek Guard Station on the Hungry Horse Reservoir. The lower 3 miles are poor-to-fair cutthroat fishing. Elk, Logan, and Betty creeks are all too small for fish.

Dolly Varden Creek. So named after the fish that are in it; Dolly Varden flows northward for 10 miles to the Middle Fork of the Flathead River at Schaefer Meadows: But only the lower 2 miles are good fishing. There are some real good holes here in which occur cutthroat as well as Dollys and some real nice spawners in the fall. A moderately popular stream, it is easily accessible by a good USFS trail.

Essex Creek. A small stream, the outlet of (barren) Almeda Lake, followed by road for a couple of miles up from the mouth and then a good USFS horse trail along its narrow, steep-walled, timbered canyon for 4 miles to the Middle Fork of the Flathead River opposite the Walton Ranger Station. Essex is crossed at the mouth by U.S. 2 and is moderately popular with the younger set for 2½ miles of poor, 6 to 8 inch cutthroat fishing.

Flotilla Lake. A lightly fished but excellent lake for 8 to 15 inch cutthroat. Flotilla is something like 146 acres in size, deep with rocky shores, in subalpine scrub-timber country reached by a *very* steep trail (a "man killer") 2 miles above Scott Lake.

A "KEEPER"
Courtesy "Doc" Newberry

Courtesy "Doc" Newberry

FLATHEAD RIVER (Middle Fork)

LOTILLA LAKE (in background)
Courtesy USFS

INQUISITIVE
Courtesy Craig Black

Gateway Creek. A most spectacular stream, Gateway heads in the "Big River Meadows" in Indian country and flows for about 3 miles through the Gateway Gorge where it is dwarfed below vertical limestone walls that rise over 1100 feet in height above it, and eventually joins Trail Creek to form the Middle Fork of the Flathead River. The lower 3 miles below the gorge are in a flat-bottomed "pastureland" valley and are good fishing for 8 to 10 inch cutthroat.

Granite Creek. A nice stream that flows to the south and is followed by a good USFS trail for 9 miles down a narrow, steep-walled, heavily timbered canyon to the Middle Fork of the Flathead River 9 miles below the Schaefer Ranger Station. It is real good fishing for eat'n size cutthroat, and medium-to-whopping sized Dolly Varden; but is closed for spawning.

Lake Creek. About 3 miles long in timber and burned over country between Scott Lake and the Middle Fork of the Flathead River at Three Forks; and followed by a good horse trail all the way. This is a pretty little stream with some 25 foot falls near the mouth. It is lightly fished but good for 8 to 10 inch cutthroat, and an occasional Dolly.

Lodgepole Creek. About 7 miles long with 5 miles of fishing, Lodgepole flows southwest down a fairly low gradient canyon to the Middle Fork of the Flathead River at Three Forks, and is followed from mouth to headwaters by a good horse trail. It is sometimes closed for spawning, but reportedly contains a good population of Dolly Varden and cutthroat trout.

Long Creek. A "short stream," about 8 miles long in a timbered canyon and followed by a USFS trail from its headwaters to mouth on the Middle Fork of the Flathead River 6 miles above the Nimrod campground. It would be good fishing for cutthroat and Dolly Varden, but is sometimes closed for spawning.

Marion Lake. A fairly deep (145 feet at maximum), 80 acre lake in a high, brushy pocket on the north side of Essex Mountain; reached by the Marion Creek trail 3 miles from the Essex road, or 4 miles from the Walton Ranger Station on the Middle Fork of the Flathead River and U.S. 2. Marion is moderately popular and excellent fishing for 7 to 14 inch rainbow-cutthroat hybrids. There are no BIG ones.

FLATHEAD RIVER (Middle Fork)

Miner Creek. A small inlet of Scott Lake, followed by a trail down its steep, timbered canyon; the lower mile is fair fishing for small cutthroat, and Dollys in the fall.

Morrison Creek. A small, southward flowing tributary of Lodgepole Creek that is followed by a trail for 11 miles up its narrow timbered canyon from mouth to headwaters west of Slippery Bill mountain. Morrison is closed for spawning but used to be good fishing for cutthroat and Dolly Varden in years gone by.

Schaefer Creek. About 8 miles long, in a narrow timbered canyon; and followed by a trail for 6 miles of fishing to the Middle Fork of the Flathead River at the Schaefer Ranger Station. The lower end is moderately popular for 8 to 10 inch cutthroat, and 3 to 10 pound Dolly Varden. The big spawners come up in the fall.

Scott Lake. Here is a really shallow (no more than 4 feet anywhere not counting the mud), 30 acre lake in good moose country; they walk right across and feed in the middle of it. It is reached by trail 3 miles from Three Forks on the Middle Fork of the Flathead River 5 miles below the Schafer Ranger Station. There are some nice springs here that keep the lake from freezing out, and it's excellent fishing for 12 to 14 inch cutthroat.

Stanton Creek. A small stream that heads in Stanton Glacier and flows 5 miles eastward through an old, brushy burn to Stanton Lake, and is followed by a good trail 1½ miles below the lake to the Middle Fork of the Flathead River 7 miles above Nyack. It is fair fishing below the lake, and used to be good for about 1 mile above in some old beaver ponds, for 8 to 10 inch cutthroat trout.

Stanton Lake. Thirteen hundred yards long by 300 yards wide, no more than 29 feet deep anywhere; in a timbered valley 1½ miles by good trail up Stanton Creek. There is some open ground at the lower end and along the west side, and there are some big swampy beaver ponds above with lots of logs and brush so that it is difficult to get around during high water. Quite a few folks hike in although the fishing is poor for rainbow, cutthroat, and rainbow-cutthroat hybrids, mostly in the 6 to 9 inch class, plus lots of whitefish and wheelbarrow loads of suckers.

Strawberry Creek. Flows due south from Badger Pass for 9 miles down a narrow, timbered canyon to Gateway Creek, and is followed by a trail the entire distance. There is considerable "traffic" by Indians and hunters up and down the trail and the creek receives a moderate amount of fishing pressure for good catches of 8 to 12 inch cutthroat.

Trail Creek. A small stream followed from headwaters by a good USFS trail for 7 miles to its junction with Gateway Creek at the head of the Middle Fork of the Flathead River. The lower 2½ miles are in open park country and, although seldom fished, are good for 8 to 10 inch cutthroat trout.

Tranquil Basin Lakes. Two, about ½ mile apart in glacial cirques in Tranquil Basin; reached to within ¼ of a mile by the Edna Creek trail 1 mile up the Middle Fork of the Flathead River from Nimrod. The upper lake is about 10 acres, the lower 8; both are deep, seldom fished but sometimes good for 2 to 3 pound cutthroat.

Tunnell Creek. A small stream flowing to the north down a narrow timbered canyon to the Middle Fork of the Flathead River 8 miles above the Nyack Ranger Station and U.S. 2. The lower 5 miles are heavily fished for good catches of 7 to 8 inch cutthroat.

Twenty Five Mile Creek. Is only 6 miles long, and followed the whole way down its steep-walled canyon by a trail to the Middle Fork of the Flathead River 12 miles below the Schafer Ranger Station. It's posted for spawning but is "reportedly" fair fishing for 8 to 12 inch cutthroat, and 14 to 22 inch Dolly Varden.

BULLS
Courtesy Meir's Studio

North Fork of the Flathead
A Float Fisherman's Paradise

The North Fork of the Flathead River originates in British Columbia, Canada, and flows some 40 miles before entering the States. Once arrived, it flows at a gradient of 15 feet per mile until it reaches the confluence of the Middle Fork some 58 miles below the international boundary. The river is the physical boundary separating Glacier National Park on the east side and the predominately public land of Flathead National Forest on the west bank.

Waters of this upper river system essentially start from glacier melt which is generally clean, clear and asthetically beautiful. However, such waters contain only small amounts of the primary nutrients that afford good fish growth as compared to most top-rated trout streams in the state. Fishing in this river differs also because it is dependent upon a migratory fish population while most other waters rely upon a resident or non-migratory fish.

The North Fork has acquired a reputation as a float fisherman's paradise and it has been included as part of the state's Recreational Waterway System. It's a large, crystal-clear stream which flows through a panoramic backdrop of Glacier Park with its snowfields and picturesque mountains. Access is no problem as it is afforded along a road that runs the entire length of the river. Still, even in the midst of such availability, seclusion is no farther than moving around the next bend. Fish found in these waters are the result of their migratory habitat that causes continual movements of fish from Flathead Lake downstream some 55 to 113 miles. Fishing for migratory fish is never constant and reliable because it changes with weather and water conditions, but su eeeu

The North Fork has acquired a reputation as a float fisherman's paradise and has been included as part of the state's Recreational Waterway System. It's a large, crystal-clear stream which flows through a panoramic backdrop of Glacier Park with its snowfields and picturespue mountains. Access is no problem as it is afforded along a road that runs the entire length of the river. Still, even in the midst of such availability, seclusion is no farther than moving around the next bend. Fish found in these waters are the result of their migratory habitat that causes continual movements of fish from Flathead Lake downstream some 55 to 113 miles. Fishing for migratory fish is never constant and reliable because it changes with weather and water conditions, but such variability affords a greater than usual challenge to the angler as he not only has to pick his fishing hole and lure, but also has to schedule his trip to fit that of the fish's travels.

Your recreational experiences here will assuredly be memorable ones.

LANEY HANZEL
Fisheries Biologist
Montana State Fish and Game Department

North Fork Drainage

Bailey Lake. On the North Fork road 7½ miles above Columbia Falls, its maybe 20 acres, was rehabilitated in 1963 and present plans are for the State Fish and Game Department to use it as a cutthroat brood lake.

Big Creek. A fair sized stream in brushy, timbered mountains; crossed at the mouth by the North Fork road at the Big Creek Ranger Station 19 miles above Columbia Falls, and followed upstream by a logging road for 13 miles of good fishing for 8 to 10 inch cutthroat; 10 to 16 inch Dolly Varden, and a few big spawners up from the main river. Its been closed for years.

Canyon Creek. A small stream in a steep-walled, brushed up, timbered canyon; it empties to the North Fork of the Flathead River 3½ miles above that stream's junction with the Middle Fork and is crossed at the mouth by the North Fork road and followed upstream by a trail for 1½ miles of fair fishing for 6 to 12 inch cutthroat and a very few Dollys; but 'twill do you no good cause its closed the year around.

Chain Lakes. See Tripple Lakes.

Coal Creek. About 20 miles long, in a narrow, timbered canyon on the eastern flanks of the Whitefish Range, Coal Creek empties to the North Fork of the Flathead River 7½ miles above the Big Creek Ranger Station and is crossed at the mouth by the North Fork Road. The lower 2½ miles are accessible by a USFS trail, the remainder by a logging road up Langford Creek from the Big Creek Station. If it wasn't permanently closed it would be good fishing for 8 to 10 inch cutthroat; and fair for 10 to 16 inch Dolly Varden, plus a few BIG spawners up from the river.

Cyclone Lake. At 4100 feet above sea level, Cyclone is about 120 acres and mostly less than 20 feet deep; in a timbered mountain pocket reached to within 300 yards (a trail here) by a good USFS all purpose (the Coal Creek) road about 9 miles above the Big Creek Ranger Station. Its not fished much but is good for 10 inch cutthroat trout and a few 14 to 16 inch Dollys. 'Twas also planted with Grayling in 1967 and there are still some left; lots of suckers too.

Diamond Lake. Is 6.4 acres, 16 foot maximum depth, in timber all around, planted with cutthroat in 1971 and nobody knows how they've made out. If you care to look you can get there on foot about a mile from the end of the Coal Creek road.

Elelehum Lake. A used-to-be barren lake that is 11.2 acres, 25 feet at the deepest, and is reached by foot trail a mile above the end of the Elelehum Creek road (which takes off from the Big Creek road about six miles above the Big Creek Ranger Station). It was stocked with cutthroat in 1971 and is darned good fishing now for 10 to 12 inchers.

Half Moon Mill Pond. The one-acre mill pond on the Half Moon sawmill, a couple of miles west of Columbia Falls. In the early 1950's it was used as a kid's pond and still contains a few fair size brookies.

Hay Creek. A small stream with about 13 miles of fishing, followed by a truck road down its timbered, brushy valley on the eastern slopes of the Whitefish Range to its mouth on the North Fork of the Flathead River 13½ miles above the Big Creek Guard Station. It is hard to fish but good for small cutthroat, especially in beaver ponds about a mile above the mouth, and now and then a Dolly.

NORTH FORK NEAR POLBRIDGE
Courtesy Ernst Peterson

DOLLY
Courtesy "Doc" Newberry

Hay Lake. A small lake in a swampy meadow, reached by a hundred yards or so of trail from the end of the Hay Creek road. This one is not too popular but is fair fishing for 10 to 12 inch cutthroat.

Hunts (or Hunchberger, Huntsberger or Hunsberger) Lake. Take the Whale Creek road about 15 miles upstream to a USFS foot trail which bears off to the left (south) for three miles to the shore of this lake which is: 5 acres, deep, in burned over mountainous country, seldom fished but reported to be fair for 8 to 10 inch cutthroat trout.

Link (or Lynx) Lake. Hooooo Boy!!! Take the Whale Creek road 9 miles upstream to the Shorty Creek (no fish here) road which bears off (in a couple of miles) to the south and follow it for 10 miles or so to a USFS trail which in turn takes you another mile east and on in to the shore. Its hot and cold fishing, 15 acres, 25 to 30 feet deep, irregularly shaped with small rocky islands here and there, ringed by steep subalpine mountains all around about a mile northwest of Red Meadow Lake and the fish are mostly cutthroat that will run around 2 to 3 pounds.

Moose Creek. A small tributary of the North Fork of the Flathead River, 3 miles south of the Ford Guard Station; crossed near the mouth by the North Fork road and followed upstream for 11 miles by logging road through dense brush and timber. It is not often fished but good for small cutthroat trout and a few fair-sized Dolly Varden.

Moose Lake. A shallow, mud-bottomed, weedy lake, about 17½ acres, marshy at the west end, in a timbered valley below Moose peak reached by traveling 13 miles up the Big Creek road from the Ranger Station and then 4½ miles up the Hallowat road and finally another five miles up the Kletomas Creek road. Wow! In spite of all this its moderately popular and fair fishing too for 10 to 12 inch cutthroat trout.

Mud Lake. A shallow lake, about 15 acres, in a timbered valley below and west of Glacier View mountain; reached by the Langford Creek-Coal Creek road 3 miles above the Big Creek Ranger Station. Mud Lake is too weedy to fish except in the spring when the water is high, and even then it's only fair for 10 to 12 inch cutthroat trout, and a few nice ones in the 18 to 20 inch bracket.

Nasukoin Lake. Is reached on foot ¼ mile cross-country from the upper end of the Moose Creek road. Once there its about 6½ acres by 27 foot maximum depth, was planted with cutthroat in 1971 and they were doing fine as of the summer of 1973. So who knows —?

Nine Mile Lake. See Spoon Lake.

Ninko Creek. A steep, brushy, headwaters tributary of Whale Creek; the lower 2 miles are fair fishing for small cutthroat trout, and Dolly Varden spawners; or would be except that its closed the year around. Anyway its crossed at the mouth by the Whale Creek road 12 miles above the Ford Guard Station and is followed by a trail for three miles to headwaters — if you're just looking for exercise.

Red Meadow Creek. A small, brushy stream about 13 miles long in a narrow timbered canyon eastward to the North Fork of the Flathead River where its crossed at the mouth by the North Fork road about 20 miles above the Big Creek Ranger Station and is followed upstream by a logging road past the headwaters. Red Meadow is hard to fish because of the dense brush but is fair in the many beaver ponds along its length for 6 to 10 inch cutthroat and 8 to 10 pound Dolly Varden.

Red Meadow Lake. A beautiful, 19 acre but shallow lake whose south side is real brushy from an old snow slide beneath high rocky peaks. The north side is timbered and followed around by the Red Meadow Creek road 12 miles west of the North Fork road, or 6 miles above Upper Whitefish Lake (in the Flathead Drainage). Red Meadow is quite popular and fair fishing for 8 to 14 inch cutthroat trout and Grayling.

South Fork Coal Creek. A small stream (only a couple of miles of fishing) that is reached at the mouth by the Coal Creek road and followed upstream by a logging road. Its real brushy, and would be hard to fish if it weren't closed for spawning....for 8 to 12 inch cutthroat and Dollys up to 10 pounds.

Spoon (or Nine Mile) Lake. A 70 acre lake, about 30 feet deep in spots, mostly brushy around the shore and marshy on the south end; reached by road 7½ miles north from Columbia Falls. It was rehabilitated in 1963 and is fair-to-middling good fishing now for cutthroat trout.

A SAMPLE

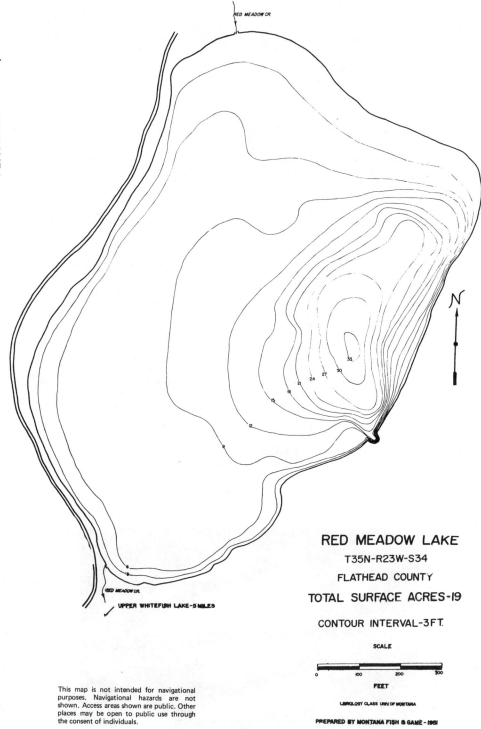

Trail Creek. See Yakinikak Creek.

Triple (or Chain) Lakes. Three nice little lakes (and a fourth that hardly qualifies) in subalpine country a mile northeast of Link Lake, or a mile north and 1200 feet above the Red Meadow Creek road. No. 1 (at the top of the string) is 4.8 acres by 50-plus feet deep and was stocked in 1969. No. 3 is 22.7 acres by more than a hundred feet deep; No. 4 (at the bottom of the string) is 17½ acres by only 14 feet maximum depth; and all three (nos. 1, 3, 4) are good fishing for 7 to 16 inch cutthroat trout.

Tuchuck Creek. Flows 4½ miles from Tuchuck Mountain down a real steep canyon above, and across a couple of miles of relatively flat timberland below the Yakinikak Creek; it is crossed at the mouth by the Yakinikak road and followed to headwaters by a USFS trail. Tuchuck is good fishing in the lower reaches for pan-sized cutthroat, especially so in some brushy beaver ponds about 1½ miles above its mouth. There's a public campground at the mouth of the creek, 9 miles above the North Fork road.

Tuchuck Lake. This is a nice little lake, about 8 acres in size by 20 feet maximum depth; was planted with cutthroat in 1971 and they are reportedly doing real good as of now. To get there you hike ¼ mile north of the Tuchuck Mountain trail about 2 miles south of the Canadian line, a couple of miles east of the Mountain.

Whale Creek. About 20 miles long, in a fairly wide, timbered valley and followed by a logging road from a point on the North Fork road ten miles above Polebridge, to headwaters. This is a good fishing stream (or would be except for the fact that its closed the year around) for 8 to 10 inch cutthroat, 10 to 16 inch Dolly Varden in spring and summer — and BIG spawners in the fall.

Whale Lake. A shallow, rock and mud-bottomed lake, about 20 acres, in timbered mountains reached to within 4 miles by the Whale Creek road at the very head of the drainage 20 miles above the North Fork of the Flathead River (except that the last four miles have washed out and will not be rebuilt). There are several large springs in this lake that keep it from freezing out and it's fair (though spotty) fishing for 10 to 16 inch cutthroat trout.

Yakinikak (or Trail) Creek. A real pretty stream below; crossed near the mouth by the North Fork of the Flathead road about 4 miles above the Ford Guard Station; the lower 6 miles are closed for spawning. The middle reaches sink for about 2 miles. The upper reaches are followed by a logging road but are steep, brushy and in a narrow, timbered canyon. The creek is, however, good fishing for 7 to 12 inch cutthroat trout, and 10 to 16 inch Dolly Varden in the summer, and sometimes (depth of water in the middle reaches permitting) for some nice BIG spawners of the fall. Its closed below the falls about 4½ miles above the mouth.

FLATHEAD RIVER (South Fork)

HUNGRY HORSE RESERVOIR
Courtesy USFS

South Fork of the Flathead
A Wilderness Paradise

Originally an integral part of the Flathead Lake and River system, the construction of Hungry Horse Dam has cut off the South Fork and isolated the fish population above the dam from that of the main river. The reservoir formed by the dam extends for 32 miles upstream and ranges from ½ to 5 miles wide. The South Fork extends for another 52 miles upstream to its remote, southern-most headwaters in Danaher Creek. Numerous tributaries offer exceptional trout spawning facilities, and fishing rates as good to excellent in the reservoir and tributaries as well. Montana westslope cutthroat trout, Dolly Varden and mountain whitefish are the main game fish, and these reside in the streams during the summer as adults. The remainder of the year, most mature fish are in the reservoir which can be fished the year around. About half of the South Fork drainage, including some of its best tributary streams and lakes, lies within the Bob Marshall Wilderness. This is a roadless area of outstanding scenic beauty from which all mechanical methods of transportation are banned. Access is solely by foot or horseback.

Numerous horse and foot trails are maintained and prominently marked by the U.S. Forest Service with suggested campsites every few miles along the main trails. The sites are primitive and help confine the scars of man's use to these small areas. Grass for pasturing horses is scarce, and horsemen are required to carry concentrated pellet food for their stock.

Fly fishing and spin casting in streams of the South Fork are unexcelled. The tranquil beauty of the region coupled with the cool, clear mountain air gives an experience of unforgettable splendor. Black spotted trout or the west-slope cutthroat trout range up to 20 inches and 3 pounds while Dolly Varden may range up to 20 pounds. Mountain whitefish are readily taken, primarily by bait fishermen.

Float trips, using rubber rafts brought into the upstream areas by pack horse, are possible down to the reservoir below the Spotted Bear Ranger Station. A very dangerous and impassable gorge near the junction of Bunker Creek necessitates some 3 miles of portage around this hazard on the South Fork in the vicinity of the Meadow Creek Ranger Station. A small airfield is located at Meadow Creek just outside the wilderness boundary.

Hundreds of high, icy cold mountain lakes are clustered throughout the drainage like jewels in cirque pockets formed by earlier glacial activity. Although most were barren of fish originally, many have populations of cutthroat trout established in them now. Outlets of these lakes form spectacular falls as they cascade down the steep and abrupt slopes.

The headwater areas are accessible from the west by trails from the Swan River road (state Highway 209), from the south by trails from forest roads joining state Highway 200, or from the east by pack trail from the head of Gibson Reservoir,* along the Chinese Wall and finally up and over the White River pass.

*See The Montanans' Fishing Guide, East.

ROBERT SCHUMACHER
Fisheries Biologist
Montana State Fish and Game Department

CAMP ALONG THE SOUTH FORK
Courtesy USFS

South Fork Drainage

Addition Creek. A nice little stream followed by a road (closed to motorised vehicles) for six miles down its steep, timbered canyon to the South Fork of the Flathead River at the Spotted Bear Ranger Station. The lower 4 miles are occasionally fished for good catches of 7 to 9 inch cutthroat trout.

Aeneas Creek. A small inlet of the Hungry Horse Reservoir; in a steep V-shaped timbered canyon; reached by the west side road 22 miles above the dam and followed by a poor (abandoned) trail upstream for 3 miles of poor-to-fair, pan-sized cutthroat fishing.

Aeneas Lake. Fairly deep, about 2 acres, in a cirque near timberline ½ mile east and over the mountains from Birch Lake (which is in the Swan River drainage). It is reported to have had fish in years gone by, but its present status is doubtful.

Arres (or Ayres) Creek. Reached at the mouth by the Danaher Creek trail 1½ miles above the Basin Creek landing field. Arres is only a couple of miles long but the lower reaches contain fair numbers of small cutthroat and, in the fall some not-so-small Dolly Varden spawners. It is almost never fished.

Babcock Creek. A seldom visited, steep-timbered-mountain stream that is crossed at the mouth by the Young Creek trail and followed upstream for 3 miles of good 8 to 10 inch cutthroat and 3 to 15 pound Dolly Varden (in the fall) fishing.

Ball Lake. A deep, 3 acre lake on a rock ledge in timbered country; reached by trail 1½ miles from the end of the Posy Creek road on the Quintonkon drainage. Ball is lightly fished but good for 10 to 14 inch cutthroat trout.

Bartlett Creek. A nice sized tributary of the South Fork of the Flathead River; crossed at the mouth by the South Fork trail 2 miles below the Big Prairie Ranger Station, and followed upstream for 5 miles of fair 8 to 10 inch cutthroat (and 3 to 10 pound Dolly Varden spawners in the fall) fishing. If it was any place else Bartlett would be heavily fished, but here it is seldom bothered.

Basin Creek. Heads in Grizzly Basin and is followed by a trail for 6 miles through a steep canyon above, and relatively flat country below, to Danaher Creek at the Basin Creek air field. The lower reaches (which are on the main trail to Benchmark) are good fishing for small cutthroat, plus a few Dollys and whitefish in the fall.

Bigelow Lake. A 33 foot deep, 24 acre lake which is not shown on most maps; in heavy timber at 6000 feet above sea level on the head of Bigelow Creek. It is most easily reached by a cross-country hike 2½ miles from the end of the Wheeler Creek road but — let's face it, there's no really easy way and its seldom if ever visited or fished although there is a small population of 16 to 18 inch cutthroat.

Big Fish Lake. See George Lake.

FLATHEAD RIVER (South Fork)

BIG SALMON LAKE
Courtesy USFS

BIG HAWK LAKES
Courtesy USFS

Big Hawk Lakes. Three small subalpine lakes in beautiful (Jewel Basin) country between 5800 and 6300 feet above sea level; at the head of Jones Creek; reached by a trail 3 miles north from the end of the Wheeler Creek road. The upper lake used to be barren; the middle one is 32 feet deep by 4 acres; the lower one is 35 feet deep by 30 acres. The upper lake was planted with cutthroat the summer of 1968. The lower lake is swarming with nice fat 5 to 19 inch cutthroat that will really take a fly. And you should know — only *hikers* are allowed. No motor bikers, skiidoers and the like.

Big Salmon Lake. A large deep lake, about 4 miles long by ½ mile wide, in a fairly high (elevation 4500 feet), timbered canyon in the Bob Marshall Wilderness area; reached by a good USFS pack trail 19 miles from the end of Holland Lake road; or 20 miles from the Spotted Bear Ranger Station at the end of the Hungry Horse Reservoir road. There's a trail all along the west side of Big Salmon and a campground at the upper end. It is hit by most everybody coming through, and provides good fishing for 12 to 17 inch cutthroat, 1 to 15 pound Dolly Varden, plus a few rainbow-cutthroat hybrids (whose ancestors probably worked their way down from Woodward Lake), and whitefish along with quite a few suckers.

Big Salmon River. Really a nice sized creek, with lots of good looking water too, on the main trail over the divide from the Swan River drainage and Holland Lake. The upper reaches, above Brarrier Falls, were once stocked with cutthroat which have apparently long since frozen out. Below the falls for 3½ miles to Big Salmon Lake, and below the lake for 1 mile to the South Fork of the Flathead River it is fair fishing for 10 to 14 inch cutthroat, and some real nice Dollys in the fall. There is a campground at the mouth, another at the upper end of the lake, and yet another about 2 miles above.

Black Bear Creek. A fairly open stream in burned over country, Black Bear is followed by a trail from its headwaters for 5 miles to its mouth on the east side of the South Fork of the Flathead River 2½ miles below the Black Bear Guard Station. It is seldom fished but excellent for 6 to 10 inch cutthroat, and an occasional good-sized Dolly Varden.

MARYOTT FALLS, BIG SALMON CREEK
Courtesy USFS

BLACK LAKE
Courtesy USFS

Blackfoot Lake. A lightly fished lake, 15.6 acres, as much as 22 feet deep in spots but with mostly a gradual dropoff, in a timbered pocket 5500 feet above sea level on the headwaters of Graves Creek. Blackfoot can be reached by a good trail 6 miles east from the end of the Noisy Creek (Flathead drainage) road, or 4 miles from the end of the Graves Creek road. It was planted with rainbow in the 1940's and is fair fishing now for 6 to 14 inchers, plus an occasional 10 to 16 inch cutthroat. And, you'd better plan on *hiking* in because the area is *closed* to all other forms of transportation. There oughta be more places just like it.

Black Lake. A very scenic 50 acre alpine lake, fairly deep with a small island in the middle, a steep dropoff at the upper end and shallow near the outlet. There are cliffs and talus around the head of the lake, and timber below. It is drained by Graves Creek and reached by a trail 7 miles up that stream from the Hungry Horse Reservoir, or 1½ miles from the end of the Noisy Creek (Flathead drainage) road. The fishing is reported to be excellent (if you hit it right and use corn for bait) for 8 to 10 inch cutthroat and a few BIG rainbow that have been reported to 15 pounds. P.S.: It was restocked with cutthroat in 1973 so should be no worse and probably even better now.

BUNKER CREEK DRAINAGE
Courtesy USFS

BLACKFOOT LAKE; You come in over the ridge
Courtesy USFS

FLATHEAD RIVER (South Fork)

Bruce Creek. A steep little stream that sometimes goes almost dry but there are lots of small falls and some good holes. It is reached at the mouth by the Addition Creek logging road and is followed upstream for a couple of miles of fair 6 to 9 inch cutthroat fishing.

Bunker Creek. A good-sized stream followed by a logging road (sometimes closed but usually open in the summer) from its mouth for 7 or 8 miles upstream and then by trail to headwaters below Thunderbolt Mountain. It flows eastward from its beginnings for 15 miles down a heavily timbered (well it *was* before the USFS sold the stumpage; it's now mostly clear cut and looks like the valley of the moon without many of the esthetics), flat-bottomed valley to the South Fork of the Flathead River 9 miles above the Spotted Bear Ranger Station. The lower 10 miles below the mouth of the Middle Fork used to be moderately popular and good fishing for 8 to 12 inch cutthroat and 2 to 10 pound Dolly Varden (especially during season in the fall). However, what with the "improved" access and all, the pressure will likely be stepped up and although the creek itself was saved from being scalped, the fishery will likely deteriorate.

Burnt Creek. The outlet of Big Knife Lakes, Burnt Creek flows for 8 miles to the South Fork of the Flathead River, 5 miles below the Big Prairie Ranger Station. It is crossed at the mouth by the South Fork trail, and followed upstream for 2 miles of good 8 to 10 inch cutthroat, and (in the fall) 3 to 15 pound Dolly Varden fishing. However most people find the main river so much more attractive that it is seldom bothered.

Camp Creek. A small tributary of Danaher Creek; crossed at the mouth by a trail 1 mile below the Basin Creek Landing Field and followed upstream for 4½ miles through its steep, narrow canyon to Camp Creek Pass. The lower reaches are fished a little by trail crews passing through, and are fair for 8 to 12 inch cutthroat, and Dollys in the fall.

Cannon Creek. About 8 miles long, in timbered country on the eastern slopes of the Swan Range, reached at the mouth by the Gorge Creek trail and followed upstream by trail for 5 miles of good 8 to 10 inch cutthroat fishing.

CLAYTON LAKE
Courtesy USFS

TRAIL RIDERS ALONG CAMP CREEK
Courtesy USFS

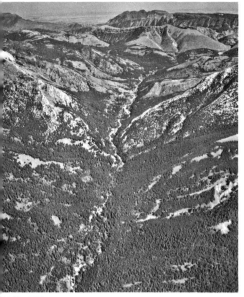

DANAHER CREEK DRAINAGE
Courtesy USFS

CLIFF LAKE left, BLACK LAKE right
Barren pothole foreground   Courtesy USFS

FLATHEAD RIVER (South Fork)

Clayton Creek. The outlet of Clayton Lake, followed by a trail along the south side for about 3 miles, and then by a road for 2½ miles to its mouth on the Hungry Horse Reservoir. Clayton is real steep above, and poor fishing at best for pan-size cutthroat plus a few lunkers to 5 pounds or better.

Clayton Lake. A really beautiful lake in a scenic (Jewel Basin) timbered pocket at 5900 feet above sea level, about 4 miles "as the crow flies" west of the Hungry Horse Reservoir and easily reached by a trail 3 miles from the end of the Clayton Creek road. It is 61.5 acres, shallow at both ends but over a hundred feet deep in the middle and mostly brushy along the shores. In it is a fair population of 8 to 18 inch (and a few that'll run up to 10 pounds) cutthroat (you can see 'em) that are plenty hard to catch but will take flies in the spring.

Cliff Lake. Hike 3 miles down the Graves Creek trail from Black Lake, or 3 miles up the Graves Creek trail from Handkerchief Lake, and then follow up the little drainage to the west, one mile to Cliff Lake. It's 5550 feet above sea level, 23.0 acres, 74 feet deep with steep dropoffs, below talus on the west and southwest with a little timber on the southeast rim. The State Fish and Game Department stocked it with cutthroat in '67 and it's good fishing now.

Conner Creek. A short, small tributary of Sullivan Creek, in logged-over country; followed by a road for about 2 miles of fair (6 to 7 inch) cutthroat fishing.

Crater Lake. A 75 foot deep, 23.5 acre cirque lake, real pretty in rocky alpine country, 5650 feet above sea level, reached by a trail 4 miles south from the end of the Noisy Creek (in the Flathead drainage) road. It was stocked with rainbow years ago but they didn't make it so — it was restocked with cutthroat in '68 and is fair-to-middling good fishing now.

Crimson Creek. The high (between 5000 and 7000 feet elevation), short, steep and rocky outlet of Crimson Lake; this stream flows for 2 miles to Marshall Creek. The lower 1½ miles are followed by a trail, and produce, or would if they were fished, fair catches of small cutthroat — and some large Dolly Varden spawners in the fall.

Damnation Creek. Heads on Pagoda Mountain and is followed by a trail for 3 miles down its very narrow, steep canyon to the South Fork of the Flathead River 4½ miles above the Black Bear Guard Station. This one is real brushy with lots of downfalls, etc., and no joy to fish. However, the lower reaches are good, though seldom bothered for 7 to 9 inch cutthroat and an occasional Dolly Varden in the fall.

Danaher Creek. A slow, meandering beaver-dammed stream (it looks more like eastern water than a high mountain creek) that follows for 10 miles through timbered canyons, and 6 miles across a broad mountain meadow, to the headwaters of the South Fork of the Flathead River 5 miles above the Big Prairie Ranger Station. Danaher is on one of the main trails to the Blackfoot country and is heavily fished for good catches of 10 to 14 inch cutthroat, and an occasional whitefish or Dolly.

Dean Creek. A small, shallow stream with a series of falls 1½ miles above its mouth, on the headwaters of the Spotted Bear River; crossed at the mouth and followed to headwaters by an unmaintained USFS trail. Dean Creek is seldom fished, but is fair below the falls for 8 to 10 inch cutthroat trout.

Doctor Creek. A small, brushy stream, the outlet of Doctor Lake; followed by a trail for its full length of 3 miles to Lick Creek and fair fishing in lots of small holes for 8 to 10 inch cutthroat, and not a few 2 to 3 pound Dolly Varden.

Doctor Lake. A real deep (almost a hundred feet in places), 80 acres, subalpine cirque lake reached by a dead end USFS trail 4 miles above the Shaw Creek Guard Station, or 11 miles over Gordon Pass from Holland Lake. It is rocky and brushy around the shore and good fishing for skinny, over-populated, 10 to 12 inch cutthroat, and an occasional 8 to 15 inch whitefish or Dolly (these last will run up to a couple of feet or so).

Doris Creek. A very small stream that is heavily fished for cutthroat, and whitefish spawners in the fall, from the Hungry Horse Reservoir. Doris is crossed at the mouth by the westside Hungry Horse road 3 miles above the Dam, and is followed by a logging road for about 5 miles upstream. There are lots of blow-downs and brush in the canyon above, but the lower mile is fairly open and easy to fish.

GEORGE LAKE
Courtesy USFS

DANAHER CREEK
Courtesy W.E. Thamarus, Jr.

Dudley (or Tent) Creek. A small stream in heavily timbered country, Dudley is real steep above but flattens out for about 2 miles below before emptying into the northeast side of the Hungry Horse Reservoir. It is crossed at the headwaters by the East Side road and is good fishing in the lower reaches for 10 to 12 inch cutthroat trout.

Emery Creek. Another small stream that is followed by a good logging road for its full length of 8 miles through timbered mountains to the northeast end of the Hungry Horse Reservoir. There are some falls about 1½ miles above its mouth but the fishing is fair both above and below for 6 to 7 inch cutthroat.

Felix Creek. Flows for 3 miles through rolling timbered country to the east side of the Hungry Horse Reservoir, 4 miles above the Betty Creek Guard Station; and is followed along by a logging road about ¼ of a mile to the north and above the stream. It is subjected to moderate fishing pressure and is just poor-to-fair for small cutthroat trout. Since the '64 flood spawners can no longer come up from the reservoir and so all of the fish that are here, are resident.

George Creek. The steep and rocky outlet of George Lake, followed by a trail for 3 miles down a narrow timbered canyon to Gordon Creek at the Shaw Creek Guard Station. The lower ½ mile is fair fishing for 8 to 10 inch cutthroat, and Dollys in the fall, but about the only one to fish it is the "Lone Ranger."

George Creek. The outlet of Sunburst, Olor and Inspiration Lakes, George Creek is reached at the headwaters by a trail 9 miles over Inspiration Pass (in the Swan Range) from the end of the Soup Creek road (in the Swan River drainage) and is followed by a trail for 11 miles down its most spectacular, steep-walled gorge to Bunker Creek and on to the South Fork of the Flathead River. The lower 8 miles are good fishing for 8 to 10 inch cutthroat trout.

George (or Big Fish) Lake. A real deep lake (over 200 feet), 128 acres in an alpine setting right on the Clearwater — South Fork divide. It is hard to get to but the best way in is reportedly by a trail from Holland Lake for 10 miles over Gordon Pass to the Shaw Creek Guard Station, then 3 miles up the George Creek trail and finally another 2 miles (horizontally plus 1000 feet vertically) cross-country to the lake which by the way, usually stays frozen over until late July and freezes up again in early October. It used to be spotty fishing for 20 inch cutthroat, was replanted in '65 and could use another replant now.

Gill Creek. About 4 miles long, with no trail but crossed at the mouth by the Little Salmon River trail 9 miles upstream from the South Fork of the Flathead River. It is steep, rocky, and poor fishing (near the mouth only) for small cutthroat and Dolly Varden in the fall. This one is seldom if ever fished.

Gordon Creek. A beautiful, wide, easily fished stream that is reached at the headwaters by a trail over Gordon Pass, 8 steep miles from Holland Lake, and is followed by a trail for 17 miles to its mouth on the South Fork of the Flathead River, 3 miles above the Big Prairie Ranger Station. There is a campground at Shirtail Park and the stream is good fishing too, for 3 to 8 pound Dolly Varden, and 8 to 10 inch cutthroat.

Gorge Creek. See *in* Sunburst Lake.

Graves Creek. A clear mountain stream that is crossed at the mouth by the west side Hungry Horse Reservoir road, about 22 miles above the dam, and is followed upstream by a logging road for 1 mile to Handkerchief Lake, and then

GRAVES CREEK FALLS
Courtesy USFS

HEADING OUT ON THE HUNGRY HORSE
Courtesy USFS

another 1½ miles beyond, and thence by a trail to headwaters. Graves is good fishing below the lake for 9 to 12 inch cutthroat and grayling, and a few 10 to 12 inch rainbow trout. There are some falls about 1 mile above the lake that are a barrier to the grayling, and above them are mostly small cutthroat and a few small rainbow.

Handkerchief Lake. A good-sized (32 acres) lake, in timber and brush; reached to within 300 yards by road a half mile from the mouth of Aeneas Creek on the Hungry Horse Reservoir. This lake is heavily fished and fair for mostly grayling plus some cutthroat. The state record for grayling (a five pounder) was caught here the summer of '74 and there are (reportedly) larger ones still around. Note: There are NO beaches and NO boat launching facilities, but there IS a USFS campground. It's good ice fishing too.

Harris Creek. A small inlet to the east side of the Hungry Horse Reservoir, crossed at the mouth by the East Side road 4 miles north of the Betty Creek Guard Station. It flows through a steep, timbered canyon above but the lower 1½ miles are good fishing for 10 to 12 inch cutthroat.

Harrison Creek. Is followed by a USFS trail for 5 miles through a very narrow, steep-walled, timbered canyon above, and across the flat-bottomed valley of the South Fork of the Flathead to that river below. It is reached at the mouth by the east-side trail 8 miles above the Spotted Bear Ranger Station and is occasionally fished for fair catches of 6 to 9 inch cutthroat trout, plus a few good-sized Dolly Varden in the fall.

Helen Creek. About 5 miles long and paralleled by one of the main "cross-over" trails through most spectacular country from the White River near Brushy Park to the South Fork of the Flathead River just above the Black Bear Guard Station. Lots of traffic goes by, but very few people stop to fish this stream for its goodly numbers of 7 to 9 inch cutthroat trout.

Highrock Creek. A tiny, brushy tributary of the Little Salmon River; crossed at the mouth by a trail and poor fishing for maybe ½ mile above for small cutthroat, and Dolly Varden spawners in the fall. Highrock is very seldom fished and just as well, too.

Hodag Creek.
A small, unimpressive tributary of the South Fork of the Flathead River, reached at the mouth by the east-side trail 1 mile below the Black Bear Guard Station. Hodag is seldom fished but the lower 2 miles are fair for 6 to 9 inch cutthroat, plus a few Dolly Varden spawners in the fall.

Holbrook Creek.
Holbrook flows to the northeast for 10 miles down a steep, deep, V-shaped canyon in partly timber and partly burned over country to the South Fork of the Flathead at the Holbrook Ranger Station. It is followed all the way to headwaters by a trail, and the lower 4 miles are good fishing for 8 to 10 inch cutthroat, and a few 3 to 15 pound Dolly Varden late in the season.

Hungry Creek.
Is reached at the mouth by the west-side South Fork of the Flathead River trail at the Black Bear air field, and is followed upstream for 5 miles to headwaters on the slopes of Marmot mountain. Hungry Creek is seldom fished but the lower couple of miles are good for 8 to 10 inch cutthroat and an occasional Dolly Varden.

Hungry Horse Creek.
About 6 miles long, on a timbered bench above, and flowing to the northeast end of the Hungry Horse Reservoir. It is paralleled by a logging road and easily fished (a real good kid's creek) for 6 to 10 inch cutthroat above the falls (3 miles above the mouth) only, plus a few Dollys below.

Hungry Horse Reservoir.
A large reservoir; 34 miles long by 1 to 4 miles wide with many coves, headlands and a most irregular shoreline, beautiful when it's full and G-d awful when it's drawn down, which is each fall and winter. Hungry Horse occupies the once densely timbered but now grossly scarred with many clear cuts (courtesy of the U.S. Forest Service), flooded-out valley of the South Fork of the Flathead River and is accessible all the way around by roads originating from U.S. 2 a few miles south of Coram. There is a resort at the upper end and 5 public campgrounds. The fishing is fair-to-good now for 10 to 14 inch cutthroat, 12 to 20 inch Dolly Varden, and 10 to 12 inch whitefish and a very few grayling near the mouth of Grave's Creek. However like most large draw down reservoirs that are managed with little or no consideration for the fishing potential, the sizes of the game fish are decreasing and the rough fish populations are exploding.

Jenny Creek.
Heads below Youngs Pass and is followed by a trail down its narrow, timbered canyon for 4 miles to the head of Youngs Creek. The lower mile contains a few small cutthroat and some Dolly Varden in the fall, but is very seldom fished by stray passers-by.

Jenny Lake.
A shallow, 4 acre pond in brushy timber just east of Blaine Mountain; most easily reached by a trail 1½ miles from the end of the Doris Creek road. Jenny is occasionally visited by Boy Scouts, and is fished somewhat for small cutthroat.

Jewel Basin Lakes.
Five little lakes in a beautiful, subalpine basin about an hour's hike by good trail, east from the end of the Noisy Creek (Flathead drainage) road. The smaller lakes are barren, but the largest two are both deep and support fish; North Jewel is about 7 acres, South Jewel about 4 acres; and the fishing is poor for 1½ to 2 pound cutthroat and rainbow trout.

Knieff Creek.
A very small stream in a heavily timbered, V-shaped canyon; crossed at the mouth by the west side Hungry Horse Reservoir road about 20 miles above the dam, and followed by a very poor trail upstream for a couple of miles of poor-to-fair fishing for pan-size cutthroat.

FLATHEAD RIVER (South Fork)

Koessler Lake. A lovely, subalpine cirque lake with a little island in its southwest end; real deep, 45 acres, a half mile northwest of Doctor Lake, no trail. Koessler was planted with cutthroat in the late 1920's and provides good fishing now for real skinny, 12 to 14 inch trout.

Lake 7 Acres. Two, Upper and Lower, 6.5 and 11.9 acres, over 100 and 76 feet deep, ¼ of a mile apart in a brushy subalpine cirque reached by an exceedingly steep and very poor (unmaintained) trail 3 miles north from the Aeneas Creek trail. Both used to be sterile (no reproduction), but had fish once and the lower one was planted with cutthroat in '67 and is now good fishing once again....for those few intrepid souls who make it in.

Lena Lake. An 84½ acre, fairly deep lake with steep dropoffs; in high, timbered mountains; reached by a good USFS trail at the head of the Big Salmon River. Lena was planted away back in the 20's with rainbow trout whose multitudinous progeny now average 10 to 20 inches in length and provide fair-to-good fishing for the few people who try it.

Lick Creek. The outlet of Lick Lake; Lick Creek flows through a narrow timbered canyon for 4 miles to Gordon Creek and is followed along the lower reaches by the Gordon Pass trail. It is a small, brushy stream, that is hard to fish but fair for "camp-eating" cutthroat.

Lion Lake. A deep, 35 acre lake in timbered rolling country right on the road from Government Town to the Hungry Horse Dam. It is very popular, and there is a USFS campground plus a nice sandy swimming beach in the southwest corner. The fishing is generally good for 12 to 16 inch cutthroat and brook trout. It gets a lot of pressure (in winter time through the ice too) but is restocked annually.

Little Salmon River. A nice sized stream, about 15 miles long and followed the whole way down its narrow, steep-walled, heavily timbered valley by a good USFS trail to the South Fork of the Flathead River 6 miles above the Black Bear Guard Station. Little Salmon is only lightly fished but is one of the best Dolly Varden streams in the state. They're hard to catch but easily visible and of good size, ranging up to a reported maximum of 18 pounds. There are some real nice 12 to 15 inch cutthroat here too.

JEWEL BASIN LAKES
Courtesy USFS

NORTH JEWEL
Courtesy USFS

LENA LAKE
Courtesy USFS

LITTLE SALMON RIVER DRAINAGE
Courtesy USFS

FLATHEAD RIVER (South Fork)

Logan (or North Fork Logan) Creek.
Empties into the east side of the Hungry Horse Reservoir near the Betty Creek Guard Station and is good fishing for mostly 10 to 12 inch cutthroat, plus an occasional rainbow, brook and whitefish. It's about 5 miles long in brush and timber, real steep above but followed closely by a logging road.

Lost Johnny Creek.
A brushy, 6 mile long inlet of the Hungry Horse Reservoir, crossed at the mouth by the west side reservoir road 4 miles above the dam and followed upstream by a logging road for about 4 miles. There are some falls about 1½ miles above the mouth and it's fair fishing above them for small resident cutthroat, and below them for good size spawners from the reservoir.

Lost Mare Creek.
A real small tributary of Hungry Horse Creek; crossed about a mile above the mouth by the east side reservoir road 6 miles south of Emery. It is steep above but fair-to-good fishing in the lower reaches for 8 to 12 inch cutthroat.

Lower Twin Creek.
A 9 mile long stream in an old brushy burn. Lower Twin flows to the South Fork of the Flathead River; is crossed by a road at its mouth about 1 mile above the Hungry Horse Reservoir; and is followed upstream for 6 miles by a USFS trail. It is not a pleasant stream to fish but the lower 4 miles are good for 10 to 12 inch cutthroat and resident Dolly Varden; plus some big spawners in the fall, come up from the reservoir.

Margaret Creek.
A small tributary of Hungry Horse Creek, about 6 miles long and followed by a logging road for 4 miles of heavily fished (but good) water for 6 to 8 inch cutthroat.

Margaret Lake.
A truly beautiful alpine lake, 5575 feet above sea level, 46 acres, by 75 feet deep, with talus slopes around the upper end and timber below. Take the west side Hungry Horse Reservoir road for about 25 miles above the dam, then a logging road up Forest Creek (which has little or no fishing potential) and finally 2½ miles of trail to the lake shore. This one is seldom visited but has a fair population of really nice 11 to 16 inch (and some recorded to 6 pounds) cutthroat that are plenty hard to catch.

FLATHEAD RIVER (South Fork)

PILGRIM LAKES
Courtesy USFS

MARGARET LAKE
Courtesy USFS

Marshall Creek. A nice looking, fairly open stream, the outlet of 5 unnamed lakes on Crescent mountain. Marshall is crossed at the mouth by the Youngs Creek trail 1 mile below the Jenny Creek Guard Station, and is followed by a trail for 4 miles of fair cutthroat, and late season Dolly Varden fishing. It is lightly-to-moderately fished by hunters and dudes passing through.

Mid Creek. Rises on the steep, timbered slopes of Silvertip mountain and is followed by a travesty of a trail, that crosses the stream a dozen times and is washed out in every instance, for 7 miles to the South Fork of the Flathead River 6 miles below the Black Bear Guard Station. It is very seldom fished but excellent for 7 to 10 inch cutthroat, and Dolly Varden spawners in the fall.

Middle Fork (Bunker) Creek. Heads on the Swan River-South Fork Flathead divide below Warrior mountain and flows for 6 miles down a steep canyon above and through fairly flat, timbered country below to Bunker Creek. It is seldom fished although easily accessible by a logging road (restricted to the public), and good in the lower reaches for 7 to 9 inch cutthroat trout.

Murray Creek. A real short, brushy stream that empties into the Hungry Horse Reservoir and is crossed at the mouth by the East Side reservoir road 14 miles above the Betty Creek Guard Station. Murray is only lightly fished but the lower two miles are good for 10 to 12 inch cutthroat.

Nanny Creek. Only about 3 miles long, in good deer and elk country; reached at the mouth by the Upper Twin Creek trail, and also by a logging road 3½ miles from the Spotted Bear River road from a point 2 miles up from the ranger station. Although small, the lower reaches are sometimes fished by hunters, for fair catches of camp-eat'n cutthroat.

Necklace Lakes. Six little (1-12 acres) lakes in high, timbered country just east of the crest of the Swan-South Fork of the Flathead drainage divide, and drained by Smoky Creek to the Big Salmon River. They're easily reached by about 5 miles of trail (horizontally, plus a couple of thousand feet vertically) eastward from Holland Lake. They range in depth from 10 to 50 feet, and all are fair-to-good fishing for 7 to 9 inch rainbow, but only those directly on the trail sustain even a light amount of fishing pressure.

North Biglow Lake. A 33 foot deep, 24 acre lake which is not shown on most maps; in heavy timber at 6000 feet above sea level on the head of Biglow Creek. It is most easily reached by a cross-country hike 2½ miles from the end of the Quintonkin Creek logging road. It's seldom if ever fished, but has a small population of 16 to 18 inch cutthroat trout.

North Creek. Small, short, and steep; a tributary of Upper Twin Creek; crossed at the mouth by the Twin Creek trail and seldom if ever fished for the few 8 to 10 inch cutthroat it contains.

North Fork Helen Creek. Is crossed at the mouth by the Helen Creek trail and followed by a trail for 4 miles up its heavily timbered, steep-walled canyon to headwaters. The lower reaches are good for small cutthroat, but are "out of the way" and very seldom fished.

North Fork Logan Creek. See Logan Creek.

Paint Creek. A fishing stream you can step across; it's about 2 miles long, in rolling timbered country, empties into the east side of the Hungry Horse Reservoir 1½ miles north of the Betty Creek Guard Station, is criss-crossed by logging roads, and provides fair fishing for 10 to 12 inch cutthroat — and a very few whitefish.

Pendant Lakes. Three of them: 9.1, 2.3 and 1.3 acres in high, timbered country just a couple of miles over the divide from Upper Holland Lake (in the Swan drainage). The outlet drops over some high falls so that there is no fish migration. However, they were planted years ago and have been poor-to-fair fishing ever since for 12 to 18 inch cutthroat trout. Not many anglers stop by.

Pilgrim Lakes. Two little subalpine lakes in a small cirque just south and below the summit of Three Eagle mountain. They're real hard to get to; cross-country for 1½ miles from the Alpine Trail about 4 miles south from the end of the Noisy Creek (Flathead drainage) road. The upper one is only 2.8 acres, but the lower one (which is 29.4 acres and a whopping 135 feet deep) is red hot for 6 to 13 inch cutthroat trout. You'll find a very few big cutthroat in the upper one too but it's not worth your while.

Posy Creek. A little tributary of Quintonken Creek in a "blow-down" logged-over area; crossed at the mouth by the Quintonken road and followed by road upstream for 1½ miles of fair, pan-sized, cutthroat fishing. This one is seldom bothered.

Prisoner Lake. A 4 acre, subalpine lake just below Shale peak at the northeast end of the Flathead Alps; reached by a steep 1½ mile cross-country scramble west from the end of the South Fork of the White River trail. Prisoner Lake reportedly has a good population of 6 to 8 inch cutthroat although it's so shallow that it looks as though it should freeze out.

Quintonken Creek. A small, fast mountain stream flowing to Sullivan Creek about a mile above the Hungry Horse Reservoir; it's about 12 miles long, in a steep-walled logged-over canyon followed by a logging road for 10 miles of good summertime fishing for small cutthroat and grayling, and fair fall-and-winter fishing for whitefish and Dolly Varden spawners.

Rapid Creek. Heads on the western slopes of the Continental Divide just north of the Trident peaks, and is followed by a good USFS pack trail for 7 miles to Danaher Creek, 4 miles below the guard station. The lower 3 miles between

Sentimental mountain and Ursus hill, are good fishing for 8 to 10 inch cutthroat, and in the fall, for Dolly Varden spawners that sometimes run as large as 10 pounds. It is, however, seldom fished — in favor of Danaher.

Riverside Creek. A small inlet of the Hungry Horse Reservoir; it's pretty steep above, but the lower reaches, which are crossed near the mouth by the East Side road 12 miles north of the Betty Creek Guard Station, are flatter and fair fishing for about 1½ miles. They contain mostly small cutthroat, and Dolly Varden that average between 8 and 10 inches, plus an occasional lunker up from the reservoir (which is also good fishing).

Shelf Lake. A deep lake, about 15 acres; in timbered lowland reached by a (restricted but generally open in the summer) USFS road just above (north) of the Spotted Bear River a mile below Silvertip Cabin. Shelf was planted with cutthroat trout years ago and was good fishing for 17 to 22 inchers, restocked in 1968 but reportedly froze out, and it is barren now.

Silver Tip Creek. A small stream followed by a trail for 8 miles down a heavily timbered, flat-bottomed canyon to the Spotted Bear River, 12 miles above the ranger station. It is not often fished but is good for 1 mile above the mouth (to the falls) for 10 to 12 inch cutthroat. Incidentally, the namesakes of this stream still abound in the near vicinity.

Slide Creek. A very small tributary of Sullivan Creek, in logged over country, paralleled by a (closed) road for a couple of miles of fair 6 to 8 inch cutthroat fishing.

Snow Creek. A short, small tributary of the South Fork of the Flathead River, crossed at the mouth by the west side trail 1 mile above the Black Bear air field, and followed by a trail for 1 mile of fair 7 to 9 inch cutthroat fishing.

South Creek. This small tributary of Upper Twin Creek is crossed about 1½ miles above the mouth by a trail 2½ miles up Bent Creek from the Spotted Bear River, 3 miles above the ranger station. It is fair cutthroat fishing for the first mile, (past an old abandoned beaver dam, "Webb Lake") but is seldom if ever bothered.

South Fork White River. A smallish stream that is excellent fishing for small cutthroat trout, and Dollys in the fall. It is followed by a trail for 5 miles to headwaters and is fished a fair amount by hunters and dudes on their way to and from the White River Pass.

Spotted Bear Lake. Reached by trail 1½ miles east from the ranger station, it is something like 18 acres, mud-bottomed and fairly deep in the middle but marshy and hard to fish from the shore. It was planted in 1973 and is excellent fishing and heavily fished now for 16 to 18 inch cutthroat trout.

Spotted Bear River. A good sized, easily fished, open stream that is followed by a road (that is closed above Beaver Creek) for 8 miles above its mouth on the South Fork of the Flathead River at the Spotted Bear Ranger Station, and a good USFS horse trail for another 22 miles to headwaters. There are no fish above Dean Falls (19 miles upstream) but it is good-to-excellent fishing below for 10 to 20 inch cutthroat, 8 to 10 inch whitefish, and 3 to 15 pound Dolly Varden. This is a good place to go for the dyed-in-the-wool fly fisherman.

Squaw Lake. Two and eight-tenths acres, in barren, rocky alpine country; reached by a trail 3½ miles south from the end of the Noisy Creek (Flathead drainage) road. This Squaw (lake) had fish years ago, but was never much and seldom is visited now. It was planted with cutthroat in '66 and '68, has been good since but may winter kill.

FORDING THE SOUTH FORK NEAR BIG PRAIRIE
Courtesy USFS

Stadium Creek. A very small tributary of Cannon Creek; reached at the mouth by a trail but seldom fished or even visited. The lower mile contains a good population of 7 to 9 inch cutthroat trout.

Sullivan Creek. A nice-sized stream, 15 miles long in a narrow, timbered canyon, and followed for 10 miles by a road to its mouth on the Hungry Horse Reservoir, 8 miles above Ranger View. Sullivan is moderately popular and good summer fishing for small cutthroat, and fall and winter fishing for large Dolly Varden spawners and whitefish.

Sunburst Lake. A beautiful, very deep (over 200 feet), 142.3 acre, subalpine lake; reached by the "all-purpose" Bunker Creek road to the mouth of Gorge Creek (where there is some limited good fishing below the gorge for about a mile) and up the Gorge Creek road for a mile and from here on in a USFS trail for 9 miles. The last couple of miles are restricted to hikers only: or — you can also come in over Inspiration Pass from the end of the Soup Creek road in the Swan River drainage. Sunburst lies below Swan Glacier and is seldom free of ice before July 1 but is excellent fishing when it does open up, for 14 to 20 inch cutthroat trout.

Tango Creek. Three miles long, steep, rocky, and crossed at the mouth by the Big Salmon River trail 2 miles above Barrier Falls. It is followed by a trail for 1½ miles of fair 7 to 9 inch cutthroat fishing but isn't bothered once in a blue moon.

Tent Creek. See Dudley Creek.

Three-Eagle Lakes. Two, at 6175 feet and 5500 feet above sea level on the northern slopes of Three Eagle Mountain reached cross-country a rugged mile over-the-hill north from Pilgrim Lakes. They're both set in talus with scrub timber around the outlets, have maximum depths of 78 and 82 feet, were stocked with cutthroat in the fall of '67 and should be excellent fishing by this year (1975). They grow slowly up there.

Tiger Creek. This small, brushy tributary of Hungry Horse Creek is followed by an old logging road for 1 mile through timbered country upstream as far as the mouth of Turmoil Creek. It is lightly fished for a couple of miles, for good catches of 8 to 12 inch cutthroat.

FLATHEAD RIVER (South Fork)

Tin Creek. A small tributary of the South Fork of the Flathead River; crossed at the mouth by the west side Hungry Horse-Spotted Bear road 1½ miles above the reservoir. There is no trail upstream but the first mile is moderately popular and fair fishing for 6 to 9 inch cutthroat.

Tom Tom Lake. A 33 feet deep, 8 acre pothole at 5700 feet above sea level in heavy timber, reached by a 1 mile cross-country hike south from the upper end of the Wheeler Creek road. It is overstocked with skinny, 7 to 13 inch cutthroat.

Trickle Creek. Heads below and east of Oreamnos peak in the Swan Range and is followed by a trail for 5 miles through steep, timbered mountains to Cannon Creek. This Trickle is almost never fished but the lower couple of miles are good for 7 to 9 inch cutthroat.

Upper Twin Creek. A heavily fished stream, especially during hunting season, that is crossed at the mouth by the South Fork of the Flathead road a mile above the Hungry Horse Reservoir, and is followed by a good USFS horse trail for 14 miles to headwaters. Upper Twin flows beneath a natural rock bridge 4 miles above its mouth (about ½ mile below the mouth of South Creek) and is pretty well log-jammed in the lower reaches. It is good fishing for 8 to 12 inch cutthroat.

Wheeler Creek. Eight miles long, in a steep-walled, timbered canyon and followed by a road for 6 miles to its mouth on the west side and 10 miles below the head of the Hungry Horse Reservoir. Wheeler is lightly fished but good in the summer for small cutthroat, and in the fall and winter for whitefish and good-sized Dolly Varden spawners.

White River. A wide, shallow stream flowing south and then west for 20 miles down its flat-bottomed valley to the South Fork of the Flathead River 1 mile below the Holbrook Guard Station. The White River is a principal thoroughfare for those going and coming from the Sun River country, and the stream is heavily fished from its mouth upstream for 8 miles to Needle Falls for 10 to 16 inch cutthroat, and 3 to 15 pound bull trout. Long stretches of this stream above the falls (which are quite high and act as a fish barrier) go dry in late summer, and the seldom fished holes that remain are jammed with 7 to 10 inch cutthroat like sardines in a can.

WOODWARD LAKE
Courtesy Jack Dollan

RAINBOW
Courtesy Jack Dollan

SPOTTED BEAR RIVER AND LAKE
Courtesy USFS

SPOTTED BEAR RIVER
Courtesy USFS

Wildcat Creek.
This very small tributary of Wounded Buck Creek is reached at the mouth by the Wounded Buck road and followed upstream by a logging road for about 1½ miles beyond a series of big falls. There are no fish above the falls but it's fair fishing below for pan-size cutthroat.

Wildcat Lake.
A real deep, steep dropoff, 38.6 acre alpine lake that is surrounded by talus and rocky cliffs and is reached by a good USFS trail 3 miles north from the end of the Noisy Creek (in the Flathead drainage) road. Wildcat was planted via packhorse many years ago and is now good fishing for cutthroat that will average 16 inches, and range up to 16 pounds. This is a good lake to hit.

Woodward Lake.
A fairly deep, 65.2 acre lake in high timbered country just east of the Swan-South Fork divide, drained by Cataract Creek and reached cross-country 3 miles up Smokey Creek from Necklace Lakes. Woodward is fished quite a bit for a "way back" lake and is fair-to-good for 10 to 12 inch rainbow trout (a very few recorded to several pounds) that are apparently the result of two plantings sometime in the 1930's, and rainbow-cutthroat hybrids.

Wounded Buck Creek.
A brushy little, low gradient stream in a heavily timbered, flat-bottomed valley; crossed ½ mile above the mouth by the Hungry Horse road 7 miles above the dam, and followed by a logging road for 4 miles upstream (to the mouth of Wildcat Creek). The middle reaches are good fishing for small cutthroat and the lower reaches are one of the best spawning areas in the entire Hungry Horse drainage.

Youngs Creek.
A nice size stream, about 15 miles of fishing in a narrow, brushy canyon followed by a trail to the South Fork of the Flathead 6 miles above the Big Prairie Ranger Station. It's good fishing for 3 to 4 pound Dolly Varden, and 1 to 3 pound cutthroat in beaver ponds along the middle reaches near the Hahn Guard Station.

GLACIER NATIONAL PARK

MANY GLACIER HOTEL
Courtesy Danny On

Glacier National Park

Glacier National Park is an area of massive mountain ranges magnificently fashioned by water and glaciers, with about 50 glistening glaciers and 200 sparkling lakes. Its 1600 square miles of picturesque landscape is enlivened by a profusion of wildflowers and abundant wildlife.

Each year the Park finds 1½ million visitors arriving mostly by car, camper or motorized home, from which they view the vistas from scenic roads. A night or two is spent at a hotel or lodge, or at one of the developed campgrounds, and boats are rented, horses and guides hired, and trails hiked to explore further the natural delights. Campfire programs and guided hikes offered by Park Naturalists are informative and popular.

Magnificent trails (700 miles of them) lead the fisherman and backpacker into the backcountry, to use the 700 remote campsites. Backcountry visits are increasing each year at a rate greater than the general park visit increase.

Glacier Park has changed little over the years and while the scenes are essentially the same as 200 years ago, subtle changes have taken place. Natural processes like gradually retreating glaciers, and evolving forests marked by fires, continually change the landscape. Changing, too are management policies, such as those toward predators, once hunted and killed to protect the "good" animals; toward bears, once passively allowed to feed in garbage dumps; toward deer and other animals once fed supplemental hay in the 1930's. Certain fires might soon be allowed to burn themselves out; research shows lightening caused fires can be a natural and necessary part of forest life.

Changing policies toward fish date from the late 1960's. For many decades Glacier National Park stocked fish, even exotic species, in many Park lakes and streams. Today the practice is discontinued. Park fishing regulations are designed to protect the native species: cutthroat, lake trout, Dolly Varden and arctic grayling. Only three of these species may be caught per day, though the limit is five if non-native species are included, such as the rainbow and kokanee. A "fishing for fun" area has been established along lower McDonald Creek, where fish caught must be released immediately to the stream. Fish in the Park are now considered not so much for sport as for their place in the natural ecosystem.

GLACIER NATIONAL PARK

GOING TO THE SUN MOUNTAIN
Courtesy Ernst Peterson

KINTLA LAKE
Courtesy Ernst Peterson

Ecosystem is the key word, and today national parks such as Glacier are managed from this viewpoint. The term describes natural units of landscape where plants, animals and the environment are interacting. Ecosystems come in many types and sizes, an example of which is the aquatic ecosystem where water, plants, insects, fish, mink, bear, otter, osprey and eagles all play a part.

Glacier National Park is a beautiful place to visit and enjoy, and to see the natural processes continue to develop and shape the scenery and maintain the plant and animal community. Here it's easy to become a little bit of botanist, geologist, zoologist, meteorologist, ecologist and philosopher.

EDWIN L. ROTHFUSS
Chief Park Naturalist

ST. MARY LAKE
Courtesy Ernst Peterson

Glacier Park Lakes

Name	Rating	Accessibility	Species	Best Season	Boat Access
Arrow	F	7.5 miles by trail from North Fork Road	cutthroat & Dolly Varden	June & July	no
Avalanche	F	2 miles by trail from Avalanche Campground	cutthroat	June & July	no
Bowman	F	by secondary road	cutthroat, Dolly Varden & whitefish	early & fall	yes
Bullhead	G	4 miles by trail from Many Glacier	eastern brook	June & July	no
Camas	F	10.5 miles by trail from North Fork Road	cutthroat	early & fall	no
Crossley	G	9 miles by trail from Chief Mtn. Customs	whitefish & mackinaw	June & July	no
Elizabeth	G	13 miles by trail from Many Glacier	grayling, rainbow — eastern brook	June & July	no
Ellen Wilson	G	10 miles by trail from Lake McDonald Lodge	cutthroat & eastern brook	July	no
Fishercap	F	5 miles by trail from Swiftcurrent Campground	brook trout — rainbow	June & July	no
Francis	G	6 miles by trail from head of Waterton	rainbow	June & July	no
Glenns	F	11 miles by trail from Chief Mtn. Customs	cutthroat, whitefish & mackinaw	June & July	no
Grinnell	F	4 miles by trail from Many Glacier	rainbow & eastern brook	July	no
Gunsight	F	6 miles by trail from Sun Road	cutthroat & rainbow	June & July	no
Harrison	G	3.1 miles by trail from No. 2 Highway	cutthroat, Dolly Varden & whitefish	June & July	no

GLACIER NATIONAL PARK

Name	Rating	Accessibility	Species	Best Season	Boat Access
Hidden	F	2 miles by trail from Logan Pass	cutthroat	July	no
Howe	G	4.7 miles by trail from North Fork Road	cutthroat	Early part of season	no
Isabel	G	16.5 miles from Walton Ranger Station by trail	cutthroat & Dolly Varden	June & July	no
Josephine	F	1.5 miles by trail from Many Glacier	rainbow, eastern brook & kokanee	June & July	no
Kintla	G	by secondary road	cutthroat, Dolly Varden, whitefish & kokanee	early & fall	yes
Kootenai	G	2½ miles by trail from head of Waterton	brook trout	June-August	no
McDonald	F	by paved road	cutthroat, rainbow, Dolly Varden, kokanee, lake trout & whitefish	early & fall	yes
Lincoln	G	8.5 miles by trail from Sun Road	cutthroat & Dolly Varden	June & July	no
Logging	G	4.5 miles by trail from North Fork Road	cutthroat & Dolly Varden	fall	no
Lower Quartz	G	4 miles by trail from Bowman Lake	cutthroat & whitefish	early & fall	no
Medicine Grizzly	G	5 miles by secondary road — 5 miles by trail	cutthroat & rainbow	July	no
Middle Quartz	G	6 miles by trail from Bowman Lake	cutthroat, Dolly Varden	early & fall	no
Moxowanis	G	16 miles by trail from Chief Mtn. Customs	brook trout	June & July	no
Old Man Lake	G	8 miles by trail from Two Medicine Campground	cutthroat	June & July	no
Poia	F	5 miles by trail from Sherburne Reservoir	grayling	June & July	no
Ptarmigan	F	3 miles by trail from Many Glacier	cutthroat & eastern brook	July & August	no
Quartz	E	6 miles by trail from Bowman Lake	cutthroat & Dolly Varden	early & fall	no

ON LOGAN PASS
Courtesy Ernst Peterson

GRINNELL GLACIER
Courtesy Ernst Peterson

GLACIER NATIONAL PARK

Name	Rating	Accessibility	Species	Best Season	Boat Access
Red Eagle	G	4 miles by trail from end road	cutthroat, Dolly Varden & rainbow	June & July	no
Red Rock	F	3 miles by trail from Many Glacier	rainbow & eastern brook	June & July	no
Rogers	F	4.5 miles by trail from North Fork Road	cutthroat & whitefish	early part of season	no
St. Mary	F	by paved road	cutthroat, mackinaw, rainbow, whitefish & Dolly Varden	early & fall	yes
Sherburne	P	by paved road	northern pike	August when lake level drops	yes
Slide	G	5 miles by trail from Chief Mtn. Highway	cutthroat & Dolly Varden	July & August	no
Snyder	F	4 miles by trail from Lake McDonald Lodge	cutthroat	early part of season	no
Swiftcurrent	F	by paved road	rainbow, eastern brook & kokanee	June & July	yes
Trout	G	4.5 miles by trail from North Fork Road	cutthroat	June & July	no
Two Medicine	G	by paved road	cutthroat, eastern brook & rainbow	June & July	yes
Upper Kintla	G	8 miles by trail	Dolly Varden	early & fall	no
Upper Two Medicine	G	5.5 miles by trail from Two Medicine Campground	eastern brook & rainbow	June & July	no
Waterton	F	by paved road	cutthroat, rainbow, eastern brook, mackinaw & whitefish	July	yes
Windmaker	F	5 miles by trail from Many Glacier	rainbow & eastern brook	June & July	no

Glacier Park Streams

Name	Rating	Accessibility	Species	Best Season	
Belly River	G	by trail from Chief Mountain Road	rainbow & grayling	early season	—
Boulder Creek	F	by trail 4 miles from Sherburne Dam	cutthroat	June & July	—
Cut Bank Creek	G	4 miles by secondary road from U.S. 89	eastern brook & rainbow	early season	—
Kennedy Creek	F	by trail 4 miles from Chief Mtn. Highway	rainbow	June & July	—
Middle Fork Flathead	G	by paved road U.S. No. 2	cutthroat, Dolly Varden & whitefish	during runs in August & fall	—
North Fork Flathead	G	by secondary road	cutthroat, Dolly Varden, whitefish & kokanee	during runs in August & fall	—
Upper McDonald Creek	P	by paved road	cutthroat	early season	—

Worms: Prohibited to dig for in the Park. Use of minnows or bait fish is prohibited.

Motorboats limited to lakes accessible by an approved road. Boat permit required.

Generally worms are better when water is high. Flies better later in season.

Success of fishermen depends on skill, boat or lure used, weather conditions, time or year and body of water fished. Generally lakes better fishing than streams. Fishing in low elevation waters is best early and late in the season. Fishing in high mountain lakes is best in July and August.

Courtesy USFS

The Kootenai "Dammed Forever"

The Kootenai River rises in Canada on the westslope of the Rockies and flows south and west through northwestern Montana and northern Idaho before turning north again and flowing back into Canada to Kootenay Lakes. The Kootenai is almost 485 miles long and is the second largest stream in Montana, with most of its water coming from the high, rugged, glaciated mountains of Canada. Prior to the completion of Libby Dam, this emerald green river and it's tributary streams provided some of the best fishing for wild cutthroat, rainbow trout, and mountain whitefish in Montana. A float trip through the narrow valley, surrounded by rugged mountains with abundant game populations, was an exciting and memorable experience. The impounding of the Kootenai River in 1972 by Libby Dam, located about 15 miles northeast of Libby, has changed 90 miles of this once beautiful and wild river into a fluctuating reservoir with an average annual drawdown of about 120 feet. Construction of the Re-Regulating Dam in the late 1970's will impound 10 miles more of the river. The fishing in the new reservoir has been good for the first two years, with 12-18 inch cutthroat and rainbow providing most of the action. However, large plants of hatchery cutthroat will be necessary in future years to maintain a satisfactory fishery in the fluctuating reservoir environment, which is most suitable for rough fish such as suckers and squawfish.

The only free-flowing part of the Kootenai River left in Montana is the 38 mile section from the re-regulatory dam to the Idaho Border. This area provides good fishing for rainbow and cutthroat trout in the 12-18 inch bracket, with rainbow over 3 pounds not uncommon. A trophy fishery for large rainbow in the 5-10 pound class is found in the spring near the mouths of Callahan Creek and the Yaak River. The only population of white sturgeon in Montana is found in the river below Kootenai Falls. The largest individual taken weighed almost 100 pounds and fish over 30 pounds are common.

Access to the river is generally good and many people fish from the bank, but a float trip is the most enjoyable way to fish this large river. Whitetail deer and Rocky Mountain Big Horn Sheep are often observed by fishermen floating the river. Floaters are cautioned to bypass Kootenai Falls which is certain death to anyone foolish enough to go over it in a raft or boat. Anglers should also be cautious about sudden increases or decreases in river flow as a result of the operation of Libby Dam Powerhouse.

The most productive flies are the Renegrade, Black gnat, Royal and Grey Wolf, Green Wooley Worm, Small Yellow May (size 18-20), Royal Coachman, Weighted Bitch Creek and the Muddler Minnow. For lure fishing, the Thomas Cycle, Meps Spinner and Red and White Daredevil are hard to beat.

Nearly all of the tributaries provide good fishing for pan-sized rainbow, cutthroat and brook trout with some of the larger streams consistently producing fish of the 10-16 inch bracket.

Numerous small and medium-sized lakes in the drainage provide excellent fishing for rainbow, westslope cutthroat and brook trout. Yellow perch and largemouth bass are also found in a few lakes.

Visitors to the area should bring their hiking shoes and plan to walk into some of the high mountain lakes in the Cabinet Mountain Wilderness Area. Good trails lead to these beautiful high, cold lakes which are in basins formed by glacial activity. Most lakes contain good populations of rainbow and cutthroat trout in the 8-12 inch bracket. The knowledgeable angler brings his skillet and a little butter and cooks the freshly caught mountain trout on the spot, a real gourmet treat. A size 18-20 Black gnat and or Renegrade will insure in most instances that the ingredient of the cookout, trout, is not missing.

BRUCE MAY
Fisheries Biologist
Montana State Fish & Game Department

P.S. If you want a stagnant water hole replete with miles and miles of gooey mud banks and winds that'll blow you off the reservoir if they don't drown you first, this is it folks, the Koocanusa Special. Think of it, all of this was prepared just for you, for the paltry sum of only $466,000,000.00 of your money including a $9,000,000.00 pittance for the Corps' bridge to glory across the north end of the mud hole and $67,500,000.00 more to build the double lane paved highways that encircle the whole d—thing "to keep the area as natural as possible." All of this conceived and constructed just for you by that great environmentally oriented agency, the United States Army Corps of Engineers. THEY CARE.

R.L.K.

Kootenai Drainage

Alkali Lake. About 107 acres by 35 feet deep! A real "homely" lake with lots of drowned timber all around; in rolling timbered country reached by a poor road 3 miles south from Eureka. Alkali was planted with rainbow in 1962, but they died out. So...it was replanted with largemouth bass in 1967 but — they didn't make it. The whole thing is barren now, too alkaline.

Alvord Lake. Fifty six acres, about 40 feet deep in the middle, with shallow dropoffs and a mucky bottom, lots of sunken logs, lilies, and rushes. Alvord lies in a heavily timbered pocket in the hills just north of Troy and is accessible by a poor road 2 miles from town. It used to have trout, but now there are only good size, 6 to 11 inch, yellow perch; 2 to 3 pound largemouth bass (at least one that went to 6.8 pounds), and small sunfish.

American Creek. A small stream that heads on Canuck peak and flows north for about 5 miles through a narrow, timbered canyon to Canada, just east of the Idaho line. It is accessible by logging roads from Idaho a "hundred miles from nowhere" and is lightly fished but excellent for small rainbow trout.

Ant Creek. See Brimstone Creek.

Arbo Lake. See Wee Lake.

Baree (or Standard) Lake. Deep, 8½ acres, in a rocky cirque accessible by trail up Baree Creek 2½ miles from the Silver Butte-Fisher River road; seldom visited and poor fishing indeed for 6 to 12 inch cutthroat. It's scheduled to be planted in '75 or so the local prognosticators say.

Barron Creek. Heads between Banfield and Blue mountains and flows east for 7 miles down a brushy-bottomed, logged over canyon to the reservoir (Koocanusa) 8 miles north by paved highway from the dam, and is followed to headwaters by a logging road. The middle reaches (for about 3 miles) are fair fishing for small brook trout, and a few cutthroat in some meadows above. If the reservoir isn't drawn to heck-and-gone down, ther should be some good size whitefish too, come up from the big water.

Basin Creek. A small, headwaters tributary of the East Fork of the Yaak River; the lower 2 miles are accessible by a logging road, the next 3 by a USFS trail. Basin Creek is lightly fished and fair-to-good for pan-size cutthroat and brook trout.

Bear Creek. The middle reaches (2 miles) of this small, beaver-dammed tributary of Libby Creek are accessible by the Cherry Creek road, the upper reaches (2½ miles) by USFS trail. The lower reaches are quite steep and seldom fished; the middle and upper reaches flatten out and are heavily fished for 8 to 14 inch cutthroat and Dolly Varden.

Bear Lakes. Three, in a rocky cirque, with some brush around the shore; accessible by the Iron Meadow Creek trail 3½ miles from the Silver-Butte-Fisher River road. Upper Bear is barren. Middle Bear has had fish in the past but is barren now. Lower Bear (10.6 acres and deep) is fair fishing for cutthroat trout that will *average* 17 inches. Not too many of them — but nice.

Beetle Creek. A very small tributary of Pete Creek, crossed at the mouth by the Pete Creek road and followed upstream for 1½ miles by a logging road. There are a few small cutthroat and brook trout right at the mouth.

Betts Lake. Five acres, shallow but with steep dropoffs from the bank; no trail, access is across posted private land ½ mile west of the Wolf Creek road and 1½ miles south of the Fairview Guard Station. Betts is good fishing for brook trout that average 8 to 10 inches with maybe a few to 14 inches.

Big Creek. Flows for 8 miles eastward from the junction of its North and South forks at an old USFS cabin, through a narrow timbered canyon to Koocanusa 7 miles below the Big bridge. The lower 3 miles flow through a steep gorge, the rest of the stream is accessible by a good gravel road and has been moderately popular and very good fishing for 4 to 10 inch rainbow.

Big Cherry Creek. A fair sized tributary of Libby Creek; accessible at the mouth by a paved road 2 miles south of Libby and followed upstream by a good gravel road for 16 miles to headwaters in steep, timbered country. The lower reaches have been mined (the bends straightened out) and the best fishing is in the middle and upper reaches, which are real good for 9 to 10 inch rainbow, cutthroat and some brook trout.

Big Hawkins Lake. See Hawkins Lakes.

Blue (or Green) Lake. Misnamed Green Lake on some maps; about 12 acres. The entire south side lies along the edge of a big rock slide and is real deep (up to 70 feet or better) straight off from shore; in rolling, timbered country accessible by poor road 1 mile south from Stryker. It's excellent fishing for 8 to 10 inch brook trout and zillions of shiners.

Blue Sky Creek. A small tributary of Grave Creek; empties in about 1 mile above the Clarence Guard Station and is accessible for 5 miles by a logging road. Blue Sky is seldom fished but has been excellent for pan-size cutthroat plus an occasional one up to a pound. Here is the kicker: it is now being managed as a migratory Cutthroat Creek and, assuming that the Corp's pride (Kookanoosa) is not drawn down too far, there should be some real nice spawners come up via Grave Creek, say along about 1975 or '76 around June or July, and later on in the fall, say in August or September, some nice 10 to 18 inch Dollys plus maybe a few up to 10 pounds.

Bobtail Creek. About 9 miles long, the upper 3 miles in a steep timbered canyon, the lower reaches in farmland. Bobtail empties to the Kootenai River 3½ miles west of Libby and is accessible for its full length by road. It is a narrow, meandering stream with some beaver ponds near the mouth, and good fishing for small rainbow and brook trout, especially in the meadows about 3 miles above the mouth.

Bootjack Lake. A private, 12 acre pond about 70 feet deep, right beside U.S. 2 in rolling timbered country at the south end of Pleasant Valley. Much of this lake is shoal with lots of logs and much brush around the margins. It froze out and was replanted with cutthroat in 1961 and is presently good fishing for 10 to 14 inchers.

Border Lake. Here is the northwesternmost lake in Montana — just east of the Idaho Line and south of Canada. It's available by trail 2 miles north of Hawkins Lake. Border is 11.9 acres, more than 40 feet deep in some areas, and although seldom fished is good fishing for a natural population of 8 to 11 inch brook trout.

Boulder Creek. The outlet of Boulder Lakes, followed to the east by a logging road for 8 miles to Koocanusa ½ mile south of the Big Bridge. It's steep and seldom fished in the lower canyon reaches but the upper reaches are available by logging roads and the entire stream is fair fishing for 6 to 12 inch cutthroat. There are a few Dolly Varden below the Boulder Falls about a mile above the mouth.

Boulder Lakes. Big (about 20 acres) and Little (4 acres) at the head of Boulder Creek, accessible by a good USFS road 13½ miles in from the west end of the Big Koocanusa Bridge 8 miles south of the new site of Rexford, to within 1½ miles of Little Boulder, and then a steep hard climb way up for another ½ mile to Big Boulder. The smaller lake is in timber all around, fairly shallow with lots of submerged logs around the edges, and good fishing for 6 to 12 inch cutthroat. The larger lake lies below rocky cliffs and talus except on the timbered north side. It was planted back in the '50s but froze out.

Bramlet Lakes. See Lower Bramlet Lake.

Brimstone (or Ant) Creek. A small tributary of Fortine Creek; accessible at the mouth by road 1 mile south of Fortine, and followed upstream by road for 4 miles through meadow and scattered timber. About 2 miles are easily fished, and fair-to-good for 7 to 10 inch brook trout, plus an occasional cutthroat at a ratio of about 9 to 1. Now hear this: a fish barrier at the lower end has recently been removed, so in a year or two there should also be some nice Dolly Varden and spawners come up from below in the fall.

KOOTENAI FALLS
Courtesy USFS

THE WAY IT USED TO BE!
Courtesy Lance Schelvan

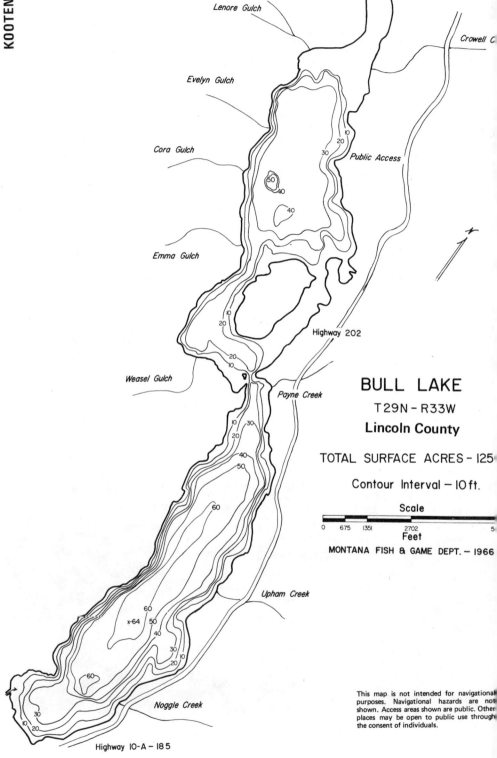

CEDAR LAKES BELOW DOME MOUNTAIN
Courtesy USFS

UPPER CEDAR LAKE
Courtesy Miller's Studio

Bristow Creek. A small stream that heads below Lost Soul Mountain and is followed by a logging road eastward down its partly logged, brushy valley to the Koocanusa reservoir 10½ miles by the west side (paved) highway north from the dam. There are about 6 miles of fair water here (in a limited sort of way) for pan-size cutthroat and brook trout.

Bull Lake. A big (5 miles long by ½ mile wide by 64 feet maximum depth) lake in a timbered valley; followed along the east shore by a paved road 15 miles south from Troy. The west shore is quite steep, the east shore a gentle slope on which are about two dozen or so cottages, a resort and a couple of USFS campgrounds. It is only poor-to-fair summer-time fishing for 10 inch to 10 pound Dolly Varden, and fair for 12 to 14 inch Kokanee. It's fished somewhat in the winter, too, through the ice for small catches in rainbow and cutthroat. The lower end is good fishing for 1 to 2 pound largemouth bass along with lots of trash fish.

Burke Lake. About 40 acres, maybe 40 feet deep, in timbered country; reached cross-country a mile north from the Hawkins Creek road a mile above Hawkins Lake. It used to be barren but is now excellent fishing for rainbow that will run upward to 3 pounds. Its over populated and seldom fished probably because no one can find it.

Burnt Creek. A small stream flowing for 8 miles through a narrow timbered canyon to the Yaak River near the Sylvanite Ranger Station; accessible by road for 6 miles of fishing. It is heavily fished near the mouth for good catches of 8 to 12 inch rainbow, cutthroat and brook trout, and fair above for 6 to 8 inchers.

Cable Creek. A real small tributary of Bear Creek in the old 1910 burn; accessible by a jeep road for 3 miles from mouth-to-headwaters in a narrow, steep-walled canyon. Cable is only occasionally fished but good for 6 to 8 inch rainbow.

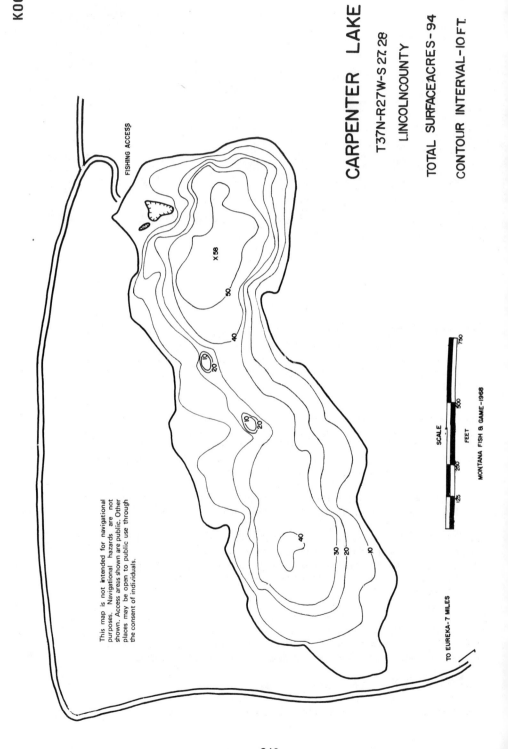

Cad Lake. Pretty small, only 3.7 acres by 30 feet deep in one place; in woods reached by a road 1½ mile east of Horseshoe Lake. It was stocked with cutthroat in 1972 and is now good fishing for 6 to 12 inches. If you're planning on using a boat, you'll have a 150 yard drag. Hey, it's on the south side of the road just east of Cibic Lake, so watch out, 'tis easy to get mixed up.

Callahan Creek. A fair-sized stream flowing to the Kootenai River at Troy. The lower reaches are in a real rough canyon (you have to literally swim through it) but it is crossed by a road a couple of times above. It's fair fishing for 8 to 10 inch rainbow, and a few nice Dolly Varden and cutthroat in the canyon.

Caribou Creek. A real small "headwaters" tributary of the East Fork Yaak River in a steep timbered canyon; accessible at the mouth by the East Fork road and followed upstream for ½ mile of poor fishing for pan-size cutthroat. There's a campground at the mouth.

Carpenter Lake. See Tetrault Lake.

Cedar Creek. Drains Cedar Lakes for 9 miles through a steep canyon to the Kootenai River 4½ miles west of Libby; the lower 2 miles are accessible by road, then take a good USFS trail on in to the lakes. This stream is seldom fished but is good for camp-fare trout.

Cedar Lakes. Upper and Lower, in an alpine cirque right below Dome Mountain in the Cabinet Range; reached by the Cedar Creek trail about 12 miles from Libby. Upper Cedar is about 54 acres by 300 feet deep; Lower Cedar about 21 acres and deep enough to keep from freezing out. Both are good fishing for 8 to 12 inch cutthroat trout.

Cibid Lake. Now, you hit this one just a few hundred yards before you come to Cad Lake, just a short bit east of Horshoe Lake. It is 11 acres, 69 feet deep and surrounded by trees. Cibid was planted with rainbow in 1973 and is pretty good fishing now for 8 to 10 inch fish, plus a fair number of largemouth bass — up to 5 pounds.

Clarence Creek. A tiny (headwater) tributary of Grave Creek; accessible at the mouth by the Grave Creek road at the Clarence Ranger Station and followed upstream through a narrow, timbered canyon for 2 miles of fair fishing. Cutthroat and a few Dolly Varden, 6 to 8 inches. Used to be, that is! It was restocked with migratory cutthroat in 1973 and by 1975 it should be better than ever for 4 to 7 inchers and by 1976 you should be getting 10 to 18 inch spawners in the spring and Dollys in the fall — up to 10 pounds or so. Always assuming that the Corps will abide by its promises and leave some water in the reservoir. Now *that* is some assumption!

Clay Creek. This little tributary of the South Fork of the Yaak River is crossed at the mouth by the South Fork road and followed to headwaters by a logging road. The lower mile is in ranchland, the upper reaches (4 miles) in a timbered canyon. The lower 2½ miles are poor fishing for 6 to 8 inch brook trout.

Cody Lakes. Three little lakes in timbered country at the head of (barren) Cody Creek; accessible by driving 8 miles up the Cody Creek road which takes off to the east from the Fisher River road at the mouth of Alder Creek, 5 miles upstream from the Kootenai River. Upper Cody is barren; Middle Cody is about 10 acres; Lower Cody is about 8 acres; and both are deep and used to be excellent fishing but froze badly and now are poor at best. The Middle lake contains a small population of rainbow that average 16 inches and range up to 20; Lower Cody contains brook trout

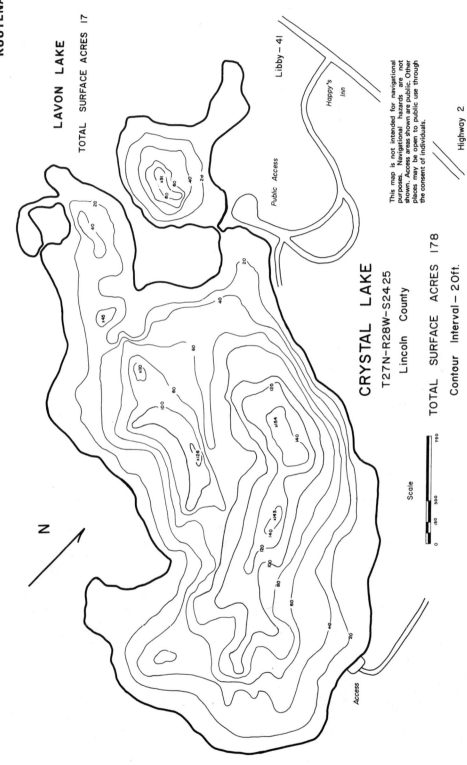

of about the same size. Neither are too well known and as a consequence of all this, are but seldom fished.

Contact Creek. This tiny outlet of Wishbone Lake flows to Granite Creek and is accessible at the mouth by the Granite Creek trail. It is only about 2 miles long, in real rough, brushy subalpine country and is almost never fished, although it reportedly has small trout.

Copper Creek. A very small tributary of Keeler Creek, accessible at the mouth by the Keeler Creek road 8 miles from Troy and crossed here and there by the Lake Creek and Iron Creek roads. The lower mile is poor fishing for 4 to 7 inch brook and cutthroat trout, but is very seldom bothered.

Costich Lake. A fairly shallow, 60 acre reservoir in open meadow approached to within ¼ of a mile by a county road a couple of miles east from Eureka. It is lightly fished by a few local folks for largemouth bass that average around 1 to 1¼ pounds, and a few good size rainbow and brook trout. It partly winterkills and summerkills too. In fact, it did better when it was a bass lake...period.

Cripple Horse Creek. Twelve miles long, flowing to Koocanusa 7 miles north of the dam where it's crossed by the "paved" double-lane east side scenic(?) highway, and is accessible by road to headwaters. The lower reaches go dry in summer but the middle 6 miles have provided some poor, tough fishing for 5 to 7 inch brook trout. It has not been the most popular creek in the district. However all of that is going to change now, courtesy of the Corps. The State Fish and Game department is going to try to manage it with a migratory population of cutthroat on the premise of minimal reservoir fluctuations (would you believe less than 200 feet?) and if all works out as hoped for, they should provide some good spring fishing for nice-size spawners, and fall fishing for 10 to 18 inch whitefish. And if the Arabs have a sudden change of heart and start *giving* us their oil, we'll all have gas to drive up and fish it.

Crystal Lake. A real deep lake (143 feet), 176 acres in size, in fairly flat country surrounded by timbered mountains right behind Happy's Inn on U.S. 2 about 40 miles southeast from Libby at the northwest end of Thompson Lakes. Crystal was poisoned out in 1960, planted with rainbow and kokanee in 1964, and is good fishing now for these (average size about 15 inches), along with LOTS of Pumpkinseeds and yellow perch.

Crystal Lake. A private, commercial fishing pond in open pastureland on the north side of U.S. 93 about ½ mile north of Fortine. It's excellent fishing, after you pay your money, for 16 to 18 inch cutthroat and rainbow trout. There is also a 9 hole golf course here, and a private landing field.

Curley Creek. Drains Perkins Lake for 5 miles through Idaho, then 1½ miles in Montana, and lastly for another 3 to the Kootenai River in Idaho again. It is reached by road 4½ miles north of Leonia, and contains mostly small (but now and then a nice one that will go to several pounds) rainbow trout.

Cyclone Creek. A really small stream, 3 miles of fishing, empties into the Yaak River 2½ miles above the Sylvanite Ranger Station and is crossed at the mouth by the Yaak River road and followed upstream by USFS trail. It is moderately popular for such a small stream, and fair fishing for 6 to 8 inch rainbow, cutthroat and brook trout.

KOOTENAI

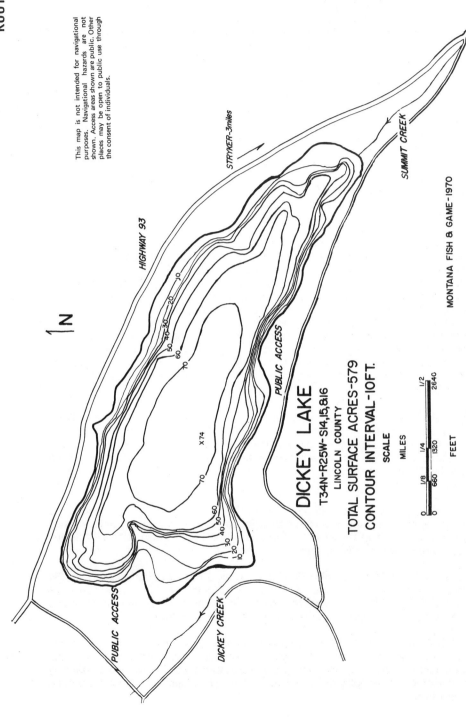

Dahl Lake. About 165 acres, like about 10 feet or less deep, an ageing lake with lots of marsh that floods in the spring but is dry by fall. It has lots of pumpkinseeds and perhaps a few largemouth bass. Is mostly on private land — seldom fished. In open prairie, sagebrush flats accessible by gravel road from the Raven Ranger Station. You can skip this one because it's not even good for ducks — it freezes up early. However, the surrounding country IS pretty good for deer and elk.

Dahl Lake Fork of Pleasant Valley Creek. A small stream about 3½ miles long above Dahl Lake, and 3½ miles below to its juncture with Pleasant Valley Creek. It flows through mostly posted meadowland; is accessible by road below the lake and a trail above, and is easily, although not often, fished for 6 to 10 inch brook trout, plus now and then a good one.

Davis Creek. A wee-small headwaters tributary of Fortine Creek; reached at the mouth near the Twin Meadows Guard Station by the Fortine road, and followed upstream by a logging road for ½ mile of tough fishing for small brook trout.

Deep Creek. The lower 5 miles of this small stream flow through rolling, timbered country; are reached at the mouth by the Fortine road 15 miles south from Eureka and are followed by a jeep trail, through mostly posted land. They are fair fishing for small cutthroat and brook trout, plus a few Dollys and rainbow near the mouth.

Deep Lake. A shallow, 6½ acre lake that has marshy shores but gets deep out a-ways (to better than 50 feet); about ½ mile south of Marl Lake in wooded hills reached by a road 5 miles west and south from Fortine. Deep Lake is private, NO fishing for the public, but fair for the owners of 8 to 11 inch brook and rainbow trout plus some whitefish and lots of suckers.

Dickey Creek. Small; flows for 2 miles from Dickey Lake through rolling, timbered hills to Murphy Lake, accessible by a truck trail and easily fished for 8 to 10 inch brook and rainbow trout.

Dickey Lake. Real deep, 579 acres, in timbered country with U.S. 93 running along the south side and a county road on around. Dickey is only poor-to-fair fishing for mostly a pound or so rainbow, plus 6 to 8 inch kokanee salmon, brook trout, cutthroat and, if you go deep, a very few lake trout. There are about a dozen private cabins here and a couple of campgrounds.

Doble's Ponds. Drive ½ mile north from Alkali Lake by jeep trail and then turn left (west) on a private ranch road for a mile along the north side of the Young Creek Meadows to 5 little private ponds in semi-open timberland above the head of the Meadows. From the bottom up they are about ¾, 4, 1, 1, and 3 acres, are quite shallow with much aquatic vegetation around the margins, and fair fishing for 6 to 12 inch brookies.

Dodge Creek. Eight miles long from the junction of its North and South Forks; the upper reaches in a canyon, the lower reaches beaver-dammed in rolling timbered country and logged over land: Dodge empties into Koocanusa 8 miles north of the Big Bridge and is accessible by a double lane road for 4 miles of tough fishing. It's not very popular but fair fishing for 4 to 8 inch brook and a few cutthroat trout. Shadow Falls are located here, 2 miles above the mouth and there are more falls about 6 miles upstream.

Double Lake. Two alpine lakes, North and South, about 50 feet apart and connected by a 10 foot channel. There is a little island in the North Lake; the South

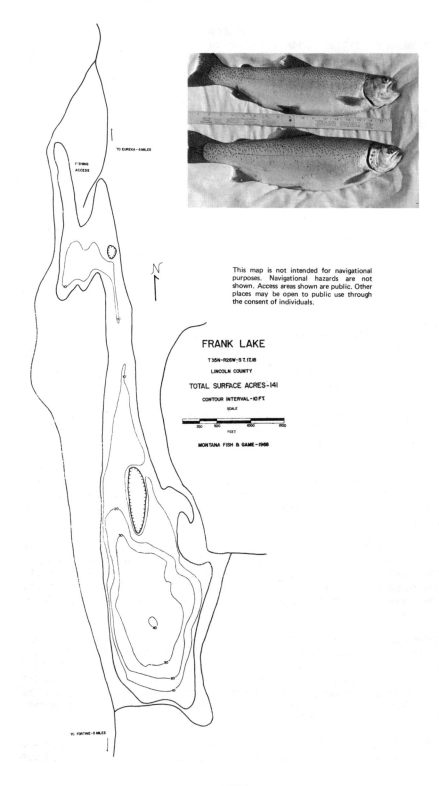

PARADISE LAKE IN BLUEBIRD BASIN
Courtesy USFS

LOWER BEAR LAKE
Courtesy Lance Schelvan

Lake is filling up with talus; both together total about 12 acres and are bordered by brush and scrub timber. To reach them take the Granite Creek trail 3 miles from the end of the road, and then fight your way for 2 miles through a jungle of "Devil's Claws" up Contact Creek to the shore. Once there you'll be all alone and will find the lakes to be excellent fishing for skinny 6 to 10 inch, overpopulated brook trout.

Double N Lake. Is a private, 100 acre 15 to 30 feet deep, long and narrow half-moon-shaped flooded meadow with aspen on the north and evergreens on around; reached a half mile west off U.S. 2 about 10 miles south from Libby. The Double N is really a "family playground" but permission to fish is usually granted, and it's truly fabulous for mostly 14 to 16 inch rainbow, plus an occasional cutthroat that will average around 10 inches but sometimes run up to as much as 5 pounds.

Dudley Creek. A tiny tributary of Edna Creek; reached at the mouth and for a mile upstream by road; seldom fished but fair for small brook trout in beaver ponds here and there along the way.

Dudley Slough. About 5 acres, in a private meadow in rolling timberland; reached by a road 2½ miles west of Trego, and another 1½ miles south from the Ant Flat Ranger Station. Dudley is good fishing for 8 to 12 inch brook trout.

Dunn Creek. A small westward flowing tributary of the Kootenai River which it joins ½ mile below the dam. Dunn Creek flows for 14 miles down a narrow timbered canyon and is followed by a logging road along the north side for 8 miles. The fishing is fair in early spring for 8 to 10 inch cutthroat in the lower reaches and 6 to 8 inchers above. The lower end dries up in late summer.

East Fisher River. A fair-sized stream in a timbered valley about 40 miles south from Libby on U.S. 2; drains Sylvan and Miller Lakes and is followed for its full length of 12 miles by a good county road. East Fisher is moderately popular and fair fishing for 5 to 10 inch brook, rainbow and a few cutthroat trout.

East Fork Pipe Creek. Small, about 8 miles long; flows through a steep, rocky canyon in the middle reaches; is crossed at the mouth by the Pipe Creek road 1 mile above the Turner Guard Station, and is followed by a logging road to headwaters. The East Fork is only lightly fished but good for 6 to 12 inch rainbow, cutthroat and some fair-sized Dolly Varden. Note! The middle reaches look like there shouldn't be any fish — but they're there.

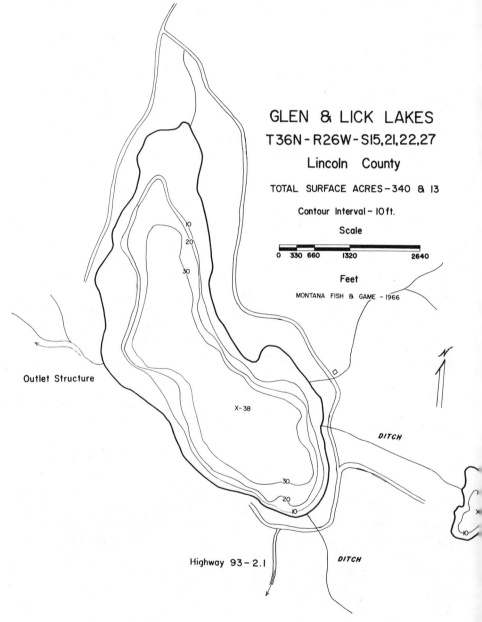

East Fork Yaak River. Reached at the mouth by the Yaak River road 4 miles above the Upper Ford Ranger Station and followed upstream past the falls, 1½ miles above the mouth, for a total of 5 miles of good fishing water in heavily timbered country just this side of Canada. It is moderately popular for 8 to 12 inch rainbow, cutthroat and brook trout that average around 10 inches and range up to 30.

Edna Creek. A small stream about 10 miles long and accessible all the way by logging roads to its junction with Fortine Creek 9 miles south of Fortine. Edna is brushy, hard-to-fish, and beaver-dammed but good fishing for about 6 miles upstream for 6 to 8 inch cutthroat, brook and some rainbow trout.

Elk Creek. A small tributary of McGinnis Creek flowing through brush and timber to its mouth 3½ miles above the Raven Work Center. There are about 3 miles of fair fishing here for 7 to 9 inch rainbow, cutthroat and brook trout. It's all accessible by road.

Fairway Lake. Maybe 2 acres, long and narrow, in timber; reached by a very poor road and ¼ mile of trail, 1½ miles west from the Spar Lake road (12 miles south from Troy). It's fair fishing for pan-size cutthroat.

Falls Creek. A tiny stream in a steep, timbered canyon; reached at the mouth by the Lake Creek road 4 miles south from Troy. A moderately popular early season stream, it is fair fishing for mostly 6 to 10 inch brook and cutthroat trout, plus a few BIG spawners. However, its almost all on private land.

Falls Creek Lakes. Tow cirque lakes at about 7000 feet elevation in rocky alpine country on the side of Dome Mountain at the head of Falls Creek 3½ miles and 4500 feet above the end of the road. NO TRAIL! The upper lake is about 2 acres, the lower 8 acres; and both "reportedly" contain fish.

Fish Lakes. See Therriault Lakes.

PE OF THE FISHER
FORE AND AFTER ASSAULT BY THE CORPS OF ENGINEERS
urtesy R. Konizeski

Fish Lakes. North, Middle and South are the only ones with fish. As a group they are five little alpine lakes in a steep, rocky canyon at the head of the Vinal and Windy Creek drainages; reached from either drainage by 3 or 1½ miles of trail. North Fish Lake is 6.9 acres by 33 feet deep, Middle Fish Lake is only 2.6 acres by 16 feet deep, and South Fish Lake is 14.7 acres by 40 feet deep, North and Middle were stocked with cutthroat in 1971 and should be good fishing now. There is a good holdover population of rainbow in South Fish Lake but they're not hit over hard because it takes a good man to get there.

Fisher River. A small river or a good size creek; flows for 30 miles from the junction of the East and Silver Butte Fisher Rivers to the Kootenai River 14 miles above Libby, and is followed by good roads all the way. The lower reaches flow through a narrow, steep-walled canyon, the middle and upper reaches through a flat-bottomed, timbered valley. Here *was* a beautiful, easily fished stream that *used to be fair* fishing for mostly 8 to 10 inch whitefish, some rainbow and a few cutthroat. Unfortunately, the U.S. Army Corps of Engineers, in its infinite wisdom, has straightened the lower 5½ miles. Although they have added some rock dams here and there it is hard to imagine this stretch ever producing many fish again. Fact is, it's little more than glorified drainage ditch now — courtesy of the Corps.

Five Mile Creek. Flows for 10 miles through steep, timbered mountains to the east side of Koocanusa a dozen miles up from the dam where its crossed by State 37. There are about 3 miles of good early season fishing here for 10 to 12 inch brook trout (especially in a passel of beaver ponds about 1½ miles upstream from the mouth), plus a few rainbow and cutthroat trout. The lower 1½ miles were recently stocked with migratory cutthroat and (given reasonable water-level fluctuations in the Corp's Pride (Koocanusa), there should be some excellent spring fishing for spawners in a few years. Anyway, there is a good road for the full length of this creek so it's readily accessible whether there's to be any fish in it or not.

GRANITE LAKE
Courtesy Miller's Studio

UPPER HANGING VALLEY LAKE
Courtesy Miller's Studio

Flattail Creek. A small stream, crossed at the mouth by the Seventeen Mile Creek road and followed for 2 miles along the middle reaches by an abandoned USFS trail 1½ miles from La Foe Lake. Flattail is reported to be good fishing for 10 to 12 inch rainbow — in beaver ponds 2½ miles above the mouth. It's very seldom bothered.

Flower Creek. The outlet of Sky Lakes, Flower flows through steep timbered mountains for 12 miles to the Kootenai River at Libby and is the municipal water supply. It's all closed except for the lakes at headwaters but good fishing if you could get it for 6 to 14 inch brook and a few rainbow trout. There is a good road along the lower reaches and a trail above to the lakes.

Flower Lake. Located on Flower Creek 3 miles above Flower Creek Reservoir (which is closed to fishing), in a closed, timbered basin but open around the shores. It is only 1.6 acres by maybe 25 feet deep, was stocked with cutthroat in 1971 and good fishing now. P.S.: It has neither inlet or outlet; a closed system.

Fortine Creek. A good, easily fished tributary of the Tobacco River; reached at the mouth by a road 3 miles north from Fortine and followed south by truck roads through rolling, glaciated country for most of its 28 mile length. It contains 8 to 10 inch cutthroat, rainbow and brook trout.

Foundation Creek. The smallish outlet of a small, barren lake on Mount Wam; flows for 2 miles to Grave Creek and is reached at the mouth by the Grave Creek road 4 miles above the Clarence Guard Station. The lower mile is fair fishing for pan-size cutthroat but the stream is too small to stand much pressure.

Fourth of July Creek. A tiny stream in a narrow canyon; followed by a logging road for the first 2 miles and crossed at the mouth by the Yaak River road 1½ miles below the Sylvanite Ranger Station. The fishing is fair-to-good for 6 to 8 inch rainbow and cutthroat trout.

Frank Lake. A hundred and thirty-seven acres, deep on the north end, in timbered hills about 6 miles south of Eureka; reached by a poor, rutted, muddy road through swampland to the south end of the lake, or along an old railroad grade and then a poor road to the north end. Frank is known locally as Little Duck Lake and is excellent fishing for 12 to 16 inch rainbow. The fishing pressure is correspondingly heavy for this part of the country.

French Creek. This little tributary of the West Fork of the Yaak River is reached at the mouth by the West Fork road and followed upstream by a logging road for about a mile of poor fishing for small rainbow, cutthroat and a very few brook trout. It is not recommended.

Frozen Lake. A long, narrow, 200 acre lake that straddles the Canadian line and is reached from our side by a trail 1 mile from the end of the Frozen Lake road, east from the Weasel Guard Station. It is fair fishing (on our side, of course) for Dolly Varden which are mostly too small to make the size limit in this country, but are legal in Canada. There are no facilities on our side but a good campsite in Canada. "Wetbacks" are welcome.

Garver Creek. A very small tributary of the West Fork of the Yaak River; reached at the mouth by a trail 1 mile from the upper end of the Pete Creek road and followed upstream by an unmaintained foot trail for about 1 mile of poor fishing. It contains small rainbow trout but is seldom fished.

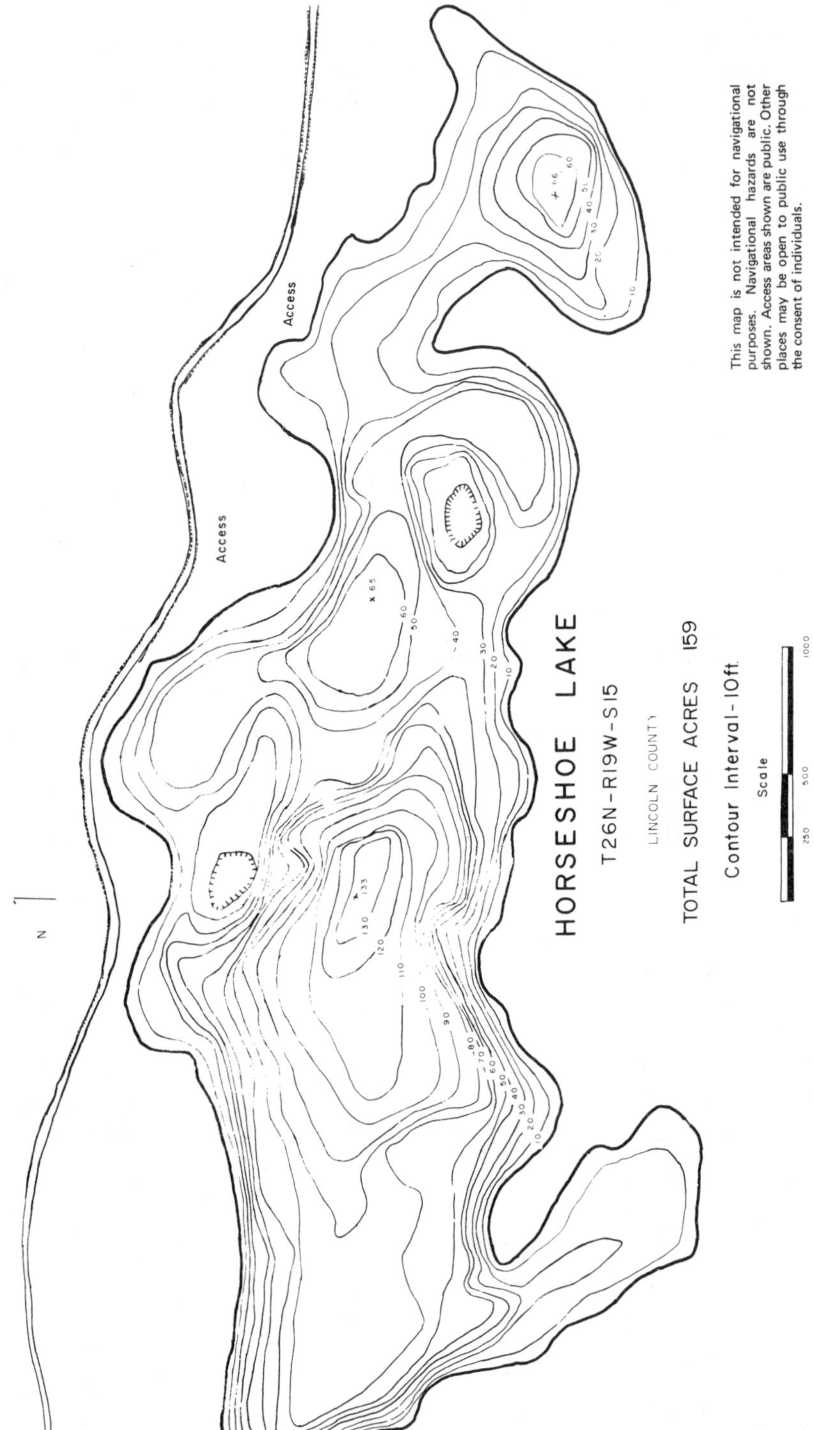

HORSESHOE LAKE
T26N-R19W-S15
LINCOLN COUNTY
TOTAL SURFACE ACRES 159
Contour Interval-10ft.
Scale
250 500 1000
Feet
N
Access
Access
This map is not intended for navigational purposes. Navigational hazards are not shown. Access areas shown are public. Other places may be open to public use through the consent of individuals.
MONTANA FISH & GAME-1967
x 65
x 133
10
20
30
40
50
60
70
80
90
100
110
120
130

PPER HAWKINS
ourtesy USFS

GRANITE CREEK
Courtesy Lance Schelvan

Geiger Lakes. Two high lakes about a mile apart in rocky country just east of the Cabinet Divide; accessible by a good USFS horse trail 4 miles from the Bramlet Creek jeep road near the West Fisher Guard Station. Lower Geiger is 35 acres, deep, in scattered timber and rock, with some brush around the shore. Upper Geiger is 15 acres, deep and perhaps 200 feet higher than Lower Geiger. Both lakes are fair fishing for cutthroat that average around 10 to 12 inches. They'll maybe average a little bigger in Upper Geiger.

Glen Lake. A dammed lake with about 17 feet of water-level fluctuation and a maximum area of around 300 acres; in hilly, timbered country accessible by road along its northeast side, 6 miles east of Eureka. It was poisoned in 1962 and planted with rainbow and kokanee in 1964. It's still good fishing for 12 to 14 inches.

Good Creek. A small mountain stream flowing through a narrow rocky canyon south to Big Creek a mile below the Big Creek Guard Station. It is crossed at the mouth by the Big Creek road, and is paralleled for 5½ miles from mouth to headwaters by a USFS trail. The lower reaches are good fishing for pan-sized rainbow, or would be if anybody bothered.

Granite Creek. Nine and a half miles long in a steep timbered canyon above and rolling hills below. The lower 5 miles are followed by the Bull Canyon road 6 miles south from Libby; the upper reaches by trail. This stream is mostly shallow but there are some nice pools in the middle reaches that are good fishing for small rainbow. It is a nice safe place to take the kids.

Granite Lake. A deep, 55 to 60 acre lake in rocky country; reached by 6 miles of easy trail from the end of the Granite Creek road. The fishing here is usually good-to-excellent for small cutthroat with now and then a lunker to maybe 16 inches or so.

Grave Creek. A fair-size tributary of the Tobacco River; reached at the mouth by road 8 miles south from Eureka, and followed upstream by a county road for 10 miles of fishing, some of which is on private land. Grave Creek is mostly fished by locals and has been good for 9 to 10 inch rainbow, plus a few that went to 2 pounds or so. There are also Dollys, an occasional brook and cutthroat trout, and whitefish. Now! The State Fish and Game Department is going to insure fish

KOOTENAI

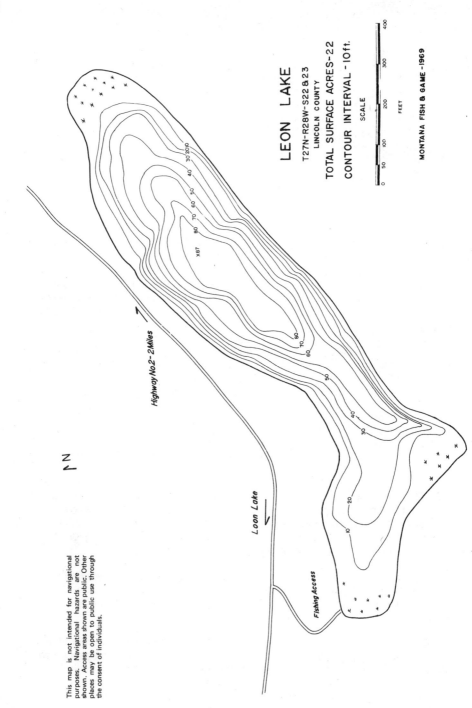

passage at the diversion dam (about 5 miles upstream from the mouth) so there will be good spawning runs of cutthroat and Dolly Varden from the reservoir...supposing there is any water in it; and also whitefish. Spring spawning for 10 to 18 inch cutthroat; August-September for the Dollys and whitefish.

Green Lake. See Blue Lake.

Grimms Meadows Lakes.
Two (PRIVATE) lakes at the head of Lake Creek; the lower one is just a beaver pond, the upper one is maybe a couple of acres and good and deep, in a swampy mountain park reached by the Lake Creek-Fortine road about 35 miles south from Eureka. They are not often fished but are fair-to-good for cutthroat and brook trout that average around 7 to 9 inches, with a few lunkers to a couple of pounds.

Grob Lake.
Take U.S. 93 one and a quarter miles north through mostly open rolling hills from Eureka, turn left (west) on State 37 for 2¾ miles, then right (north) on a county road for a mile where you jog left around the Iowa Flats School and continue on for ¾ of a mile to a jeep road on your left that takes you to within 150 yards of the shore. It's 25 acres mostly in a steep little open grassy pocket but with some timber on the southeast end, is weedy around the margins and chuck-a-block full of sunfish and bullheads.

Grouse Lake.
Seven acres, with a gradual dropoff and lots of marsh around the edges; on Grouse Mountain and reached by trail a mile from the Keeler Creek road. It's planted with brook trout every third year and the fishing potential varies from fair-to-good to real good.

Hanging Valley Lakes.
Two little alpine lakes, about ½ mile apart at the head of Hanging Valley; reached by a trail 2½ miles above the end of the Flower Creek road and then 1¼ miles of steep cross-country climbing to the lower lake. They are not fished much but are fair for skinny 6 to 13 inch cutthroat.

Hawkins Creek.
Drains Hawkins Lakes for 2½ miles through a steep, timbered canyon to Canada, and is accessible for 1½ miles of poor fishing for small cutthroat, by a logging road from the head of Pete Creek.

Hawkins Lakes.
Two beautiful alpine lakes about 1½ miles apart: 12.9 acres by 38 feet deep and 4.2 acres by 20 feet deep. The lower lake (or Big Hawkins) is reached to within ½ mile by a logging road north from the headwaters of Pete Creek. Little Hawkins is (mostly) reached by trail a mile from the end of the road. They are both good fishing for 6 to 14 inch cutthroat. Both were planted in 1971.

Hellroaring Creek.
A small, straight stream in a narrow canyon, flowing to the Yaak River 5 miles above the Sylvanite Ranger Station; it's crossed by the Yaak River road at the mouth and is followed upstream to headwaters by a logging road for 7 miles of fair fishing. This seldom fished Hellroaring stream contains real small rainbow, cutthroat and brook trout.

Helmer Lake. See Vinal Lake.

Hidden Lake.
An 8 acre by 65 foot deep lake in timber and brush in the old 1931 burn at the head of Spread Creek; seldom fished because nobody can find it although it's easy enough, right up the creek about a mile above the end of the road. It used to be barren but was planted with cutthroat in 1969 and may still have some left. The only way to find out is to go there and try it.

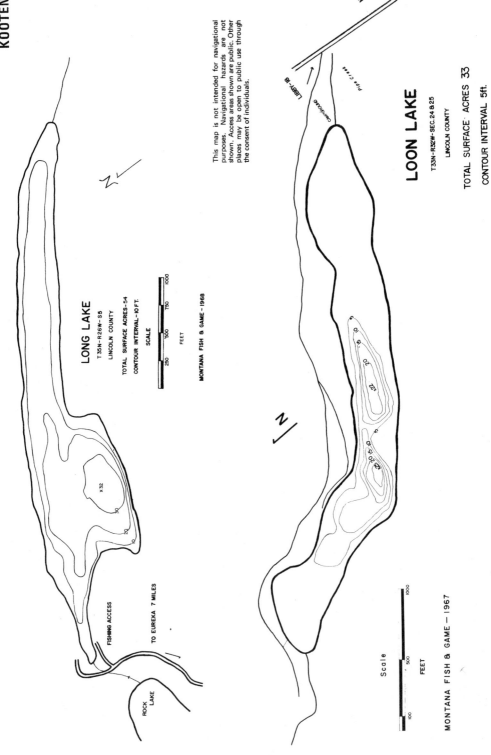

LEIGH LAKE
Courtesy Miller's Studio

LEIGH CREEK FALLS
Courtesy USFS

Hidden Lake. A fairly deep, 10 acre lake reached by a very poor road, in rolling, timbered hills 2½ miles south of Dickey Lake and 2 miles west of Bull Lake. This one is seldom fished but fair for 6 to 10 inch cutthroat and brook trout.

Horseshoe Lake. Between Loon and Crystal lakes in rolling timbered country accessible by U.S. 2 eight miles east of the Raven Work Center. Horseshoe is 159 acres in size, deep and full of trash fish along with a few whitefish and some largemouth bass. Rumor has it that it's due to be rehabilitated soon.

Hoskin Lakes. Upper Hoskin is small, shallow and either barren or close to it. Lower Hoskin is 33 acres and 30 feet deep, in timbered country reached by a good USFS trail ½ mile from the end of the Pipe Creek Beaver Creek road 2½ miles north of the Dirty Shame Saloon in the town of Yaak. It is good-to-excellent fishing for 12 to 14 inch cutthroat trout.

Howard Creek. The real small outlet of Howard Lake, it flows for 2 miles through heavily timbered mountains to Libby Creek 2½ miles above Old Town. Howard is not normally fished but is good for rainbow that average 6 to 8 inches and sometimes set records of up to a pound.

Howard Lake. A 33.9 acre by 58 feet maximum depth lake in heavy timber at the head of Howard Creek; reached by the Libby Creek road about 30 miles south of Libby and then the Howard Creek road upstream for 2 miles to a public campground. It is excellent early spring and fall trolling, summer fly-fishing, and winter ice fishing for 6 to 16 inch rainbow.

Hudson Lake. A private, posted, 3 acre pothole about 40 feet deep in the middle with shallow dropoffs; in open rangeland a half mile west of Sophie Lake and 2½ miles south of Gateway. It's excellent fishing for 12 to 18 inch largemouth bass, brook and rainbow trout.

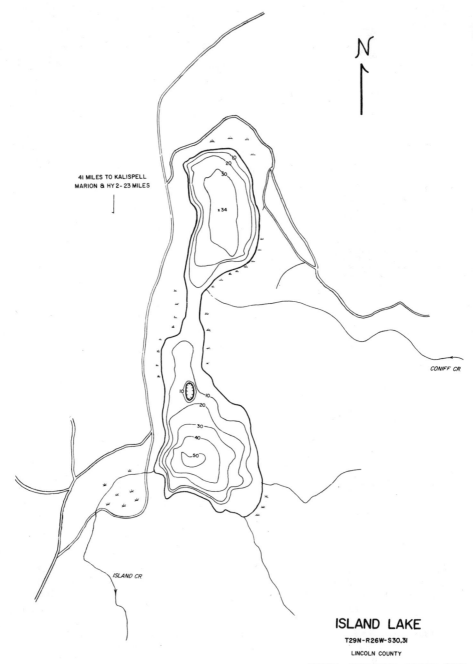

Indian Creek. Flows west through a little timbered canyon to the open rolling hills of the Eureka valley, and then southwest for about 4 miles to the Tobacco River 1½ miles west of town on the Great Northern Railroad tracks. It gets quite low in late summer but the lower reaches are fair kid's fishing for 6 to 10 inch brook and cutthroat.

Island Lake. This is a warm water lake with lots of weeds around the edges, in a wooded-valley in partly open logged-over country 8 miles from Marion, 23 miles from U.S. 2, and 41 miles from Kalispell. It is 225 acres by 50 feet deep (see fig.), is seldom fished and the reason why is because it supports mostly rough fish (suckers, squawfish, and pumpkinseed), along with an occasional largemouth bass.

Ivan Creek. A very small (2 miles long) tributary of Edna Creek which is crossed by a truck road at the mouth and followed upstream by a poor jeep road. It is real brushy and hard to fish, but fair for 7 to 8 inch brook trout. It's not often fished.

Jackson Creek. About 2 miles long from the junction of the North and South Forks to its mouth on Koocanusa where it is crossed by a USFS development road 3 miles above the dam. Jackson flows through brushy, logged over country and provides perhaps a half mile of heavily fished water for poor catches of small brook trout.

Jumbo Lake. Just a series of beaver ponds of about 1.8 acres total area by 15 feet deep, in timber and brush near Hidden Lake; reached to within a half mile by an old washed out road 3½ miles southeast from Dickey Lake. Jumbo is lightly fished but fair for 8 to 10 inch brook trout along with zillions of suckers.

Keeler Creek. Crossed at the mouth by the Lake Creek road 8 miles south of Troy and followed by a logging road for 10 miles to headwaters. The lower reaches are in more-or-less bottomland, the upper reaches in a steep-walled, heavily timbered canyon. Keeler provides fair fishing for mostly 6 to 8 inch cutthroat, and a few brook, rainbow and Dolly Varden. It was badly scoured in the January of '74 flood.

Kilbrennan Creek. Drains Kilbrennan Lake for 3½ miles down a timbered canyon to the Yaak River 3 miles above the Kootenai River and is reached near the mouth and also at the lake by logging roads 10 miles north from Troy. Kilbrennan goes dry in the middle reaches but is fair fishing for 6 to 10 inch brook trout in big beaver ponds above and below.

Kilbrennan Lake. Fifty-nine acres, about 50 feet deep with a shallow dropoff at each end; reached by a logging road 10 miles north from Troy, and moderately popular with a public campground and boat ramp. This lake is excellent fishing for brook and a very few rainbow trout. It really gets it on opening day—which coincides with the general (stream) season.

Klatawa Lake. About 30 acres, in a timbered pocket at the head of the West Fork of Granite Creek; reached by a steep trail 13 miles from the end of the Granite Creek road. Klatawa is seldom fished and do you know why? Because there is nothing in it!

LaFoe Lake. An old broken-out beaver pond on LaFoe Creek 1 mile north of Loon Lake. It is good fishing but seldom bothered for small cutthroat, and is reached by a trail 1 mile from the Seventeenmile Creek road near the mouth of (barren) Big Foot Creek.

KOOTENAI

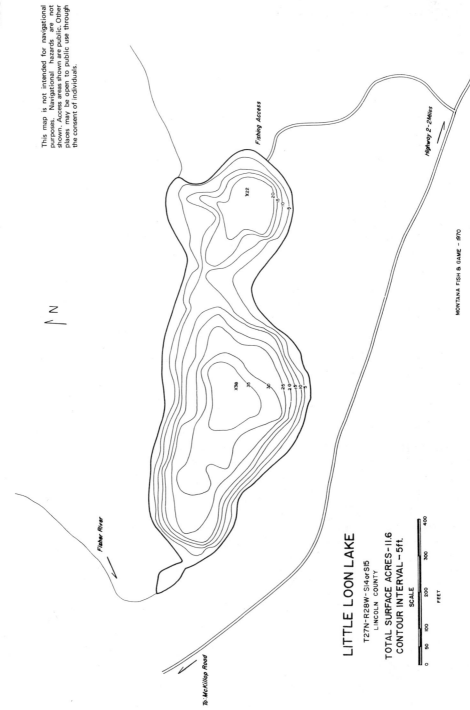

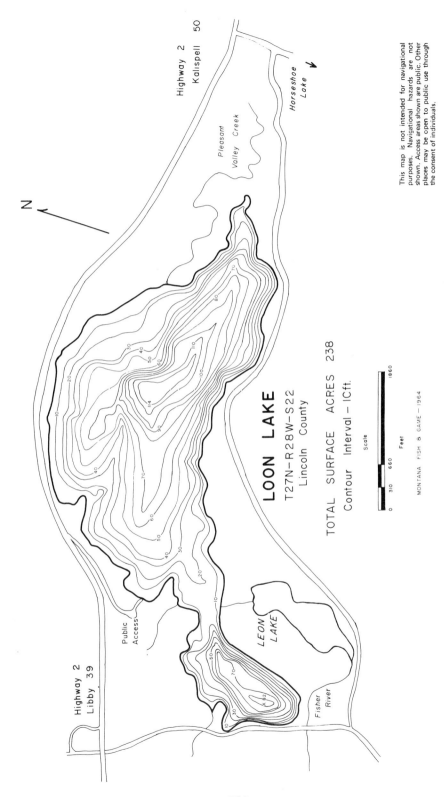

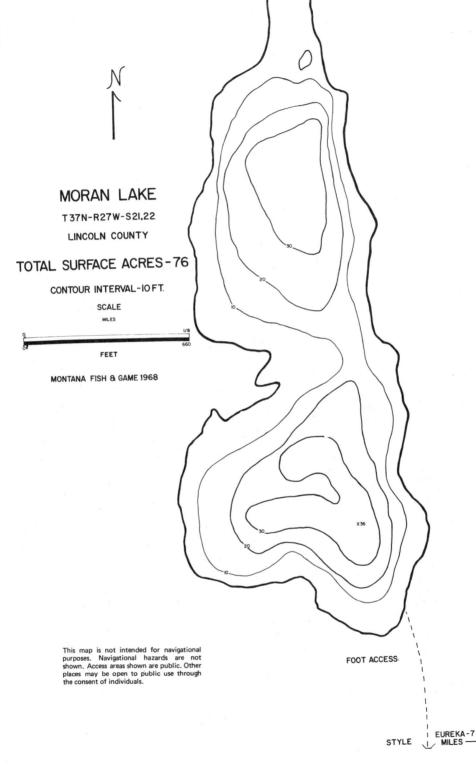

Lake Creek. A 5 mile long tributary of Swamp Creek following by a logging road from its mouth upstream past headwaters to the Five Mile Creek drainage. This is a brushy stream that is hard to fish but fair for small cutthroat trout.

Lake Creek. A brushy little stream that flows for 4 miles through heavy timber to the West Fisher River at the Guard Station. It is crossed at the mouth by the West Fisher road and followed along the lower reaches by a trail; and is seldom fished for the few 6 to 10 inch cutthroat it contains.

Lake Creek. One of the very few bottomland streams in the Kootenai drainage, Lake Creek flows northward from Bull Lake for 15 miles down its wide, open valley to the main river at Troy, and is easily accessible by good roads all the way. It provides excellent fishing for 9 to 12 inch rainbow (and some up to 20 inches), 10 to 15 inch cutthroat, 6 to 12 inch brook trout, a very few 10 to 12 inch whitefish and Dolly Varden that range all the way from 7 inches or so to maybe 7 pounds. This is one of the best creeks in the area.

Lavon Lake. Seventeen acres, 91 feet deep, Lavon connects to Crystal Lake at the west end of Pleasant Valley. Both were poisoned in 1960 and were planted with rainbow and kokanee in 1964. It's good fishing now for 12 to 14 inchers.

Leigh Creek. The outlet of Leigh Lake flowing for about 5 miles down a steep glaciated canyon to Big Cherry Creek and followed all the way by logging road and trail. It is excellent fishing for 8 to 10 inch brook and some rainbow trout, especially in pools across a big flat just below the lake.

Leigh Lake. A pretty, deep, 133 acre lake in a cirque; reached by the Leigh Creek horse trail 3 miles from a logging road up the north side of Big Cherry Creek. It is reported to be a real spotty lake but sometimes very good for 4 to 12 inch brook trout. The outlet is also good just below the lake.

Leon Lake. Deep, 18½ acres, in rolling timberland a couple of hundred yards below Loon Lake from U.S. 2. Leon was rehabilitated in 1969, planted with cutthroat in 1970 and is good-to-excellent fishing now for 8 to 14 inchers, plus a very few grayling.

Libby Creek. A little stream followed by road for 15 miles through rolling timberland to the Kootenai River at Libby. This creek has been badly messed up by mining operations but is planted each year and good fishing for 8 to 10 inch rainbow, cutthroat, a few 10 to 12 inch whitefish and fewer still Dollys.

Lick Lake. Fifteen acres, 40 feet deep, with logs around the shores, open meadows to the east and the rest in timber. Lick Lake is reached by a county road 7 miles east from Eureka, ½ mile east of Glen Lake. It is a private "fishing hole," poisoned in 1962, restocked with rainbow in 1963 or 1964 and provides good fishing now for rainbow, brookies and Dollys to 3 pounds or so.

Lime Creek. This brushy little stream flows through rolling timberland 4 miles south of Trego east of the Fortine road. It provides about 2 miles of lightly fished water for fair catches of pan-size cutthroat.

Little Cherry Creek. Reached at the mouth by the Libby Creek road, 19 miles south from Libby. There are a couple of miles of fair fishing here for 6 to 8 inch cutthroat and Dolly Varden. No Trail.

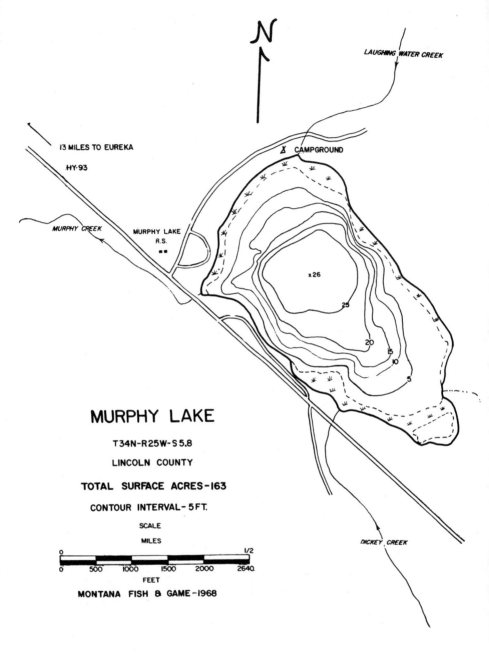

Little Creek. Proceed up the Yaak River road a couple of miles beyond the Sylvanite Ranger Station, and then ford the river to the mouth of Little Creek. There is no trail and it's seldom fished, but the first half mile is fair for 6 to 8 inch rainbow and cutthroat trout.

Little Loon Lake. This shallow, ten acre lake is really the beginning of the Fisher River. It is reached by U.S. 2 near Loon Lake, is full of rough fish and quite a few nice (3 to 6 pound) largemouth bass (some smallmouth too), and some 12 to 14 inch rainbow trout.

Little (or Upper) McGregor Lake. One-third of a mile north of McGregor Lake (and U.S. 2) in timbered hills; 33 acres, 30 feet deep with a muddy bottom, steep dropoffs, and a mostly marshy shoreline. It was rehabilitated and stocked with brook trout in 1961 and is good fishing now for 12 to 18 inchers.

Little North Fork Big Creek. A small stream with limited recreational potential, crossed at the mouth by the Big Creek road 3 miles up from the Koocanusa mud flats, and followed upstream (north) by pack trail to headwaters. There are some nice falls ½ mile up from the mouth and from here on down you can catch a meal of small rainbow — most any time.

Little Spar Lake. An alpine lake in barren rocky country at the head of Spar Creek; reached by a poor trail 4 miles from Spar Lake. Little Spar is 50.6 acres, over 90 feet deep, brushy around the shores and good fishing for 8 to 10 inch, bit headed, snaky brook trout. It could stand a lot of fishing but, for one thing, it is often iced in until mid-July.

Long Lake. A 48 acre lake with shallow dropoffs and lots of water-killed trees around the edges, in rolling hills between Lost Lake and Rock Lake; reached to within a few hundred yards by road 5 miles south from Eureka. It was planted with cutthroat in 1972 and was good fishing for awhile. However it went Kapoot...too much alkali.

Loon Creek. A heavily fished stream that is closely followed by a logging road for 5 miles from its mouth on Pipe Creek (near the Turner Guard Station) up a narrow gorge though the old 1910 burn to Loon Lake. It's fair fishing for 10 to 12 inch brook and rainbow trout in beaver ponds along the lower mile.

Loon Lake. Two hundred and thirty eight acres, fairly deep but with shallow drop offs, in timbered hill land ¼ mile west of Horseshoe Lake, reached by U.S. 2, seven miles east of the Raven Work Center. This lake is full of trash fish, plus a few sunfish, brookies, fair populations of both large and small mouth bass and a good population of 10-15 inch rainbow trout.

Loon Lake. A long narrow lake, 25.7 acres, deep in the middle but with shallow dropoffs, in a timbered valley reached by the Loon Lake road 5 miles from the Turner Guard Station on Pipe Creek. This is a real spotty lake for 10 to 12 inch (and a few to 20 inches) brook and a few cutthroat trout. It's good ice fishing too, and there is a public campground.

Loon Lake. Thirty-three acres with some deep holes but mostly shallow with marshy shores, in timbered hills 1 mile east of Mud Lake and reached by a poor road 3 miles to southwest from Fortine. It's not very popular but is good for one pound brookies.

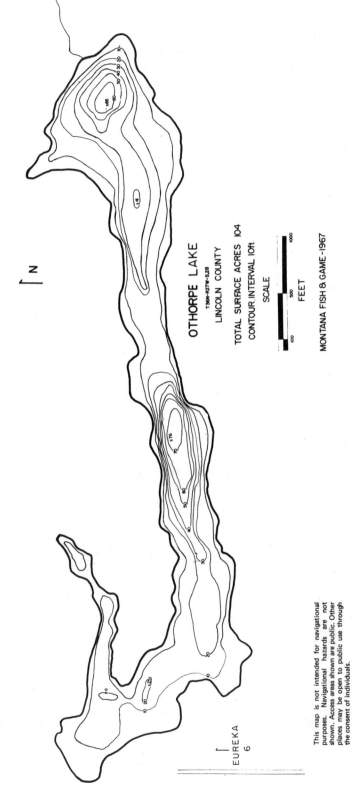

cGUIRE CREEK
ourtesy USFS

MINOR LAKE
Courtesy Miller's Studio

Lost Fork Creek. Two and a half miles of fishing down a narrow timbered canyon; and reached at the mouth by a logging road 9 miles up Seventeenmile Creek. It's not very popular but has quite a few nice holes that are good fishing for small cutthroat trout.

Lost Lake. Four miles south of Eureka in hilly timberland and followed all around the shore by a poor road. This marshy, 16.2 acre lake was planted in 1961, '62, and '67. It was moderately popular for awhile but the water dropped so low, the fish all died out.

Lost Lake. A moderately popular 4 acre lake with shallow dropoffs and mostly marshy shores, in logged-over hill country, reached by a logging road 3 miles north of Happy's Inn on U.S. 2 (or Crystal or Horseshoe Lakes). It is fair fishing both summer and winter for 8 to 10 inch brook trout.

Louis Lake. Take the Sunday Creek road 7 miles south from Stryker to the Louis Creek road, then 4 more miles west to the lake in timber all around. It is 15½ acres, 22 feet deep in the middle but marshy all around the edges, was stocked with cutthroat in 1973 and should be very good fishing for 10 to 12 inchers in a year or so.

Lower Bramlet Lake. A 40 acre alpine lake in a barren rock-and-scattered timber cirque reached by a trail 1½ miles above the end of the Bramlet Creek road, or 4½ miles from the West Fisher Guard Station. Lower Bramlet is only lightly fished but fair for 10 to 14 inch rainbow. Note! Upper Bramlet, ¼ of a mile above, is barren.

Marl Lake. Take the Meadow Creek road 3 miles west from Fortine, and then turn south for 1¼ miles to this roughly sub-circular, 106 acre lake in rolling, timbered hills; a very good spot to get lost in. Marl Lake was planted with cutthroat in 1972 and is good fishing now for 12 to 14 inchers, plus a fair number of 8 to 12 inch whitefish.

Martin Lake. Shallow, 15 acres, and marshy; in timbered country accessible by a fair road 3½ miles east from Fortine. Martin isn't fished much but is fair-to-good for 8 to 10 inch brook trout.

Martin Lake. An alpine lake at the head of Big Cherry Creek, reached by a steep 2¼ mile scramble across-country from the end of the trail 2¼ miles above the end of the Big Cherry road. Here is a lake that is seldom fished but reportedly good for small cutthroat, rainbow and brook trout.

Mary Rennels Lake. A private, fairly shallow, 100 acre lake on Clay Creek; reached by road 2½ miles above its mouth, or 4 miles from the South Fork Guard Station. It is not available to the public but is stocked, and good ice fishing for 8 to 14 inch cutthroat.

McGinnis Creek. About 5 miles of brush-and-meadowland fishing in this small tributary of Elk Creek; reached at the mouth by a road 3½ miles east from the Raven Ranger Station and followed upstream to headwaters. It's lightly fished but fair for 8 to 10 inch brook trout.

McGuire Creek. Small, steep, and followed by a trail through its rocky canyon to the Kootenai River about 5 miles below the Stonehill Guard Station. There isn't much fishing here except right near the mouth in early season for 7 to 10 inch cutthroat and brook trout.

Meadow Creek. Less than a mile long from the junction of its North and South Forks, to its mouth on the Yaak River 5 miles above the Sylvanite Ranger Station. This stream is followed along its entire length by a logging road and is fair fishing for 4 to 7 inch rainbow, cutthroat, and brook trout; if you want 'em.

Meadow Creek. A small, brushy, hard-to-fish stream in rolling timbered hill-land that is accessible by logging roads for most of its 8½ mile length from headwaters to its mouth on Fortine Creek, 1¼ miles west of town. There are lots of beaver ponds along the lower 3 or 4 miles and they're fair fishing for 8 to 10 inch brook trout and a few rainbow.

Midas Lake. See Standard Lake.

Middle Fish Lake. See Fish Lakes.

YOUNG BULL
Courtesy Cary Hull

CLARKS NUTCRACKER
Courtesy Cary Hull

Milnor Lake. Twenty-nine acres, 60 feet deep, within a hundred yards of the Bull Lake Highway about 5 miles south of Troy and ⅛ mile from Savage Lake. Milnor was rehabilitated and stocked with rainbow in 1959, was very good fishing for awhile but is now chock full of little bitty bass with once-in-awhile-a-good-one, and zillions of sunfish. Anyhow, it's all private.

Minor Lake. Deep, 20 acres, in an alpine cirque reached by pack trail up Parmenter Creek and the South Fork about 10 miles west of Libby. Here is a moderately popular "dude" lake that is excellent July, August and September fishing for 10 to 12 inch cutthroat.

Moran Lake. Shallow, 39 acres and shaped like a cow's horn; mostly in private timberland ½ mile east of Sophie Lake. Moran was rehabilitated in 1958 and is excellent fishing now for 12 to 18 inch brook trout. Access is via a stile over-the-fence and across USFS land.

Mount Henry Lake. Eight acres and a little better, just shy of 30 feet deep, a little east of the summit of Mount Henry; reached by a good USFS trail 4 miles above the end of the Hudson Creek road (Hudson is a tributary of Solo Joe Creek which in turn empties to the Yaak River 10 miles above the Upper Ford Guard Station). It's too far back to attract many people but is good fishing for 6 to 8 inch cutthroat. It was stocked in 1971. P.S.: There are also a few real nice big ones here, remnants from an early plant.

Mud Lake. A shallow, 12 acre lake with marshy shores; about ½ mile south of Marl Lake in wooded hills, reached by a road 7 miles west and south from Fortine. Mud Lake isn't bothered much but is fair fishing for 8 to 11 inch brook trout and whitefish.

Murphy Creek. Flows from Murphy Lake for 2 miles to Fortine Creek, about 1 mile south of Fortine and is easily accessible by road. A swampy meadow and beaver-ponded stream, it is hard to fish with any luck but does contain a good population of 8 to 10 inch rainbow and brook trout.

WASTELAND
courtesy R. Konizeski

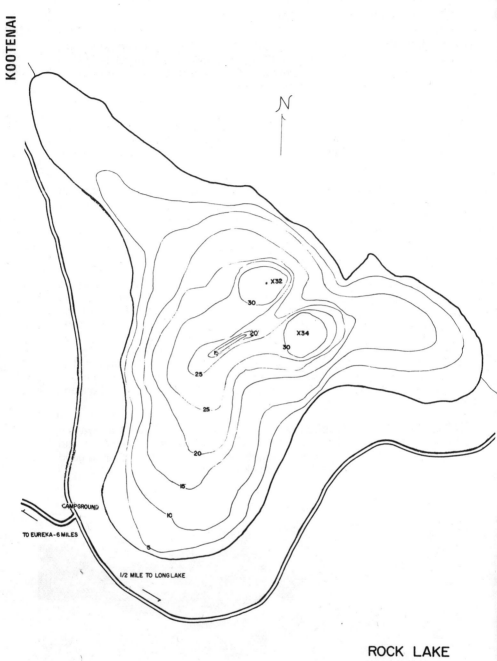

Murphy Lake. A 163 acre lake in timbered hills right by U.S. 93 a couple of miles south of Fortine. Murphy is deep at the north end, mostly marshy around the shore, and real good fishing for 10 to 14 inch (and a few reported to 8 or 9 pounds) Largemouth Bass, and fair fishing for 10 to 12 inch brook trout and whitefish. It also contains a very few rainbow and cutthroat and lots of sunfish and suckers.

Murray Creek. This little, privately owned, stream rises in a series of springs and flows for about a mile through hilly country to the Kootenai River 3½ miles south of Gateway. It's posted but good fishing especially in the beaver ponds along the upper reaches for 6 to 10 inch brookies and cutthroat.

Myron Lake. Three miles west of Happy's Inn on U.S. 2 and then a mile north by jeep road. It is 5 acres, 30 feet deep, in timber; was stocked with cutthroat in 1973 and should be good fishing for 10-12 inchers in a couple of years.

North Fish Lake. See Fish Lakes.

North Fork Big Creek. Four miles of fair fishing here, for camp-eating rainbow, in a timbered canyon accessible by a good USFS road (the Big Creek road) at the mouth and by the Boulder Divide road at headwaters. Very seldom fished.

North (Fork) Callahan Creek. Rises and flows for the most part through Idaho, but there is about 1 mile of Montana fishing in a timbered canyon near the mouth which is reached by the Callahan Creek logging road 7 miles west from Troy. It's fair for 8 to 10 inch rainbow, and a very few cutthroat.

North Fork Meadow Creek. This little stream provides about 3 miles of marginal fishing for 4 to 7 inch cutthroat and brook trout. It's followed by a logging road down its narrow, timbered canyon to Meadow Creek about a half mile above the Yaak River (5 miles above the Sylvanite Ranger Station).

North Fork Yaak River. Is paralleled by an old logging road for 3½ miles along its timbered valley to the Canadian line. The North Fork is "miles from nowhere" and very lightly fished although excellent for 6 to 8 inch rainbow and brook trout (especially in some big beaver ponds right at the line).

PARADISE LAKE
Courtesy Miller's Studio

ROCK LAKE
Courtesy USFS

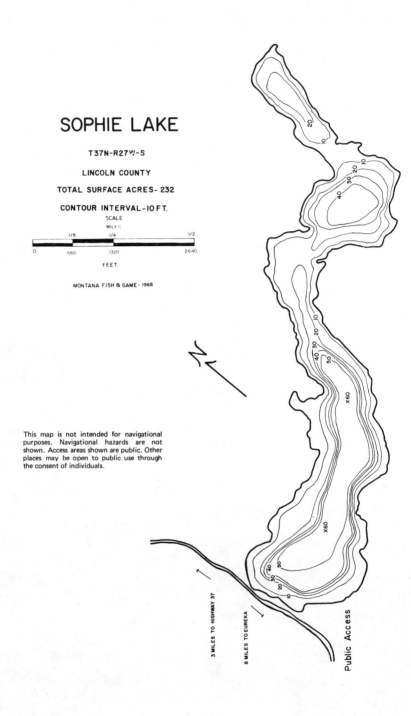

O'Brien Creek. A little mountain stream, the municipal water supply of Troy. It's mostly accessible by road for about 5 miles of fishing all told, for 6 to 10 inch cutthroat and brook trout.

Okaga Lake. See Phillip Lake.

Overdale Lodge Pond. This private 5 acre pond on Lang Creek is reached by a road 2 miles from the Yaak River road, 20 miles above the Sylvanite Ranger Station. It is stocked with rainbow and is good fishing for 8 to 16 inchers.

Owaisa Lakes. See Sky Lakes.

Parmenter Lake. Take a good USFS trail for 7½ miles above the end of the Parmenter Creek road, and then the North Fork trail 1¼ miles to within ⅛ mile of the shore. It's about 5 acres, in rocky country; not often fished but reported to contain trout.

Parmenter Lake. Take a good USFS trail for 7½ miles above the end of the Parmenter Creek road, and then the North Fork trail 1¼ miles to within ⅛ mile of the shore. It's a small lake in rocky country; not often fished but reported to contain trout.

Pete Creek. A little stream in timbered mountains; it flows for about 1 mile through numerous beaver ponds in the Pete Creek meadows, then for 2 miles through a narrow canyon, another ½ mile of beaver ponds, and lastly for about 6 miles down a little flat-bottomed valley where it splits, comes together and resplits into several channels before finally emptying to the Yaak River about 15 miles above the Sylvanite Ranger Station. The Yaak River road crosses this stream at its mouth and a logging road follows it to headwaters. It is moderately popular and fair fishing for 8 to 10 inch brook and some cutthroat trout.

Phillip (or Okaga) Lake. Is a mushroom-shaped private hatchery about a mile long by a quarter of a mile wide, at the mouth of Windy Creek maybe 5 miles or so above the Ford Ranger Station. It's advertised as the "finest rainbow fishing in the State." Boats and cabins are available.

Phillips Creek. A small, meandering stream that rises in Canada but flows for about 3 miles through U.S. farmland to Sophie Lake and eventually the Koocanusa reservoir 2 miles south of Gateway. The lower reaches dry up in summer but the upper (U.S.A.) reaches contain a few small brook and cutthroat trout. It's seldom fished.

Phill's Lake. Sixteen and a half acres but shallow; in timbered country midway between the Big Bridge and Rexford and reached by a poor jeep road. Phill's was planted with rainbow in 1961 and has been fair fishing for middling-sized rainbow, but someone slipped some bullheads in and now it's good for them too — up to a pound or so, if you want 'em.

Pine Creek. A little stream flowing from Windy Pass on Cross Mountain down a narrow, flat-bottomed valley for 14 miles to the Kootenai River 10 miles below Troy. The lower reaches (about 5 miles) are readily accessible by road and are fair early-season fishing for pan-size brook trout. Quite a few local fishermen hit this one.

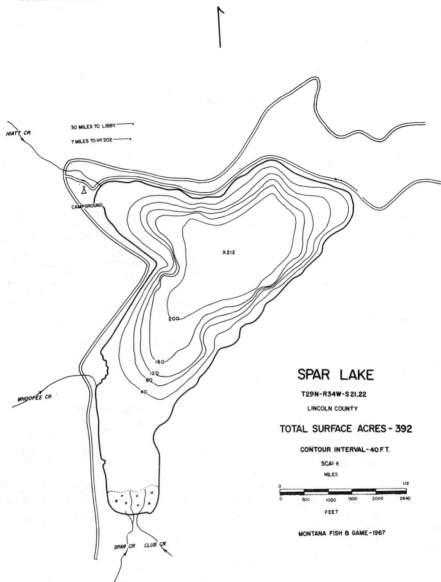

Pinkham Creek. Twenty miles long and followed by county roads westward through timbered hills from headwaters to its mouth on the Koocanusa Reservoir a mile north of the Big Bridge. Pinkham is lightly but easily(?) fished and there are lots of log jams, beaver dams and cover. From the falls down to the mouth (about 5 miles) it is all cutthroat water. There are lots of smallish rainbow (4 to 7 inches) and brookies, plus a few cutthroat too, above the falls. The upper reaches, away from the road, are the best. Excellent fishing there — for the little ones. Ooooops, almost forgot...there are some nice cutthroat spawners come up in the lower reaches in the spring. Some sections go dry in summer.

Pipe Creek. Followed for 21 miles by a good USFS oiled road from its headwaters in steep mountains through timbered hills to its mouth on the Kootenai River 3 miles west of Libby. It is one of the most popular streams in the entire Libby area inasmuch as it is readily available, easily fished, and good for pan-size rainbow, 8 to 12 inch cutthroat, and a few brook trout. The best fishing is in beaver ponds and timbered areas away from the road.

Pleasant Valley Creek. Flows from headwaters just west of Little Bitterroot Lake into-and-out-of Dahl Lake through timber and ranchland for 12 miles to the Pleasant Valley Fisher River. It's easily accessible by road although there is some posted land, which is no great loss because this creek is poor fishing at best, for small rainbow and maybe a few cutthroat and brookies.

Pleasant Valley Fisher River. Forty miles long, mostly on posted ranchland in Pleasant Valley, and timber below. The lower 11 miles are followed by U.S. 2, the next 12 miles by a good logging road. This is a moderately popular, easily fished stream that produces fair catches of 9 to 11 inch or so rainbow and brook trout, plus a few good sized ones.

Poorman Creek. Flows from Ramsey Lake for 5½ miles down a steep canyon above, and a narrow valley below through the old 1910 burn to Libby Creek at Old Town. It is mostly accessible (except for the lower mile) by a logging road and is good fishing for 6 to 8 inch rainbow.

Quartz Creek. A small stream in a steep timbered canyon, followed by road for 2½ miles from the junction of its East and West Forks to its mouth on the Kootenai River 5 miles west of Libby. Quartz Creek is only fair fishing for pan-size cutthroat, both forks are better.

Rainbow Lake. A fairly deep, 10 acre lake that drains ¼ mile to, and lies between, Upper Thompson and U.S. 2. It is seldom fished but good for Largemouth Bass, and also contains a very few good-size rainbow and lots of sunfish.

Rainbow Lake. See in Ten Lakes.

Ramsey Creek. A popular little stream, crossed at the mouth by the Libby Creek road ¾ of a mile above Old Town, and followed by a logging road and trail for about 3 miles of fishing. It used to be real good but is now only fair for 8 to 10 inch rainbow and a few Dolly Varden.

Rattlebone Lake. Is 35 feet maximum depth, 15 acres, in rolling timberland at the head of Cripple Creek, reached by road about 4 miles southwest from the Ant Flat Ranger Station. Rattlebone is a good Largemouth Bass lake that produces lots of ½ to ¾ pounders and a few that will go to 8 pounds or better, plus too many sunfish.

KOOTENAI (NOT STANDARD) WHITEFISH!
Courtesy Ernst Peterson

Courtesy E. Bridenstine Studio

Red Top Creek. Flows for 5 miles through a steep timbered canyon to the Yaak River 4 miles above the Sylvanite Ranger Station. It's crossed at the mouth by the Yaak River road and followed upstream for about 3 miles of fair fishing for 6 to 8 inch rainbow, cutthroat and brook trout.

Rock Lake. Five miles southeast of Eureka by crow-flite, fairly deep, about 34 acres with 30 foot cliffs around the east side and timber on around. There is a poor road to the south end. Rock was planted in the early '60s and was good fishing for one pound rainbow trout, but froze out the winter of '74 and is probably barren now. There's a USFS campground with boat launching facilities and whole shebang here. Even if there are no fish, you can always occupy yourself with camping and water-skiing.

Ross Creek. Flows east for 8 miles from its headwaters down a fairly flat, timbered valley to the south end of Bull Lake. The middle reaches go low, but there are about 5 miles of fishing water above and below; reached by 3 miles of road and then a fair trail. It is lightly fished bu fair for small cutthroat, and some brook trout right near the lake. There is a nice campground (one of the largest in the state) at headwaters.

Round Lake. Really just a 4 acre swamp-hole at the head of a small tributary of Summit Creek; reached by a ½ mile hike up the power line of the Sunday Creek road about 3 miles south of Dickey Lake. Round Lake is (reportedly) fair fishing at times for brookies that come in from the creek.

St. Clair Creek. A very small, lightly fished tributary of the Tobacco River, crossed at the mouth by U.S. 93 one mile southeast of Eureka and followed by a country road for 5 miles of fair fishing for pan-size cutthroat, and a very few brook trout in the lower reaches.

St. Clair Lake. A mile southwest of Big Therriault Lake; best way in is by 4 miles of trail from the end of the Stahl Creek road. It is only 3.2 acres in size with 17 feet maximum depth but was planted with cutthroat in 1971. They did OK the first few years...let's hope they're still there.

Savage Lake. Is 73 acres with about 30 in aquatic vegetation, over 60 feet deep, in wooded hills reached by the Bull Lake road a couple of miles south of U.S. 2 and about 5 miles from Troy. Savage is a popular summer home lake with three dozen or so cottages as well as public access areas. It has a fish barrier at the outlet, was rehabilitated in the late '50s and is fair fishing now for cutthroat and brook trout plus an occasional largemouth bass.

School House Lake. About 14 acres by 30 feet deep, in hilly (mostly wooded but some meadow) land; reached by the Lake Creek road 4 miles south from Troy. It has been pretty well taken over by trash fish and lots of bullheads and sunfish, but still has a few rainbow. Has gone private, is no longer managed and just as well.

Schrieber Creek. Flows to Schrieber Lake and out again for 1½ miles down a heavily timbered canyon where it is followed by U.S. 2 to its junction with the Fisher River a couple of miles south of the Swamp Creek Guard Station. It is lightly fished by local kids for a few small brookies.

Schrieber Lake. A shallow, marshy, 20 acre lake in timberland close to U.S. 2 about a mile south of the Swamp Creek Guard Station. It's seldom fished but has a few brook trout in it, average size 8 to 10 inches and maximum up to maybe 12.

Seventeen Mile Creek. Reached at the mouth by the Yaak River road 3 miles above the Syvanite Ranger Station and followed by a logging road upstream through a flat-bottomed canyon for 7 miles of good early season fishing for 7 to 12 inch cutthroat (including some spawners up from the Yaak River) and rainbow trout.

Shannon Lake. A shade under two acres, in a rocky canyon (no trees around the lake) a short half mile straight up (south) from old U.S. 2 six miles west of Troy. Shannon used to be good fishing for brookies but has been ruined by an over-large population of sunfish. It's not frequented much anymore — is in the works for a rehab.

Silver Butte Fisher River. A brushy stream 7½ miles long and followed by a good USFS road down its timbered valley to junction with East Fisher, 4 miles from U.S. 2. There are about 5 miles of fair fishing here for 8 to 9 inch rainbow, cutthroat and brook trout in that order of plentitude.

Sky (or Owaisa) Lakes. Upper (4.2 acres) and Lower (24 acres) a half mile apart in steep timbered mountains just below and south of the summit of Sugarloaf Mountain; reached by a fair USFS horse trail 5 miles from the end of the Flow Creek logging road. Both used to be fair fishing for goldens but there was no reproduction and none are left, so — the State Fish and Game Department replanted them with cutthroat in the late '60s and they're good fishing now for 13 to 14 inch trout.

Slee Lake. Private, 16 acres, in a timbered pocket reached by road 2 miles northeast of Troy, and very good fishing for largemouth bass, yellow perch and sunfish.

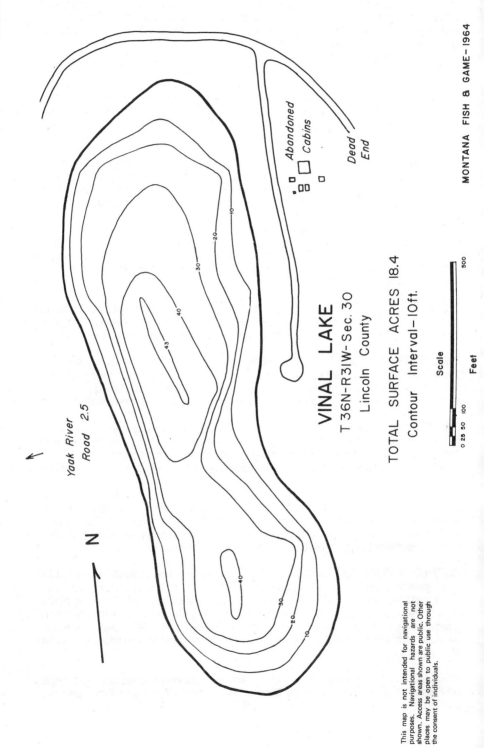

Snowshoe Creek. A tiny tributary of Big Cherry Creek, followed by a jeep road for 4 miles from its mouth to some beaver ponds at the headwaters. It is seldom fished but the ponds are fair for 10 to 12 inch rainbow.

Sophie Lake. Is 232 acres, maximum depth around 60 feet but with shallow dropoffs, in rolling grassland reached by an oiled county road 6 miles north from Eureka. Sophie was poisoned in the late '50s and subsequently planted with cutthroat; more recently with rainbow and the latter are now providing poor-to-fair fishing for 12 to 14 inchers. There are also a very few Dollys and largemouth bass, plus a large and burgeoning population of trash fish.

South Fork Big Creek. A little stream followed by an unimproved road from the headwaters for 15 miles to Big Creek at the guard station. The lower reaches are very brushy with lots of log jams. Some sections dry up in late summer. It was recently rehabilitated and stocked with migratory cutthroat that should — supposing Kookynoosy isn't drawn down too far — in a few years provided a good fishery for nice-sized spawners in the spring, and lots of little fellows the year around.

South Fork Callahan Creek. A small lightly fished stream that rises in Idaho and is followed by a logging road for about 2½ miles of Montana fishing, for 8 to 10 inch rainbow and a very few cutthroat.

South Fork Meadow Creek. A tiny, marginal stream followed by a logging road from its mouth for 6 miles to headwaters. There are about 3 miles of fair fishing for pan-size-and-under rainbow, cutthroat and brook trout.

South Fork Yaak River. A mixed-up stream, the lower reaches in a steep, inaccessible canyon, the middle reaches (about 5 miles) in a flat-bottomed open-meadow valley easily accessible by road. The canyon is best fished by going up the road for a couple miles to its head, and then fishing down. The South Fork is moderately popular and fair fishing as a whole for 5 to 9 inch cutthroat, rainbow and some brook trout.

South Fork Young Creek. Flows to Young Creek about 8 miles above the mouth. This small tributary is seldom bothered but has pan-size cutthroat in the lower ¾ of a mile.

Spar Lake. Is over 200 feet deep with steep dropoffs all around 392 acres, with no outlet and as much as 50 feet of water-level fluctuations. Spar lies in a wooded pocket among steep mountains, and is reached by the Lake Creek road 25 miles south from Troy. It is moderately popular, there's a campground, and fair-to-good fishing for kokanee salmon that occasionally run up to as much as 7 pounds, 10 to 20 inch lake trout and a scattering of brookies.

Spread Creek. A nice little stream (the outlet of Hidden Lake) in a narrow, timbered valley, easily reached by 6 miles of logging road from its mouth on the Yaak River 11 miles above the Sylvanite Ranger Station. It is moderately popular and fair-to-good fishing for small cutthroat, eastern brook and rainbow trout.

Stahl Creek. A tiny, relatively open stream in a steep, timbered valley; followed by a logging road from its mouth on Grave Creek at the Clarence Guard Station upstream for 3 miles. Like Clarence and Grave Creeks, it should provide some good spring fishing for nice-sized cutthroat spawners, plus lots of little ones.

Standard Lake. See Barer Lake.

Standard Creek. Very small, flows for 3 miles to the West Fisher River 2½ miles above the guard station; is reached at the mouth by the West Fisher road and followed by logging road and trail to headwaters. Standard is chock-full of criss-crossed logs and very good fishing in spots for 9 to 11 inch brookies plus an occasional cutthroat.

Standard (or Midas) Lake. About 10 acres, shallow, and marshy around the shore; on the Standard Creek drainage (don't confuse with Standard or Baree Lake at the headwaters of Baree Creek); reached by a one mile cross-country hike through the timber from the old Midas Mine 4 miles above the West Fisher Guard Station. It's fair fishing for camp-eating cutthroat and brook trout.

Star Creek. Empties to the Kootenai River 11 miles north of Troy; is crossed by a logging road a couple of miles above its mouth and followed by road and trail for 7 miles on upstream to headwaters. This stream is seldom bothered and only poor fishing for small rainbow and a very few Dolly Varden near the mouth plus some good sized spawners up from the river.

Stewart Creek. About 2 miles of fishable water; crossed at the mouth by the Fortine Creek road and followed upstream for 1½ miles through rolling, timbered hills. Stewart is quite brushy, but fair fishing for 4 to 8 inch cutthroat.

Sullivan Creek. About 6 miles long; accessible near the mouth by road 3 miles north of the Big Bridge and followed its full length by a good USFS oiled road. Hopefully, it will be good fishing in the lower 2 or 3 miles for nice cutthroat spawners; and above for 6 to 8 inch resident cutthroat.

Summit Creek. A tiny, brushy stream followed by the Eureka-Whitefish highway for 2 miles to the east end of Dickey Lake. It's not fished much but the lower reaches are fairly open and contain a few 6 to 8 inch brookies.

Sutton Creek. Ten miles long, the lower 7 followed by a good logging road westward down a narrow timbered canyon to Koocanusa opposite Cliff Mountain. Sutton is seldom bothered but fair early season fishing for rainbow and cutthroat in the lower reaches from 4 miles upstream to Four Mile Falls, and brook trout and small Dolly Varden near the mouth.

Swamp Creek. Reached at the mouth by the Fortine Creek road 18 miles south of Eureka, and followed upstream by a logging road for about 5 miles of fishing. Some sections are brushy, others are posted, but taken as a whole it is fair fishing for little fish; cutthroat, brook and a few rainbow trout.

Swamp Creek. A slow, meandering stream flowing through private ranch-and-timberland for 6 miles beside U.S. 2 to Libby Creek, 14 miles south of town. It is heavily fished by mostly summer people, for 6 to 8 inch brook and a scattering of rainbow and cutthroat trout.

Ten Lakes. Eight are barren, sub-alpine, 1½ miles south of the Canadian line. One (which one?) is reported to have fish, and another (Rainbow Lake) IS good fishing for 12 to 14 inch cutthroat. It is in an old burn, about 17 acres, deep, with rocks on the south and east sides. There is no trail or road; perhaps the best way in is by a 2 mile cross-country hike over the mountains west from a point on the Wigwam Creek road 3 miles above its mouth. This lake is VERY seldom fished.

IMMY LAKES IN THE CABINET RANGE
ourtesy USFS

TERRIAULT LAKE
Courtesy Lance Schelvan

Tetrault (or Carpenter) Lake.
Is 115 acres, the east end deep and the west end shallow; in open meadows reached by a good county road 7 miles north of Eureka. Tetrault was rehabilitated with rainbow in 1964 and is excellent fishing now for 1 to 3 pounders, summer and winter.

Therriault Creek.
A short stream that flows from steep timbered mountains through rolling hills a couple of miles to Griffith Creek. The lower reaches are reached by a road a couple of miles east from Glen Lake and are fair fishing for small cutthroat, brook and rainbow trout.

Therriault (or Fish) Lakes.
Two — Big and Little, at 5000 feet elevation in timbered mountains 9 miles east of Eureka by goose-flight; reached by first the Grave Creek, then the Weasel Creek, and finally the Wigwam Creek road 30 miles around from Eureka. Big Therriault has a public campground, is 55 acres, over 65 feet deep and fair fishing for 10 inch cutthroat. Little Therriault is 26.9 acres, 60 feet deep, very scenic and fair fishing for 10 to 12 inch cutthroat, they maybe run a little larger here.

Throops Lake.
A very pretty little private, 40 acre lake in some rock ledges on the north side of U.S. 2 a few miles east of the junction with Montana 22 (east of Troy). There are quite a few lilies here and the water is real black, deep in spots, and excellent fishing for ¾ to 1 pound largemouth bass, plus an occasional 8 to 10 inch brook trout.

Timber Lake.
Is 27.5 acres, maximum depth 40 feet, in timbered hills but open around the shore; reached by a poor road 5 miles south from Eureka. It was planted with rainbow in 1961 and is good fishing now for trout that will average between 1 and 1½ pounds, but you can't eat 'em in the summer.

Tobacco River.
A nice open stream flowing through (mostly private) meadow and timber for 16 miles past Eureka to Koocanusa ½ mile above Rexford. It is crossed by roads along its course and (supposing the Corps keeps its word and doesn't jerk the bottom out of Koocanusa) should be good year around fishing for

YAAK FALLS
Courtesy USFS

small (4 to 7 inch) cutthroat, 10 to 18 inch spawners in the spring, and Dollys in the fall. It should (all else being equal) be good fly fishing up to the mouth of Graves Creek. There's a boat launching ramp (and sometimes water) at the mouth.

Tom Poole Lake. Take the Yaak River road 15 miles north from Libby to about a mile south of the Turner Guard Station, then a good but steep trail for a couple of hundred yards east to this marshy lake which is 6.8 acres by 15 feet deep, in dense timber and brush, full of fresh-water shrimp and has provided some of the very best fishing (at times) in the district for 12 to 16 inch cutthroat. At other times it is a good place to go — to get skunked. As of this writing its been pretty well fished down though and is due for a replant.

Topless Lake. Lies in a logged area just east of Horseshoe lake, across the road from Cibid and Cad lakes. It is only about 9 acres, 25 feet deep, and good fishing for 10 to 14 inch grayling and cutthroat too. How it got its name is beyond me.

Trail Creek. A very small, seldom fished tributary of the West Fisher River; crossed at the mouth by the West Fisher road at the guard station and followed upstream by a trail for a couple of miles of poor fishing for 6 to 8 inch rainbow.

Upper Bluebird Lake. A 2½ acre by 17 foot maximum depth alpine lake in rock and timber country at the head of Bluebird Creek; reached by a trail a couple of miles from the end of the Bluebird-Wigwam Creek road. Upper Bluebird is right against steep, rocky slopes to the west and there is much snow and winter kill. It used to be very poor fishing for 10 to 14 inch cutthroat but was planted with cutthroat in 1966 was good for awhile and just might still be now.

Upper McGregor Lake. See Little McGregor Lake.

Vimy Lakes. Two little mountain-top potholes (3 acres in toto) a couple of hundred yards apart, almost impossible to get to but reached once in a blue moon ¼ of a mile south and 600 feet up cross-country from Double Lake; in rocky, scrub timber north of Vimy Ridge. They used to support good populations of rainbow years ago, but could be frozen out now.

Vinal Creek. Is 18½ acres, over 40 feet deep in some places with mostly shallow

Vinal Creek. In a flat-bottomed, timbered canyon; followed by a trail from its mouth on the Yaak River 3 miles below the Upper Ford Guard Station, for 4 miles upstream to some 50 foot falls. There are no fish above the falls but it's good below, and in a series of potholes below Fish Lakes (on the north branch of the creek).

Vinal (or Helmer) Lake. Is 18½ acres, over 40 feet deep in some places with mostly shallow dropoffs and open around the shore; Vinal is reached by a truck road (a half mile east of the Yaak River road) about 4 miles below the Upper Ford Ranger Station. It used to be heavily fished when the nearby radar base was active. It was rehabilitated in 1967 and planted with cutthroat that should be doing well now.

Weasel Creek. The outlet of Weasel Lake, followed north by the Graves Creek logging road for 4 miles down its timbered canyon to Canada; seldom fished and poor-at-best for small cutthroat trout.

Weasel Lake. Ten acres, fairly shallow, in a spruce-timbered basin reached by trail about a mile from the end of a marked logging road off Weasel Creek. The east side of this lake is full of debris from a big snow slide. It's not fished much but is good for small cutthroat and a few Dolly Varden.

Wee (or Arbo) Lake. A 27 foot deep, 5 acre alpine lake in timber, brush and talus; reached cross-country (half a mile up and another half down) from the end of the Hemlock Creek road which takes off to the south from the head of Seventeen Mile Creek or you can also get in by way of the Arbo Creek (barren, too small) road 3½ miles from the East Side Yaak River road 10 miles up from Troy. Wee Lake was planted years ago and is now good fishing for rainbow that will average 10 to 12 inches and go to 24. It could stand lots more pressure.

WOLVERINE LAKES
Courtesy USFS

KOOTENAI RIVER
Courtesy Miller's Studio

West Fisher River. Twelve miles long, in a narrow timbered valley; crossed at the mouth by U.S. 2, thirty miles south from Libby, and followed upstream by a truck road to headwaters. It is easily fished but only fair for 7 to 9 inch cutthroat, and a very few brook and rainbow trout.

West Fork Quartz Creek. Small, seldom fished, in a narrow timbered canyon; reached at the mouth by the Quartz Creek road and followed by a trail for 3 miles of good cutthroat fishing (average size 6 to 8 inches).

West Fork Yaak River. Flows from its headwaters on the high, alpine slopes of Rock Candy Mountain down a fairly wide canyon for 8 miles to Canada and for many miles beyond before it eventually recrosses to the States and flows for another 10 miles through a steep-walled canyon above, and a wide, timbered valley below to the main river 2½ miles above the Upper Ford Ranger Station. The upper reaches are reached by the Pete Creek road and followed all the way by logging road and trail. They are fair fishing for 6 inch cutthroat. The lower reaches are followed by a logging road for 3 miles, and then by trail to the line. They're very good fishing for 8 to 10 inch cutthroat, especially below the falls which are 3 miles upstream from the mouth.

Wigwam Creek. Empties to Weasel Creek just over the border into Canada and is followed upstream by a logging road for 6 miles of fishing. It's not very popular but fair for 6 to 8 inch cutthroat.

Williams Creek. Flows down a narrow, steep canyon to Grave Creek 1½ miles below the Clarence Guard Station; and is followed by a logging road for 5 miles to headwaters. Williams is quite small and practically never fished, but fair for 7 to 8 inch cutthroat below the falls at the lower end of the canyon, and maybe a few that will go a pound. In the upper canyon area there are only small, 6 to 9 inch resident Dolly Varden.

Windy Creek. A very small stream flowing north to the Yaak River about 5 miles above the Upper Ford Ranger Station. There is a private fish hatchery at the mouth and it's too small to fish above.

Wishbone Lake. Is 18 acres, fairly deep, in rocky cliff, talus and scrub timber on the north side of Vimy Ridge at the head of the East Fork of Contact Creek; reached cross-country ½ mile east from Double Lake. It is very seldom fished and is overpopulated with 6 to 16 inch brook. The USFS is reportedly contemplating a trail to this one.

Wolf Creek. Is 17 miles long, rocky above; slow, meandering and brushy below; followed the entire length by a logging road to its junction with the Fisher River at the Wolf Creek Guard Station. This stream is easily fished but no more than fair for small cutthroat, a few whitefish, rainbow and brook trout, plus lots of trash fish.

Wolverine Lakes. Three lakes ⅛ of a mile apart in subalpine, timbered country reached by a USFS trail 2½ miles from the end of the Wigwam Creek road. Upper Wolverine is about 15 acres and 30 feet deep; Middle Wolverine is about 25 acres and also 30 feet deep; and Lower Wolverine is only about 4 acres and maybe 7 or 8 feet deep. All three are good fishing for 9 to 11 inch cutthroat but have been better in years gone by. Lower Wolverine is unique in that it has a good inlet and a big spring area that extends out about 100 feet from shore — but there is no surface outlet.

Yaak River. Flows from within a few miles of the Canadian line, for 28 miles south to the Kootenai River a few miles east of the Idaho line and is followed by a good gravel road all the way. It is a good sized stream, mostly in a flat-bottomed valley above and a narrow, steep-walled canyon below. The lower 8 miles (below Red Top Creek) are fairly fast, and heavily fished for good catches of 10 to 12 inch rainbow, plus a few cutthroat, brook trout (these reported to 27 inches), and whitefish. The stream is much slower above Red Top Creek with lots of still, deep water that is good fishing for 8 to 10 inch rainbow, quite a few brook, and an occasional cutthroat.

Young Creek. Heads just this side of Canada and flows southeastward for 7 miles down a narrow timbered valley above, along the south side of some open meadows for about a mile (northwest of Alkali Lake), and then down a tight little brushy draw in the lower reaches to the reservoir 10½ miles north of the Big Koocanusa bridge. It's accessible at the mouth, about a mile upstream (near the Tooley Lake School), and from the meadow reaches on up by jeep and logging roads. This stream has been rehabilitated with migratory cutthroat which should provide good year around fishing for small (6 to 7 inch) trout, early spring fishing for 10 to 18 inch spawners, and fall fishing for Dolly Varden spawners that will run (hopefully) upwards to ten pounds or so. Providing always of course that the reservoir isn't sucked dry.

Courtesy Danny On

HOLLAND LAKE
Courtesy USFS

Swan River
Gateway to the
Mission Mountain Wild Area
and Bob Marshall Wilderness

High in the Mission Mountains, at an elevation of 6,600 feet, the Swan River arises from the outlet streams of Crystal and Greywolf Lakes. It begins as a roaring stream, cascading down steep, rocky slopes, falling at the rate of 450 feet per mile, until leveling out at Lindberg Lake on the valley floor. Below Lindberg Lake, the gradient diminishes and the stream is characterized by a series of alternating pools and riffles flowing through a partially covered, timbered valley to the mouth of Swan Lake. As it flows in a northward course, the volume flow increases as a network of 17 major tributaries enter the mainstem of the river. Below Swan Lake the river widens out and forms a slow meandering course to the backwater of the Pacific Power and Light Diversion Dam, 2 miles east of Bigfork.

The Swan River offers good rainbow, cutthroat and Dolly Varden trout fishing from the time spring flows begin to recede in late June through October. Rainbow trout up to 5 pounds are caught on natural bait and spoons. In mid-summer, when the water levels drop and fly hatches appear, fly fishing becomes increasingly popular and continues through late fall.

The tributary streams, especially in some of the more remote areas, provide excellent fishing for small, resident, pan-sized cutthroat and brook trout after flows recede in late June. Brook trout are predominant in the lower, warmer sections of the streams while cutthroat trout inhabit primarily the headwaters of the tributary streams. On occasion, one might hook a large Dolly Varden on its annual spawning migration from Swan Lake. Some of the more important fishing streams include Beaver, Elk, Glacier, Lion, Goat, Jim, Piper and Lost Creeks. Forest Service roads provide good access to these waters.

Swan Lake, the largest lake in the drainage, affords some good fishing for large Dolly Varden in early spring after ice-out and in the fall just before freeze-up. Trolling the shoal areas with large plugs and spoons is most effective for these lunkers.

SWAN

Several valley floor lakes which have little or no spawning facilities are maintained by stocking at various intervals with trout species. These lakes include Pierce, Lindberg, Holland, Van, Shay, Fran, Russ and Peck Lakes, and provide both summer and winter fisheries for a variety of fish. These include westslope cutthroat trout, brook trout, grayling and kokanee.

In July, August and September, the high mountain lakes open up and become a popular retreat for the back-pack and horse-back angler. Most of these remote lakes are inhabited by self-sustaining populations of westslope cutthroat trout. Trail access into the lakes varies from short, one or two miles of well-beaten trails for the novice, to steep rugged climbs for the expert.

ROBERT DOMROSE
Fisheries Biologist
Montana State Fish and Game Department

SWAN RIVER NEAR SALMON PRAIRIE
Courtesy USFS

Swan River Drainage

Barber Creek. Flows from the junction of its North and South forks (a couple of miles north of Holland Lake) for 3 miles west through rolling timber and pastureland in the Swan Valley to the river where it is crossed near the mouth by the Swan highway. The lower reaches are fair fishing for pan-size brook trout along with a few Dolly Varden and rainbow trout. The upper reaches, below the forks, support a few cutthroat trout.

Beaver Creek. This stream flows to the northeast for 12 miles through heavily timbered mountains along the Swan River-Clearwater River drainage divide to the Swan River 2½ miles below Lindbergh Lake. It is followed upstream for about 5 miles by a good logging road (west from the Swan-Clearwater highway at the summit). The lower 3 to 4 miles are good brook trout water (6 to 14 inches). The upper 6 miles are good for 6 to 8 inch cutthroat and a few brook trout.

Birch Lake. A 49.6 acre alpine lake, 105 feet deep with steep dropoffs, in a cirque near the top of the Swan Range; at 6000 feet above sea level; reached by good county roads east of Kalispell to the Echo Guard Station, then a steep 3 miles to the end of the Noisy Creek road, and finally 2½ miles by easy trail around the west side of Mt. Aeneas to the shore. This one is quite popular for a high mountain lake, and fair fishing too for small cutthroat and some rainbow-cutthroat hybrids that range from 7 to 14 inches.

Bond Creek. Drains Bond Lake for 5 miles west to the south end of Swan Lake; crossed near the mouth by the Swan highway and followed to headwaters by a good USFS trail. The upper reaches are in a steep timbered canyon but the lower mile is fairly flat, swampy in spots, and fair fishing for 6 to 8 inch brook and some cutthroat trout.

Bond Lake. Almost five acres, fairly deep, in a timbered pocket on the northern slopes of Spring Slide mountain about 6 miles east as the crow flies from the community of Swan Lake; accessible by 5 miles of good USFS (the Bond Creek) trail. This lake is fair fishing for cutthroat trout that will average around 10 inches and range up to 14.

Buck Creek. A small westward-flowing stream about 6 miles long from its headwaters on the steep western slopes of the Swan Range, to its mouth on the Swan River. It's crossed near the mouth by the Swan Valley highway 4½ miles south of the Condon Ranger Station and followed for 3½ miles upstream by a logging road and trail. The lower reaches are fair fishing for 6 to 10 inch cutthroat, rainbow and brook trout.

Bunyan Lake. A 9 acre lake, deep in spots but mostly shallow, and marshy at the east end; it lies in wooded country a mile west of the upper end of Lindbergh Lake and is reached by a logging road 7 miles from the lower end. Bunyan is moderately popular and good fishing for 6 to 13 inch cutthroat trout.

Carney Creek. A small stream, only a couple of miles long in timber and open farmland; crossed at the mouth by the Swan River road 2½ miles east of Bigfork, and several times along its course by logging roads. It is fair fishing for small rainbow and brook trout, but is generally passed up for bigger water.

SWAN

Cat Lake. In steep mountainous terraine at the head of the Cat Creek drainage a mile above Pony Lake (up the creek). It's 14 acres, 50 feet deep, was stocked with cutthroat in 1969 and is now good fishing for 12 to 14 inchers.

Cedar Creek. The outlet of Cedar Lake; flows for 11 miles down the east side of the Mission Range and eventually spreads out into a swamp adjacent to the Swan River near the Goat Creek Guard Station 15 miles south of Swan Lake on the Seeley Lake highway. The lower 3 miles are accessible by a logging road and are fair fishing (it used to be excellent) for mostly 3 to 8 inch brook trout and some good sized Dolly Varden spawners up from the main river.

Cedar Lake. An 45.9 acre alpine lake that is good and deep but with mostly gradual dropoffs; reached by a trail 5 miles from the end of the Fatty Creek road near the crest of the Mission Range. Cedar Lake is moderately popular and good fishing for 8 to 12 inch cutthroat trout.

Cilly Creek. A little stream flowing west through steep, timbered mountains to the Swan River and crossed near the mouth by the Seeley Lake highway 4½ miles south of Swan Lake. The lower mile is fair fishing for pan-size brookies and a very few cutthroat trout.

Cold Creek. From the junction of its forks, flows 3 miles to the Swan River 3 miles north of Condon. Moderately popular and fair fishing for 6 to 10 inch cutthroat, brook and some Dolly Varden that'll run up to ten pounds or so.

Cold Lakes. Two, about 80 acres each and deep ¼ of a mile apart in high alpine country, rocky above but heavily timbered around the outlets; reached by a good USFS trail 1¼ miles from the end of the Cold Lakes road. Both are good fishing for 14 to 16 inch above, maybe 9 to 12 inch below, cutthroat trout. The upper one is also known as Frigid Lake. Some folks just don't seem to get around to fishing it. Best time is in late July, August and September.

Condon Creek. This small stream heads at the crest of the Swan Range and flows for 7 miles westward through steep timbered mountains to the Swan River where it is crossed by the highway 1½ miles south of the Condon Ranger Station. The lower 3 miles are fair fishing for camp-eating brook; the headwaters are good for cutthroat trout.

Cooley Creek. Just a trickle between Rumble Creek and Buck Creek. It dries up in late summer but a few beaver dams in the headwaters are good fishing for small cutthroat trout. You gotta walk and they aren't bothered but once in half a blue moon.

Cooney Creek. A little stream flowing west from the Swan Range to the river where it is crossed a half mile upstream by the Swan highway 1½ miles above the Condon air field. Cooney Creek gets quite low in late summer but the short stretch between the road and river is fair fishing for small cutthroat and brook trout. There are some "lost" beaver dams in the headwaters.

Crazy Horse Creek. A high, mountain stream flowing southeastward to Glacier Creek right at the end of the road about 1 mile below Glacier Lake. It is seldom fished but contains a few small cutthroat trout in the lower mile.

CRYSTAL LAKE
Courtesy USFS

HEART (OR HART) LAKE
Courtesy USFS

Crescent Lake. Reached by trail a mile and a lot of switchback climbing way up west from Glacier Lake. Crescent is 23½ acres by 35 feet maximum depth, in rugged, rocky alpine country on the eastern slopes of the Mission Range. It used to be barren, but was planted with golden trout in the early '60s and was excellent fishing for awhile, but it's less than mediocre now, for ½ to 3 pounders and they're not too fat.

Crystal Lake. A deep lake, about 186.4 acres, in heavily timbered mountains; reached by a trail 2½ miles above Lindbergh Lake, or 2 miles from Meadow Lake, or 6 miles from the start of the Sunset Road in the headwaters of Beaver Creek. There is a nice camping spot on the north side of the lower end, and the lake is moderately fished for good catches of 9 to 10 inch cutthroat trout.

Deer Creek. A small inlet of Mud Lake, the upper reaches are on the western slopes of the Swan Range and are too steep for fish, but the lower reaches are beaver-dammed about a half mile above the lake and are fair fishing for pan-size brookies. To reach it go 3 miles north from Bigfork, then 3 miles east to the lake and follow around its south end to the creek.

Dog Creek. Seven miles long in heavy timber; it empties from the east to the Swan River and is crossed near the mouth by the Swan highway about 6 miles north of the Condon Ranger Station. The lower 3 miles are occasionally fished for fair catches of 6 to 10 inch brook trout.

Ducharme Lake. See Upper Piper Lake.

East (or Lost) Lake. A deep, 150 acre, alpine lake in scrub timber and talus between Goat and Daughter of the Sun mountains near the crest of the Mission Range; reached by a poor hunters' trail 1½ miles above and west of Crystal Lake. It is almost never visited but the few that do get there find it to be good fishing at times for 9 to 18 inch cutthroat (in the 3 pound class).

Elk Creek. A popular stream that is good fishing for cutthroat and brook trout and (in the fall) large Dolly Varden spawners for 4 miles above its mouth on the Swan River just west of the Condon air field. It is reached near the mouth and here and there for about 3 miles upstream by logging roads, and is followed by a trail along the east side on past the falls away up into the canyon.

Elk Lake. A seldom visited alpine lake in barren country near the crest of the Mission Range; reached by a 3 mile hike up the South Fork of Elk Creek from the end of an old pack trail up the Elk Creek Ridge, or by a good trail by way of Spook Lake, 3 miles from Mollman Lake. This one is about 200 acres, deep and poor-to-fair fishing for good-size cutthroat and rainbow trout.

Fatty Creek. Drains Fatty Lake and is followed by logging road and trail for 4 miles eastward through heavily timbered mountains to Cedar Creek. The lower reaches contain a few small cutthroat trout.

Fatty Lake. An alpine lake, about 21 acres, fairly deep with a mud bottom and brushy shores; reached by a poor hunters' trail 1½ miles northeast from a point near the end of Fatty Creek road. It is lightly fished but was planted with cutthroat long ago and is good fishing now, for 10 to 18 inches.

Fish Lakes. See Lower, Upper.

Fran Lake. The second of 3 small, undrained lakes in the timbered Swan River Valley flats about 1¾ miles by a USFS logging road westward from the Swan highway opposite the old Condon Ranger District headwaters. All 3, including

Fran Lake. The second of 3 small, undrained lakes in the timbered Swan River Valley flats about 1¾ miles by a USFS logging road westward from the Swan highway opposite the old Condon Ranger District headquarters. Fran is 8.8 acres, the lower one is about 5 acres. They were all thought to be almost barren except for a very few 16 to 18 inch brook trout. Sooooo, they were stocked in 1963 with cutthroat trout, and as of now Fran and Russ (the uppermost of the three and about 10 acres) contain good populations of 12 to 15 inchers.

Frigid Lake. See Cold Lakes.

Gilbert (or Gildart) Creek. A small stream that flows for about a mile down a steep canyon on the eastern slopes of the Mission Range, and then meanders across the timbered bottomland in the Swan River valley to ultimate junction with the river about 2 miles above the lake. Gilbert is accessible by logging road to within ½ mile above the mouth) are fair fishing for cutthroat, Dolly Varden and brook trout.

Glacier Creek. A clear, snow-fed tributary of the Swan River; reached here and there by a logging road for 6 miles from its mouth near the Condon air field to near Loon Lake, from whence it is followed south by a logging road for 7 miles to the mouth of Crazy Horse Creek, 1 mile below Glacier Lake. Glacier Creek is moderately popular and fair fishing for 8 to 10 inch cutthroat, and some rainbow and brook trout and Dolly Varden spawners.

Glacier Lake. A deep, 103.8 acre lake in heavily timbered mountains 4 miles due west of Lindbergh Lake; reached by trail a mile from the end of the Glacier Creek road. It is heavily fished in summer and good too for mostly 9 to 12 inch (plus a few lunkers) cutthroat trout. There is a campground near the outlet.

Glacier Sloughs. See Notlimah Lake.

Goat Creek. A small stream flowing down a steep canyon above and through timbered hills below to the Swan River about 14 miles above the lake; crossed at the mouth by the Swan River highway and followed upstream for 6 miles by a logging road. There are about 3 miles of fishing. Goat Creek is moderately

ELK LAKE (on right) & Barren SUMMIT LAKE
Courtesy USFS

GLACIER FALLS
Courtesy USFS

popular and provides fair catches of 6 to 8 inch cutthroat and brook trout, plus Dollys of all shapes and sizes.

Gray Wolf Lake. A deep, 300 acre alpine lake in barren, rocky (with some scrub timber) country below Gray Wolf Glacier near the crest of the Mission Range. Gray Wolf is reached by trail 8 steep miles above the upper end of Lindbergh Lake, — or, 9 miles from the end of the Beaver Creek road and is not exactly overrun with fishermen. It was planted years ago with cutthroat trout and is fair fishing at times for 12 to 18 inchers.

Hall Lake. A deep, 12 acre subalpine lake with lots of brush and talus around the north end; reached by a good USFS trail 5½ miles horizontally (and 3000 vertically) from the community of Swan Lake. It is good fishing for mostly rainbow trout, plus a few cutthroat that average around 12 inches and range up to 4 or 5 pounds.

Heart (or Hart) Lake. Reached by trail in alpine country ½ mile above and upstream (west) from Crescent Lake. It's 22.9 acres, 55 feet deep with mostly steep dropoffs, was originally barren but planted with golden trout in the early '60s and red hot for awhile, but is now only so-so for 12 to 18 inch trout. This was the best of the Crescent-Hart-Island Lake chain and was fished accordingly.

Hemlock Lake. Lies at the head of Hemlock Creek in timbered mountains; is reached by an old hunters' trail ¼ mile west from a USFS trail 4 miles north from the end of the Red Butte Creek logging road (takes off to the west from the Kraft Creek-Glacier Creek road). Hemlock is about 30 acres, deep and moderately popular for good catches of 10 to 12 inch cutthroat trout.

High Park Lake. A deep, 225 acre alpine lake on the east side of the Mission Range just below Goat Mountain and reached by a trail 2 miles above Crystal Lake. High Park is not heavily fished but is very good at times for 9 to 14 inch cutthroat trout. The scenery alone is worth the trip.

Holland Creek. Heads in Upper Holland Lake and is followed by a USFS trail for 3 miles down a steep timbered canyon to the main Holland Lake, and thence by road for 3½ miles through timbered bottomland to the Swan River about 4½ miles

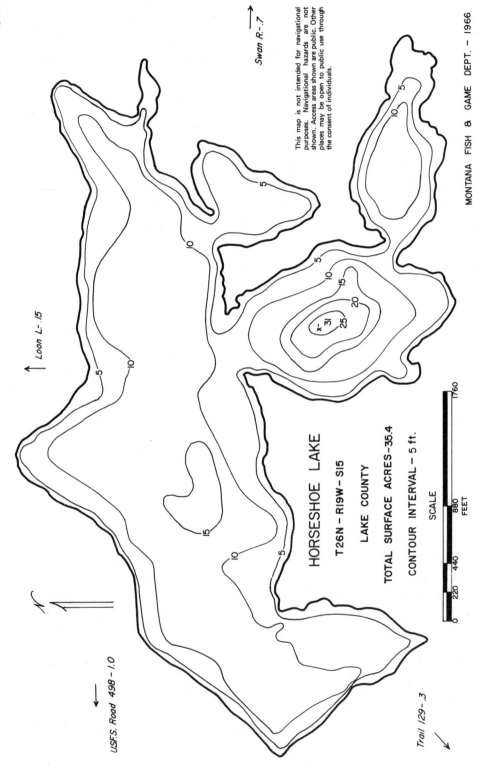

GRAY WOLF
Courtesy USFS

A GOOD CAMP SITE ON UPPER HEMLOCK
Courtesy USFS

north of the Clearwater-Swan River drainage divide. The lower reaches (below Holland Lake) are fair fishing for 6 to 8 inch rainbow, cutthroat and brook trout, along with too many trash fish.

Holland Lake. A nice, big (408 acres), recreational summer home lake at the foot of the Swan Range, reached by road a couple of miles east from the Clearwater-Swan River highway 4 miles north of the drainage divide. Holland used to be fabulous fishing in the late 1800's for whopping big trout. It now contains rainbow trout, kokanee, cutthroat trout and a few nice-size Dolly Varden, plus increasing numbers of trash fish. There are both public and private campgrounds.

Horseshoe Lake. One of the few lakes in Montana with smallmouth bass. It is a little better than 50 acres, marshy and swampy in spots with lots of lilies along the south shore; in timbered NPRR land on the south side of the Loon Lake road about ½ mile west of the north end of Swan Lake and 9 miles east of Bigfork. Quite a few people fish it now for fair catches of 1 to 4 pounders. However, the subdividers have moved in so here we go.....

Island Lake. The uppermost of the Crescent-Hart-Island Lake chain, in alpine country about ½ mile up the drainage from Hart Lake. It's 25 acres by 40 feet maximum depth, with a 1 acre rock island in the southeast end. Island Lake used to be barren, was planted with golden trout in the early '60s, was good for awhile for 10 to 16 inches but not many are left now.

Jim Creek. Drains the Jim Creek Basin Lakes (14 of them) eastward for 6 miles to the Swan River just west of the Condon Ranger Station. The lower 3 miles are right down in lodgepole bottomland in the Swan Valley; are crossed here and there by logging roads and trails; and are fair fishing for Dolly Varden, brook and a few cutthroat trout.

Jim Creek Lakes. Now here this! Jim Lake is about 20 acres, is full of fresh-water shrimp, was planted with cutthroat and rainbow trout in years gone by and is now good early season fishing for nice, fat 12 to 14 inches. It's difficult to fish from shore though because of trees and brush all around, coupled with shallow dropoffs.

Now if you scramble up the drainage (no trail) for about 1¼ miles you'll pass 3

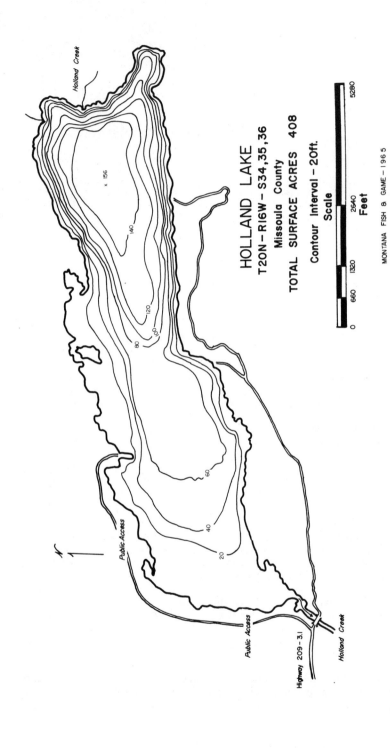

KELLY RESERVOIR
Courtesy Phyllis Marsh

ACROSS HOLLAND LAKE TO THE MISSIONS
Courtesy USFS

barren lakes and then come to Jim Creek Lake No. 6, which is 5 acres and good fishing for 12 to 14 inch cutthroat. Next follow up the "northern" inlet for ½ mile to Lake No. 7 and another ¼ mile above up its inlet to Lake No. 8. No. 7 is about 12 acres, all in timber and brush; No. 8 is 16½ acres mostly in a great rock-bound cirque but with timber around the east end. Both are good fishing for foot-long cutthroat trout.

If you're still of a mind to explore, retrace your steps to Lake No. 6 and follow up the west inlet ¾s of a mile past two small (barren) lakes to Lake No. 12. This one is 26½ acres, below rock headwalls to the north and west but bounded by timber on around. It too is good fishing for 12 to 14 inch cutthroat.

Should you be young, eager and full of beans and you want to try them all (the lakes, not the beans), there are still two more of the Jim Creek chain with trout. But they're hard to find. Lets try. Go back to the "first" lake below Lake No. 6, follow around the south side to the "second" inlet and follow it up for ½ mile to a long narrow lake (it's barren), work around it (the immediate shores will be mostly open), and keep on up its inlet for a couple of hundred yards to Lake No. 19. Now, from where you stand hike west along the north end of this lake a few hundred yards to an inlet coming in from the "north" which you should follow up for maybe 5 or 600 yards to Lake No. 20. Both of these lakes (nos. 19 and 20) are in timber, both have lotsof shallow water, and both were planted in 1969 and are good fishing now for 12 to 14 inch cutthroat. All of the rest of the Jim Creek Lake chain are barren. If you don't believe me, just spend a summer some time trying them out. I hope you make it back to the car. Oh yes, this is good bear country.

Johnson Creek.
A small stream that flows west through timbered bottomland across the Swan highway and empties into the river just above the north end of the lake; accessible by logging road and trail for about 3 miles of fair fishing. Johnson Creek contains mostly 7 to 10 inch brook trout and whitefish in the lower reaches, and brookies in the few beaver ponds above.

Kelly (or Wyman Creek) Reservoir.
Float or swim across Swan Lake at a point about ¼ of the way down from the south end (perhaps ¼ mile north of the Rock-House-On-The-Point on the west side of the lake) to the mouth of Wyman Creek (too small for fish), and hike ½ mile up it on a good trail to the old Kelly Reservoir which was built away-back-when he was one of the "copper kings" in Butte. It is 18.6 acres, in timber all around, has the remains of an old dam (with water

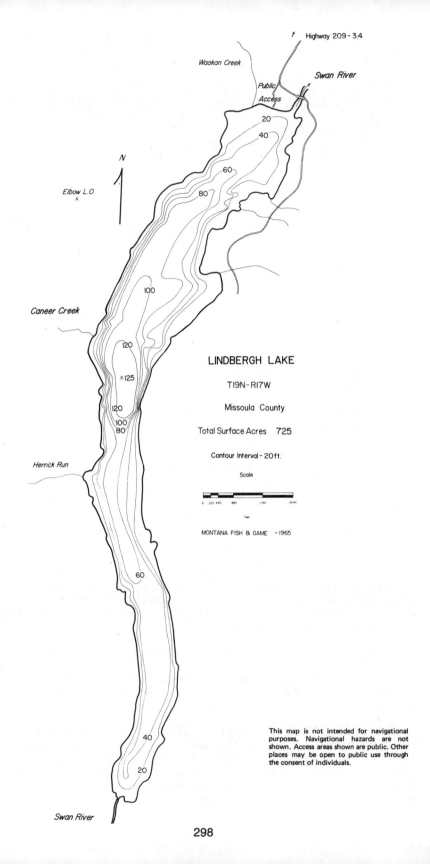

LINDBERGH LAKE
Courtesy Ernst Peterson

STILL A LONG WAY DOWN TO THE WATER
Courtesy USFS

wheel) at the outlet, lots of moss on the bottom and springs too, and is loaded with small brook trout. A private lake; not many people fish it but there are no signs telling you not to.

Kraft Creek. A small stream, crossed at headwaters by the Kraft Creek (Glacier Lake) road 7 miles up from the Swan highway and here and there by logging roads along its course to Glacier Creek. It contains a few small cutthroat and brook for the few folks who try it.

Kraft Creek. A small stream, crossed at headwaters by the Kraft Creek (Glacier Lake) road 7 miles up from the Swan highway and here and there by logging roads along its course to Glacier Creek. It contains a few small cutthroat and brook trout for the few folks who try it.

Lace Lake. About 18½ acres, in high, rocky scrub-timber country near the crest of the Mission Range; reached by trail 1½ miles above Glacier Lake. Do not confuse Lace with Jewel and Lagoon Lakes which are barren, or Turquoise Lake which lies 100 yards up the inlet to the west, and drains to it. Lace Lake is moderately popular in late summer and good fishing for 9 to 14 inch cutthroat trout.

Lily Lake. A shallow, 7.3 acre lake ¼ mile above the west side of Swan Lake near the mouth of the Swan River where it empties into the lake; a ten minute hike in heavy timber. It was stocked with grayling in 1970 and is good fishing now for 10 to 14 inchers but is seldom bothered. People just don't know it's there.

Lindbergh Lake. A long (4 miles), deep (125 feet), narrow (⅛ to ½ mile) summer home and resort lake with a USFS campground on the north end; reached by a road 3½ miles from the Swan-Clearwater highway 3 miles north of the drainage divide. It is fair-to-good spring fishing for Bulls but otherwise no better than poor-to-fair for kokanee salmon, plus a few rainbow trout and whitefish, plus increasing numbers of suckers and squawfish. It is, however, the jumping-off place for quite a few good fishing lakes and creeks in the high-up-and-away-back country.

Lion Creek. A fair size stream, followed by a trail from headwaters below Lion Creek Pass at the crest of the Swan Range, westward for 8 steep miles down the slopes of the range, and for 6 miles across the flat timbered bottoms of the Swan Valley to the river 6 miles south of Goat Creek Ranger Station. The lower 8 miles are

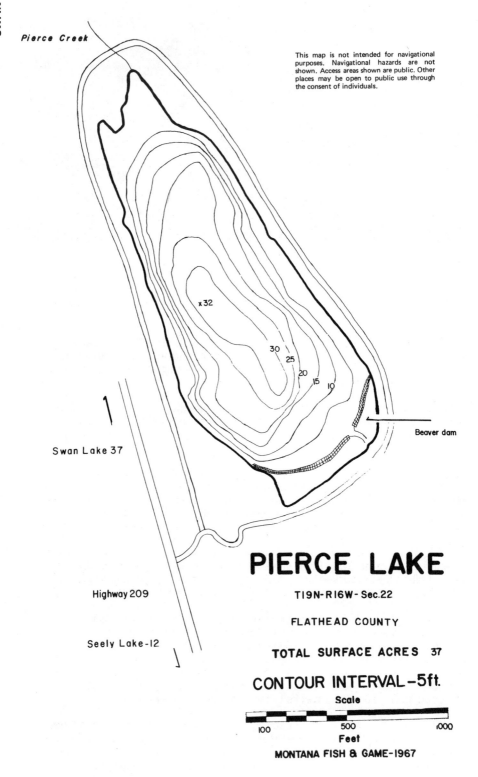

fair fishing for 6 to 10 inch brook and cutthroat trout and a few whitefish near the mouth, plus some good sized Dolly Varden spawners in the fall.

Loon Lake. Forty-four acres, 50 feet deep with mostly steep dropoffs but marshy on one end, in hilly, timber-meadow-and-marsh country a half mile west of the lower end of Swan Lake and just north of Horseshoe Lake; reached by a gravel road 1½ miles west from the Swan Lake highway. Loon was rehabilitated in 1959 and used to be good fishing for 12 to 16 inch cutthroat trout but is now about taken by sunfish. It's still fair though for largemouth bass up to 4 or 5 pounds. There's a fishing access.

Lost Creek. Really two creeks, North and South, that join west of the Swan River highway about 3 miles south of the lake. The lower reaches sink about ¼ of a mile above the river. The South Fork is followed by a logging road and trail for 4 miles of fair fishing; the North Fork by 5 miles of logging road but there is only about 3 miles of it pretty good fishing. Both forks contain 10 to 12 inch Dolly Varden plus some big spawners up from the river, plus resident populations of 6 to 7 inch brook trout in the lower reaches and a very few cutthroat trout. However there is good cutthroat fishing in the upper reaches of both streams (for pan-sizers).

Lost Lake. See East Lake.

Lower Fish Lake. Twenty acres, in subalpine country on the North Fork of Cedar Creek drainage; reached by a cross-country hike downstream for ¾ of a mile below Upper Fish Lake. It is seldom found, to say nothing of the fishing, but is reported to be excellent for 6 to 14 inch cutthroat trout.

Lower Piper Lake. See Piper Lake.

Meadow Lake. A shallow, marshy, 14.7 acre lake with a logging road right to it, ¾s of a mile south from Bunyan Lake. It is lightly fished but very good for 10 to 14 inch cutthroat trout.

Metcalf Lake. Is 13.4 acres, in timbered bottomland of the Swan Valley right on the edge of the mountains, reached by a jeep trail ½ mile from the Woodward lookout road about 14 miles south of Swan Lake. Metcalf is lightly fished and fair for undersized, 10 inches or so, overpopulated largemouth bass, along with a few MONSTERS.

Mollman Lakes. Two really "high up" alpine lakes and a pothole right at the top of the Mission Range in barren rocky country; reached (once in a blue moon) from the Flathead Valley by a good but steep trail 3½ miles over the crest of the Missions from the end of a logging road (right past the north end of the Kicking Horse Reservoir) 4 miles due east from U.S. 93, three and a half miles south of Ronan. The larger (upper and westernmost) lake used to be good fishing for large rainbow trout, but is now apparently barren. The lower lake, which is long and narrow in a north-south direction, is fairly deep below steep mountain slopes along its east side, and — in company with the shallow pothole just below its northeastern end — is fair-to-good fishing for 6 to 18 inch rainbow and cutthroat trout.

Moore Lake. Ten acres, deep, in timbered mountains on the Moore Creek drainage; reached by a trail 5 miles from the end of a logging road a couple of miles south and west from Salmon Prairie, and there is also a new (Piper Creek) road which takes off from State 209 seventeen miles south of Sean Lake that takes you 5 miles in to a USFS trail and then you hike another 4 miles to the shore. What with all of this hiking Moore is seldom visited but good fishing nevertheless for cutthroat that will run mostly between 8 and 12 inches and range up to 20 inches or more.

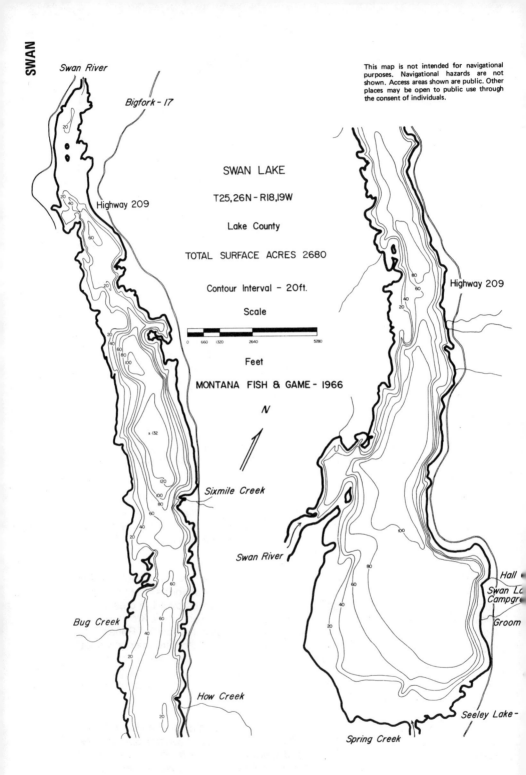

Mud Lake. From Bigfork take the Echo Lake road 3 miles north and then 3¼ miles east to Mud Lake which is about 150 acres, in farmland but timbered on the east side, real shallow and with such a soupy, muddy bottom that you need a boat to fish it. It provides, however, some fair-to-good morning and late evening (and wintertime through the ice) fishing for 6 inch to 2 pound yellow perch, largemouth bass, and 8 to 12 inch brook trout.

North Fork Cold Creek. A small stream, about 3½ miles long, the outlet of Cold Lakes, reached at the mouth by a logging road 5 miles west from the Swan highway near the Condon Ranger Station. The lower reaches are available by road and are fair fishing for small cutthroat, and a few Dollys in the fall, water permitting because Cold Creek itself gets real low at times.

North Hemlock Lake. See unnamed lake at the head of the North Fork of Hemlock Creek.

Notlimah Lake (or Glacier Sloughs). Really just a big slough about 1¼ miles long by a 100 yards wide, in heavy timber on Glacier Creek; reached by a trail a couple of miles west from the lower end of Lindbergh Lake or cross-country from the switchbacks on the Kraft Creek road. It is moderately popular and fair fishing for 6 to 12 inch cutthroat and brook trout at all seasons of the year.

Pierce Lake. A 37 acre lake with a mucky bottom and many drowned trees around the shore; in rolling timbered country reached by a road ¼ of a mile east from the Seeley Lake-Swan highway just north of the drainage divide. This is a "summer home" lake that is fair fishing for 10 to 12 inch cutthroat trout.

Piper (or Lower Piper) Lake. A beautiful, deep, 82½ acre subalpine lake in rocky, scrub timber country; reached by a USFS trail 4 miles west of Moore Lake. Piper is seldom fished, it's just too far back for most, but is good for 9 to 16 inch cutthroat trout, and a few lunkers up to 30 inches or more have been reported.

Pony Creek. The small outlet of Pony Lake flowing westward across the Swan Valley to the river near Salmon Prairie and crossed by the highway just south of the Condon Ranger Station. The lower mile is fair fishing for pan-size brook trout.

Pony Lake. A deep, 17.4 acre alpine lake on the western slopes of the Swan Range; reached by a 2½ mile cross-country hike from the end of the Pony Creek trail. It sustains a light-to-moderate fishing pressure and is reportedly fair for 8 to 16 inch cutthroat trout and an occasional lunker (1 to 3 pounds or so). Note: Don't confuse with "Cat Lake" a mile to the south.

Rainbow Lake. About 8 acres, 25 or so feet deep, with an island in the middle in rock, brush and scrub timber near the crest of the Mission Range. Rainbow is reached by an old pack trail 1½ miles south and over the crest of the range from the end of the Indian Service, Hell Roaring Creek trail which is 3 miles from the end of the Hell Roaring logging road 6 miles east from Polson in the Flathead Valley. Rainbow is seldom fished and just as well because it's barren.

Rumble Creek. A small stream, the outlet of Rumble Creek Lake; the lower reaches flow westward across the timbered bottoms of the Swan Valley to the river about 2½ miles south of the Condon Air field and there's about 3½ miles of access road east from State 209 from a point a couple of miles south of the air strip. There are a few beaver ponds here that are good fishing for pan-size cutthroat. The lower reaches are good for small brookies.

Rumble Creek Lakes. Two, in alpine country just below and west of Holland peak at the crest of the Swan Range. They are seldom visited as there is no trail. The best way in is up the South Fork of the creek a couple of miles above the end of the Cooney Lookout road. The upper lake is just a shade over 27 acres, 55 feet deep, was stocked with cutthroat in 1969 and is now good fishing. The lower lake is only about 11 acres, but real deep (over 95 feet), and has been good fishing for 6 to 12 inch cutthroat trout.

Russ Lake. See in Fran Lake.

Shay Lake. Seventeen acres, moderately deep, in timberland reached to within ½ mile by the Fatty Creek logging road 2 miles from the Swan highway, 5 miles south of the Condon Ranger Station. This is a good but temperamental winter lake that is moderately popular for 10 to 14 inch grayling and a few cutthroat trout.

Skylark Lake. A 6½ acre pond reached by an old trail 1 mile from the end of the Bunyan Lake-Meadow Lake road. This one is pretty shallow and could freeze out, but has (up to now) supported a very few cutthroat trout.

Smith Creek. A small stream with lots of beaver dams; its lower reaches flow right through the old Condon Ranger Station in swampy meadow and timberland. It is heavily fished here in early season for fair catches of 6 to 9 inch brook trout. The upper reaches have some cutthroat.

Snow Lake. Really a part of the swamp land at the head of Carney Creek; about 2 acres, shallow on private land reached to within a couple of hundred yards of the shore by a road 5 miles east and south from Bigfork. It was rehabilitated a long time ago and is good fishing for 8 to 12 inch cutthroat trout.

Soup Creek. A pretty small stream flowing west from the Swan Range to the Swan Valley and crossed by the Swan River road a mile above its mouth. Soup Creek is followed upstream by a logging road for a couple of miles of fair brook and some cutthroat trout fishing.

South Fork Cold Creek Lakes. Thirteen in all, ranging in size from 3 to 20 acres and from fairly shallow to deep, in an old burn mostly, although there is some nice timber too. The lowest one is reached to within a mile by the South Fork road, the rest just string out for about 3 miles up the canyon and north over the ridge in real rough country with no trail to connect them. They're very seldom fished but the first three you come to of any size (they are 5, 13 and 7.3 acres, a couple of hundred yards apart and all in timber except for the north end of the uppermost one) have fish. The two lower ones are good fishing for 8 to 12 inch rainbow below, and maybe 8 to 14 inches in the next one up. The upper lake has a small population of nice large rainbow (up to 16 inches). Hike on up its inlet for another half mile (past a little one acre pothole) and you will arrive at the last fishing lake in the chain which lies just below a big swamp and has a small population of nice-size rainbow trout.

Spruce Lake. See Unnamed Lake ⅛ mile southwest of Cedar Lake.

Square Lake. The site of the Fred Wagner plane crash years ago, on the headwaters of the Fatty Creek drainage but reached over the top of the Mission Range from the Flathead Valley, by an old "blazed trail" ¼ of a mile northeast and straight down over cliffs and through scrub timber from Rainbow Lake. Square Lake is about 12 acres, 25 feet deep, and was planted by airdrop with cutthroat trout way back in 1956. It hasn't been fished a dozen times since but the trout are reported to be plentiful now and run mostly from 10 to 16 inches.

RUMBLE CREEK LAKES
Courtesy USFS

LACE LAKE, with Glacier and Tourquoise
Courtesy USFS

Stoner Lake. A 31 acre pothole with about 20 in vegetation, a maximum depth of 15 feet and gradual dropoffs; in dense timber reached by a poor truck road 3 miles west from a point on the Swan highway about 1½ miles north of the Holland Creek crossing. Stoner used to be good trout fishing but winterkilled and now has only Yellow Perch.

Swan Lake. A real big lake; about 10 miles long by 1 mile wide, and followed along its east side by a good blacktop highway down the flat-bottomed, heavily timbered Swan River Valley. Swan Lake is a moderately popular resort area; there is a USFS campground on the north end, quite a few summer homes and several resorts. It used to be good fishing for 10 to 12 inch kokanee, but is now only fair, although heavily fished, for 8 to 11 inch cutthroat, rainbow, Dolly Varden (that range up to 15 pounds) and whitefish, plus largemouth bass and yellow perch in the upper end.

Trinkus Lake. Really 2 subalpine lakes about 75 yards apart; reached by the Bond Creek trail 5 easy and 1 steep mile east from the community of Swan Lake. The upper lake is barren; the lower is about 12 acres by 65 feet deep or a little less, and fair fishing for 9 to 14 inch rainbow and a few cutthroat trout. This is good elk country and the lake is fished mostly by hunters.

Turquois Lake. A 184 acre alpine lake just below and east of Glacier Peaks at the crest of the Mission Range a few hundred yards above Lace Lake. Turquoise used to be good cutthroat fishing but the reproduction was nil and only a very few lunkers are left now. How about a replant? It's happened — but Turquoise is still not so hot. Pretty tough!

Unnamed Lake at the head of the North Fork Hemlock Creek.
About 10 acres, deep, in scrub timber and talus; reached by trail ⅛ of a mile below the Hemlock lookout, 4 miles west from the junction of Red Butte and Kraft Creeks on the Glacier Creek logging road. This one was planted years ago, and is good fishing now for 10 to 12 inch cutthroat trout.

SWAN

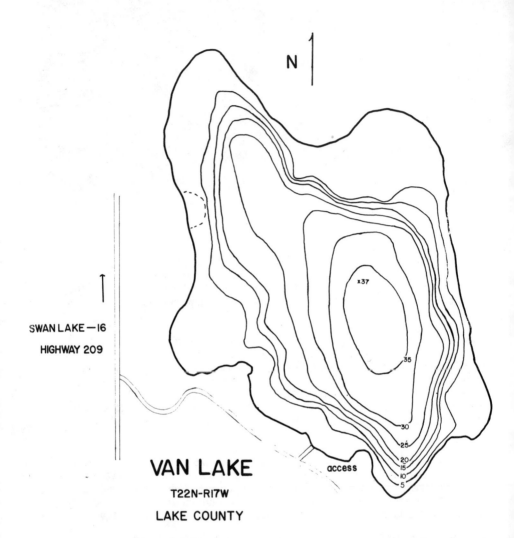

SWAN LAKE—16
HIGHWAY 209

VAN LAKE

T22N-RI7W

LAKE COUNTY

TOTAL SURFACE ACRES 58

Contour Interval - 5 ft.

This map is not intended for navigational purposes. Navigational hazards are not shown. Access areas shown are public. Other places may be open to public use through the consent of individuals.

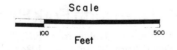

MONTANA FISH & GAME — 1967

Unnamed Lakes at the head of Fatty Creek.
Four, from ½ to 3 acres each ¼ to ½ mile apart in alpine country on the eastern slopes of the Mission Range. The lowest one lies ¼ of a mile west of the Fatty Creek trail, 2 miles above the end of the road. The others can be reached only by cross-country hikes. All four of these little ponds are excellent fishing (if you want it) for skinny, 6 to 8 inch cutthroat trout.

Unnamed Lake one-eighth mile southwest of Cedar Lake.
A 5 acre, alpine lake reached by a cross-country hike and very seldom fished but reported to be fair for 12 to 14 inch cutthroat trout.

Unnamed Lakes at the head of Woodward Creek and Tributaries.
Of these 8 unnamed lakes in an old burn at the head of Woodward Creek, 6 are reportedly barren; the remaining two (the largest of the lot) are about 10 acres each, real deep, and were stocked in 1958. They used to be excellent fishing for 10 to 16 inch cutthroat trout in past years, but may well be barren now. Moreover, they are almost inaccessible. The best way in is NOT up Woodward Creek as it is too steep, but rather up the Fatty Creek trail, 2 miles from the end of the road to Fatty Lake and then north cross-country over the mountain for 2 miles through good elk hunting and huckleberry country. If you're going to try it, you'd better avail yourself of the USGS Porcupine Creek quadrangle map but don't look for the lakes because they're not shown on it.

Upper Fish Lake.
Take a foot trail northeast for ½ mile from the north end of Cedar Lake to the old Cedar Point lookout; then another ½ mile south to Upper Fish Lake; which is about 7 acres with brushy, rocky shores in subalpine country, and is reportedly good fishing for 6 to 10 inch cutthroat trout.

Upper Holland Lake.
This one is good and deep, about 50 acres, in subalpine country below the crest of the Swan Range; reached by a USFS trail 2½ miles east and above Holland Lake. It contains a goodly number of 12 to 18 inch cutthroat trout, but they're real temperamental and more anglers get skunked than not.

Upper Piper (or Ducharme) Lake.
Unnamed on most maps, about ⅛ mile on up the drainage (¼ mile by trail) above Lower Piper, maybe 20 acres, in an alpine cirque just east of the Swan-Flathead divide. It's not stocked often but is good at times for cutthroat in the 14 to 18 inch class.

Van Lake.
Take the Swan River road 15 miles south from Swan Lake, and then a logging road 2 miles east to Van Lake in mountainous, timbered country. It is 58 acres, maybe 40 feet deep in the middle with gradual dropoffs and some marsh around the shore. The fishing has been good for 15 to 18 inch (9 pounds the reported maximum) rainbow, but it was stocked with cutthroat in 1974 and will be managed for cutthroat from here on in.

Whelp Lake.
On the drainage a few hundred yards below Gray Wolf Lake; about 5 acres, and fair-to-midling fishing for 10 to 16 inch cutthroat trout.

Wolf Creek.
A real small stream with about 1½ miles of fishing water, in timbered country right at the mouth of the Swan Valley, reached by good country roads 2 miles north and then 4 miles east from Bigfork. Wolf Creek used to be excellent fishing for small cutthroat above and brook trout in the lower end, but is now only fair in the lower end and good above.

Woodward Creek. This northward flowing stream empties into the Swan River 10 miles above the lake, and is more or less accessible by logging road and trail for about 1½ miles of fishing. It is moderately popular for fair catches of 9 to 12 inch brook and cutthroat trout, plus some Dollys up from the lake.

Wyman Creek Reservoir. See Kelley Reservoir.

HIGH PARK LAKE AND GREY WOLF LAKE IN THE MISSIONS
Courtesy USFS